PRACTICAL THIN-LAYER CHROMATOGRAPHY
A Multidisciplinary Approach

PRACTICAL THIN-LAYER CHROMATOGRAPHY
A Multidisciplinary Approach

Edited by
Bernard Fried
Joseph Sherma

CRC Press
Boca Raton   New York   London   Tokyo

Acquiring Editor: Mark Licker
Project Editor: Les Kaplan
Marketing Manager: Arline Massey
Direct Marketing Manager: Becky McEldowney
Cover design: Dawn Boyd
PrePress: Kevin Luong
Manufacturing: Sheri Schwartz

Library of Congress Cataloging-in-Publication Data

Practical thin-layer chromatography : a multidisciplinary approach ; edited by Bernard Fried, Joseph Sherma
    p. cm.
    Includes bibliographical references and index.
    ISBN 0-8493-2660-5
    1. Thin-layer chromatography. I. Fried, Bernard, 1933– . II. Sherma, Joseph. H.
QD79.C8P73 1996
543′.08956—dc20
DNLM/DLC   95-52348
                                                                                                                                 CIP

    This book contains information obtained from authentic and highly regarded sources. Reprinted material is quoted with permission, and sources are indicated. A wide variety of references are listed. Reasonable efforts have been made to publish reliable data and information, but the author and the publisher cannot assume responsibility for the validity of all materials or for the consequences of their use.

    Neither this book nor any part may be reproduced or transmitted in any form or by any means, electronic or mechanical, including photocopying, microfilming, and recording, or by any information storage or retrieval system, without prior permission in writing from the publisher.

    All rights reserved. Authorization to photocopy items for internal or personal use, or the personal or internal use of specific clients, may be granted by CRC Press, Inc., provided that $.50 per page photocopied is paid directly to Copyright Clearance Center, 27 Congress Street, Salem, MA 01970 USA. The fee code for users of the Transactional Reporting Service is ISBN 0-8493-2660-5/96/$0.00+$.50. The fee is subject to change without notice. For organizations that have been granted a photocopy license by the CCC, a separate system of payment has been arranged.

    The consent of CRC Press does not extend to copying for general distribution, for promotion, for creating new works, or for resale. Specific permission must be obtained in writing from CRC Press for such copying.

    Direct all inquiries to CRC Press, Inc., 2000 Corporate Blvd., N.W., Boca Raton, Florida 33431.

© 1996 by CRC Press, Inc.

No claim to original U.S. Government works
International Standard Book Number 0-8493-2660-5
Library of Congress Card Number 95-52348
Printed in the United States of America 1 2 3 4 5 6 7 8 9 0
Printed on acid-free paper

PREFACE

After an introductory chapter on principles, techniques, and instrumentation, this book contains chapters that describe the use of thin-layer chromatography (TLC) in eleven important scientific disciplines. This approach is novel for books on TLC; most of them have a number of introductory chapters dealing with theory, practice, and instrumentation, followed by applications chapters organized according to compound (analyte) type; contain a series of chapters on practice or applications; or are a random collection of papers from a symposium. The chapter authors comprise an international roster of expert practitioners of TLC in the most important application areas.

We proposed this new disciplinary approach not only to avoid redundancy with existing books in the field but because we believe it provides thorough coverage of the principles, practices, and applications of TLC for important sample and compound types directed specifically at workers in the most active scientific fields.

We attempted to make this book as practical as possible. Authors were encouraged to include the following topics in each chapter: an introduction that discusses the importance of TLC in the field and compares TLC to other analytical techniques; sample collection and preparation; chromatographic systems (layers and mobile phases); development techniques; detection, identification, and quantification of zones; validation of results; specific protocols for important analyses; and references through 1994.

Chapters 2 to 6 cover fields of greatest interest to biologists and biochemists and Chapters 7 to 12 to chemists and analysts. Although the chapters were written by specialists in particular disciplines, the information presented is also applicable to analyses in related fields. The book should be useful to all scientists interested in the principles and practices of TLC. It will be of greatest interest to students, teachers, researchers, clinicians, and technicians who wish to apply TLC in the disciplines covered in the chapters and in related disciplines.

We encourage readers to suggest ways to improve the book, including additional topics and authors, for the second edition.

Bernard Fried
Joseph Sherma
Lafayette College,
Easton, PA
June, 1995

ERRATA

Practical Thin-Layer Chromatography: A Multidisciplinary Approach
Bernard Fried and Joseph Sherma
(ISBN 0-8493-2660-5)

The following author affiliation should be included in the List of Contributors:

Jacques Pothier, Ph.D.
Laboratory of Pharmacology
Faculty of Pharmacology
Tours, France

In the Table of Contents, **M. A. Haseeb** should be added to the author credit 1 for Chapter 4.

Our apologies to Drs. Pothier and Haseeb for the oversight.

LIST OF CONTRIBUTORS

Jan Bladek, Ph.D.
Institute of Chemistry
Military University of Technology
Warsaw, Poland

Giuseppe Ceramelli, Ph.D.
Dipartimento Farmaco Chimico Technologico
Universita Degli Studi di Siena, Italy

Pietro Corti, Ph.D.
Dipartimento Farmaco Chimico Technologico
Universita Degli Studi di Siena, Italy

Tibot Cserháti, Ph.D.
Central Research Institute of Chemistry
Hungarian Academy of Sciences
Budapest, Hungary

Elena Dreassi, Ph.D.
Dipartimento Farmaco Chimico Technologico
Universita Degli Studi di Siena, Italy

Richard D. Fell, Ph.D.
Department of Entomology
Virginia Polytechnic Institute and State University
Blacksburg, Virginia

Esther Forgács, Ph.D.
Central Research Institute of Chemistry
Hungarian Academy of Sciences
Budapest, Hungary

Bernard Fried, Ph.D.
Department of Biology
Lafayette University
Easton, Pennsylvania

M. A. Haseeb, Ph.D.
Department of Microbiology, Immunology and Medicine
State University of New York
Health Services Center
Brooklyn, New York

Raka Jain, Ph.D.
Drug Dependence Treatment Centre
All India Institute of Medical Sciences
New Delhi, India

Mark Maloney, Ph.D.
Biology Department
Spelman College
Atlanta, Georgia

Illka Ojanperä, Ph.D.
Forensic Toxicology Division
Department of Forensic Medicine
University of Helsinki
Helsinki, Finland

Ali R. Shalaby, Ph.D.
Food Technology and Dairy Sciences Department
National Research Centre
Cairo, Egypt

H. M. Stahr, Ph.D.
College of Veterinary Medicine
University of Iowa
Ames, Iowa

Paul J. Weldon, Ph.D.
Conservation and Research Center
Smithsonian Institute
Front Royal, Virginia

CONTENTS

Chapter 1
Introduction to Techniques and Instrumentation 1
Tibor Cserháti and Ester Forgács

Chapter 2
Thin-Layer Chromatography in Bacteriology 19
Mark Maloney

Chapter 3
Thin-Layer Chromatography in Plant Sciences 33
Jacques Pothier

Chapter 4
Thin-Layer Chromatography in Parasitology 51
Bernard Fried

Chapter 5
Thin-Layer Chromatography in Studies in Entomology 71
Richard D. Fell

Chapter 6
Thin-Layer Chromatography of Skin Secretions in Vertebrates 105
Paul Weldon

Chapter 7
Thin-Layer Chromatography in Clinical Chemistry 131
Raka Jain

Chapter 8
Thin-Layer Chromatography in Environmental Analysis 153
Jan Bladek

Chapter 9
Thin-Layer Chromatography in Food Analysis 169
Ali R. Shalaby

Chapter 10
Thin-Layer Chromatography in Forensic Toxicology 193
Ilkka Ojanperä

Chapter 11
Thin-Layer Chromatography in Pharmaceutical Analysis 231
Elena Dreassi, Giuseppe Ceramelli, and Pietro Corti

Chapter 12
Thin-Layer Chromatography in Veterinary Toxicology 249
H. M. Stahr

Index 265

Chapter 1

INTRODUCTION TO TECHNIQUES AND INSTRUMENTATION

Tibor Cserháti and Esther Forgács

CONTENTS

I. Introduction 1

II. Theoretical Aspects of Thin-Layer Chromatography 2
    A. Stationary Phases 2
        1. Silica and Surface Modified Silicas 2
        2. Other TLC Sorbents 3
        3. Impregnation of Sorbents 4
    B. Mobile Phases 4
    C. Retention Parameters in TLC 5
    D. Retention Mechanism 6

III. Practical Considerations 6
    A. Sample Preparation 6
    B. Sample Application 7
    C. Development and Developing Chambers 8
        1. Linear Ascending Development 8
        2. Linear Horizontal Development 10
        3. Circular Development Methods 12
        4. Anticircular Development 12
    D. Detection of Solutes 12
        1. Derivatization 12
    E. Qualitative and Quantitative Evaluation 13
    F. Sensitivity and Reproducibility in TLC 14
    G. TLC as a Pilot Method for HPLC 15
    H. Determination of Molecular Parameters with Adsorptive and Reversed-Phase TLC 15

References 16

I. INTRODUCTION

Thin-layer chromatography (TLC) was developed more than 30 years ago for the separation and semi-quantitative determination of the individual components of more or less complicated mixtures. In the last decade, the application of various TLC methods for the separation and quantitative determination of a wide variety of organic and inorganic substances has considerably increased. This increase is probably due to the improved instrumentation and automation of the various steps of TLC analysis (gradient and forced flow methods, centrifugal development, circular rotation planar chromatography, high-pressure planar liquid chromatography, densitometry, among others. In addition, coupled spectroscopic methods (TLC-UV-VIS, TLC-MS, and TLC-FTIR) have been developed, considerably enhancing the reproducibility of TLC. TLC methods have been successfully used in many fields of research and development such as in clinical medicine,[1] forensic chemistry, biochemistry, pharmaceutical analysis for the estimation of impurity profiles of drugs and related materials,[2] drug screening, toxicology, environmental pollution studies,[3,4] as well as – to a lesser extent – in cosmetology, foodstuff analysis,[5] in the analysis of metal ions,[6] explosives and their biodegradation products,[7] and in petroleum analysis.[8]

The widespread application of TLC is also due to the fact that it is a simple and rapid analytical procedure; it is extremely flexible because a great variety of mobile and stationary phases and detection reagents can be tested and used for the solution of any separation problem.

The general aspects and practical applications of TLC have been discussed in detail in excellent books[9] and in reviews[10] dealing with the various fields of application such as the analysis of inorganic pollutants[11] and pesticide residues.[12] The new advances in the theoretical and practical applications of various TLC methods are presented and discussed every 2 years in *Analytical Chemistry*.[13,14]

The objectives of this chapter are to give a general overview of the theory and practice of TLC and to offer a brief summary of the new advances in this rapidly developing field of chromatographic separation science for researchers and serious students interested in any of the topics covered.

II. THEORETICAL ASPECTS OF THIN-LAYER CHROMATOGRAPHY

TLC separation includes the interaction of the compounds to be separated (solutes) with both a stationary and a mobile phase.

A. STATIONARY PHASES

Stationary phases are generally chemically well-defined inorganic (sometimes organic) materials with porous structures and with relatively high specific surface areas. TLC plates are prepared from these materials by binding them to a support (glass, aluminium, plastic foil) with the help of various organic (polyvinylalcohols with various molecular masses) or inorganic (gypsum) binders. The presence of binders modifies, only to a small extent, the retention capacity and selectivity of the original stationary phase. The performance of a TLC sorbent depends considerably on its specific surface area and pore volume, the mean pore diameter and the pore size distribution, and the particle size and particle size distribution. Smaller particle size and narrower particle size distribution enhances separation efficiency and improves resolution, decreases analysis time, and increases detection sensitivity. The average particle size of the traditional TLC sorbents is between 10 and 50 µm, with a fairly wide size distribution. High performance TLC (HPTLC) sorbents have an average particle size of about 5 µm, with a narrower particle size distribution. It has been proven many times that the performance of HPTLC plates is superior to that of normal TLC plates. However, in the analysis of dye components contained in hair color formulas the performance of HPTLC layers sometimes was inferior to that of normal TLC layers.[15]

1. Silica and Surface Modified Silicas

Many sorbents have been tested for TLC applications; however, silicas with various surface characteristics and silicas with covalently bonded organic ligands (amino, diol, cyano, chiral phases, silanized silica, C_2-, C_8-, and C_{18}-alkyl bonded silicas) on the surface are used the most frequently for TLC separations. A wide variety of TLC and HPTLC ready-made plates are available, facilitating the solution of many separation problems. The polarity order of the ready-made TLC plates based on silica is approximately silica > amino silica > cyano silica > octadecyl silica. It must be noted that the polarity order listed above is only of limited value, depending considerably on the chemical character of the solutes and the composition of the eluent system. The stability and wetting of the old octadecylsilica plates were insufficient in eluents containing a considerable ratio of water. The mechanical stability of the layers can be increased by adding sodium chloride or any other neutral salt to the eluent; however, the salts can influence the retention of solutes (salting-in or salting-out effects), sometimes reducing separation efficiency. The effect of salt is higher when the solute has one or more dissociable polar substructures. A new generation of RP-18 plates overcomes this difficulty; they can be used in aqueous systems, conserving their mechanical stability and providing good mobility, even for water. These plates are generally marked as "W."

The medium-polar amino, diol, and cyano layers can be used both in the adsorption and reversed-phase separation mode, depending on the composition (polarity) of the mobile phase. Although these supports exhibit good separation capacitiy, they are not frequently used in practice. Diol layers have been used for the enantiomeric separation of metoprolol, propanolol, and alprenolol using N-benzyloxycarbonylglycil-L-proline as chiral selector,[16] as well as for the separation of glycine, sulfate, and glucuronide conjugates.[17] Amino-bonded layers have been used for the detection of catecholamine in biological materials,[18] for the separation of all substrates of the 7,8-dihydropteroate synthase in the reversed-phase mode,[19] for the quantitation of 5-methylcytosine,[20] and for the separation of some steroid hormones.[21] Silanized silica plates were used for quantitative analysis of quaternary ammonium antiseptics.[22] Until recently, chiral plates (reversed-phase silica modified with Cu^{2+} and a chiral agent) have found only limited application for enantiomer separation. The overwhelming majority of chiral separations have been carried out by adding various chiral selectors to the eluent. Silica plates with concentrating zones allow the application of a relatively high volume of a dilute sample. Their application is preferable when the concentration of the solute or solutes in the sample cannot be enhanced by other methods.

2. Other TLC Sorbents

Other inorganic and organic sorbents such as alumina (neutral and acidic), magnesium silicate, diatomaceous earth (kieselguhr), celluloses and cellulose derivatives, and polyamides have found only limited application in TLC, although their separation capacities differ from those of traditional adsorption and reversed-phase layers based on silica. Alumina layers have been used for the separation of nonylphenyl ethylene oxide oligomers using acetonitrile–chloroform mixtures of various compositions as eluents,[23] and for the separation of a new type of plant regulator, jasmonates.[24] Cellulose layers have been used recently for the separation of some organomercurial antiseptics in 1M NaCl as eluent.[25] An excellent separation of alpha-carotene was achieved on magnesium oxide layers. The pigments were extracted with acetone, then transferred into petroleum ether by adding saturated NaCl solution.[26] Phenyldimethylsiloxane-treated high performance thin-layer chromatographic plates were used for the separation of flavonoids (baicalein, baicalin, wogonin, and oroxylin-A) in *Scutellaria radix*. It was found that the separation was better than on C_{18} plates.[27] Zirconia is being used more and more frequently in HPLC; however, until now it has found only limited applications in TLC.[28] Water-insoluble β-cyclodextrin polymer beads have been used as TLC sorbent.[29] It has been established that this sorbent shows retention characteristics deviating from those of traditional adsorptive and reversed-phase sorbents. The retention of solutes on this sorbent mainly depended on their steric parameters.

The separation efficiency of the various organic and inorganic sorbents has been frequently compared. Neutral aluminum oxide, silica, diatomaceous earth, silica–diatomaceous earth 1:1 (w/w), cellulose, polyamide, cyano, diol, and amino layers were tested in the adsorption separation mode for the separation of the color pigment of paprika (*Capsicum annum*) powders. For reversed-phase chromatography, polyamide, cyano, diol, and amino plates were used as received; the others were impregnated by overnight predevelopment in *n*-hexane–paraffin oil, (95:5). It was established that the best separations can be achieved with eluents comprising mixtures of tetrahydrofuran, acetone, and water, or acetone and water on layers of impregnated diatomaceous earth.[30] The separation capacity of the various sorbents depended considerably on the character of the solutes to be separated.[31] Thus, polyacrylnitrile showed a better separation of mixed minocarboxylato cobalt(III) complexes than did silica and cellulose,[32] whereas the separation of some azolidines was better on cellulose than on silica and C_{18} layers.[33] The separation capacity of cellulose was higher than that of silica for the diastereoisomers of the ethyl ester of caffeic, ferulic, and isoferulic acids.[34]

The efficiency of separation can be enhanced by mixing two sorbents. This effect was exploited for the determination of Mg in Al alloys using a mixture of cellulose and a strong acid cation exchanger in H^+ form.[35]

3. Impregnation of Sorbents

When the separation cannot be successfully carried out on any of the sorbents commercially available, or the aim of the chromatographer is the study of various molecular interactions between the solutes and other compounds adsorbed on the surface of the sorbent, the sorbents have to be modified by various additives that are not miscible with the mobile phase. These additives can be adsorbed on the support, either by predevelopment of the plates with a solution of the additive or by immersing the plate into the solution of the modifying agent. Predevelopment requires more time than the immersion; however, its application is highly recommended because the covering of the original sorbent by the modifying agent is more uniform in this instance. Impregnated sorbents (mainly silica) have been frequently used in TLC both for theoretical studies and practical separations. An efficient separation of isomeric triacetylglycerols was achieved by adding silver ion to the sorbent.[36] Silica layers impregnated with tributylamine or with tributylphosphate[37] have been successfully used for the separation and quantitative determination of Pb^{2+} from Hg_2^{2+}, $Hg_,^{2+}$ $Tl,^+$ $Bi,^{3+}$ $Sn,^{4+}$ and $Sb.^{3+}$ Silica plates impregnated with tricaprylammonium chloride were used for the separation of various bioactive compounds such as minoxidil and its intermediates,[38] natural and synthetic penicillins,[39] barbituric acid derivatives,[40] tetracyclines,[41] aromatic-N-glycosides,[42] benzodiazepine derivatives,[43] and 2,3-dihydro-1,5-benzodiazepines.[44] The presence of salts and the pH of the eluent did not influence the retention on the silica plates impregnated with tricaprylmethyl–ammonium chloride, indicating that the interaction between the surface of the TLC layer and the solutes is of a hydrophobic character. However, the retention order was different on layers impregnated with paraffin oil or with tricaprylammonium chloride, suggesting that the mechanism of retention may be different on both layers. Fast atom bombardment mass spectrometry and silica plates impregnated with 3-glycidooxipropyl were used for the determination of midazolam in blood.[45] Thorium nitrate-impregnated silica layers were used for the analysis of toxic metals in seawater and industrial wastewater.[46] The complex formation between humic acids and Fe(III) was studied by mixing Fe(III) into the silica stationary phase and by comparing the retention behavior of humic acids on the plain silica and Fe(III)-impregnated silica layers using tap water as eluent.[47]

TLC layers can also be modified by mixing the additive in the eluent used for the separation of the solute molecule (dynamic modification). Many enantiomeric separations have been carried out in this manner by adding a chiral discriminator to the eluent. Thus, dansylated amino acids were separated by using RP-18W/UV_{258} plates (Macherey-Nagel, Dürren, Germany) and adding bovine serum albumin to the eluent.[48,49] The enantiomers of dansylated amino acids were also separated by using the inclusion forming agent β-cyclodextrin as a mobile phase additive.[50] Trp, methyl Trps, and fluoro Trps enantiomers have been successfully separated on a cellulose layer with the aqueous solution of alpha-cyclodextrin.[51] Separation improved at higher temperature.[52] The macrocyclic antibiotic vancomycin has also been successfully used for the enantiomeric separation of derivatized amino acids and racemic drugs in the reversed-phase separation mode on diphenyl-F plates.[53]

B. MOBILE PHASES

The application of mobile phases in high performance liquid chromatography (HPLC) is hampered by the fact that the detection is generally carried out in the presence of the mobile phase. Solvents with UV absorbance cannot be used for the UV detection of solutes. One of the main advantages of TLC is that the mobile phase is evaporated before detection, and it does not interfere with the determination of the position of solute spots. This fact increases enormously the choice of solvents to be used in TLC as the mobile phase. The elution capacity (solvent strength) of solvents is defined by their ability to move a given solute or a given series of solutes. The elution strength is higher when the mobility of solute is higher. However, the elution strength depends both on the sorbent and on the chemical character of solutes used for its determination. The order of elution strength on silica is n-heptane > n-hexane > n-pentane > cyclohexane > carbontetrachloride > toluene > dichloromethane > diisopropyl ether > tertiary butanol > diethyl ether > nitromethane > acetonitrile > 1-butanol > 2-propanol > ethyl acetate > panol > acetone > ethanol > dioxane > tetrahydrofuran

> methanol > pyridine > water. The elution order is nearly reversed for reversed-phase TLC. Solvents with low boiling point, low viscosity, and low toxicity are preferable for TLC application. Low boiling point facilitates the evaporation of the mobile phase from the surface of the layer, whereas the use of highly viscous mobile phases results in increased development time and more diffuse spot shape. The uncontrolled adsorption of water or any other organic (solvent vapors) and inorganic (acids and bases) compounds markedly modifies the retention capacity and selectivity of layers. It is important to avoid the contamination of layers with any gases or, when it is favorable, saturate the layer with the gas phase deliberately. Demixing of multicomponent mobile phases may result in multiple mobile phase fronts. When each of them is past the solute spots, their importance is negligible; however, the spots situated between the various mobile phase fronts are generally distorted. Many efforts have been devoted to the optimization of the mobile phase composition.[54] The efficiency of the optimization of mobile phase composition for stepwise gradient HPTLC can be enhanced by computer-assisted methods using the R_f values as selection criterion.[55] Another computer-assisted optimization method was developed for stepwise gradient and multiple development TLC.[56,57] The optimization methods used in TLC have been recently compared and critically evaluated.[58]

C. RETENTION PARAMETERS IN TLC

The position of any solute spot in TLC is characterized by the retention factor R_f. It is equal to the distance between the center of spot (z_s) and the start line divided by the distance of the mobile phase front from the start line (z_f):

$$R_f = z_s/z_f \tag{1.1}$$

The value of the retention factor is between 0 (solute remains on the start) and 1 (solutes move with the front of the mobile phase). To increase the reproducibility of the determination of the solute retention, the R_f value is sometimes related to the R_f value of standard solute.

As there is a logarithmic relationship between the R_f value of solutes and the concentration of the stronger component in the eluent mixture (C), the R_M value was introduced to describe the correlation between the mobility of solutes and the composition of the mobile phase:

$$R_M = \log(1/R_f - 1) \tag{1.2}$$

$$R_M = R_{M0} + b \cdot C \tag{1.3}$$

where R_{M0} is the hypothetical R_M value of a solute extrapolated to zero concentration of the stronger component in the eluent.

The velocity of the development can be characterized by the rate coefficient (k):

$$k = (z_f)^2/t \tag{1.4}$$

where t is the development time. The capillary forces between the pores of the sorbent particles and the molecules of the mobile phase and the surface tension and the viscosity of the mobile phase have a considerable impact on the k value. Various TLC scanners have been developed for the exact determination of spot characteristics (spot position, peak area, and peak height) that are related to the quality and quantity of compound in the spot. The peak area and peak height of a densitometer scan are defined similarly to the peak area and peak height in HPLC. Peak height is the distance between the peak maximum and the baseline whereas peak area is the area of the curve between the start and the end of the peak and the baseline.

Resolution (R_s) characterizing the goodness of the separation of two neighboring spots on the TLC plate can be calculated as in HPLC:

$$R_s = 2 \cdot (z_2 - z_1)/(w_1 + w_2) \tag{1.5}$$

where z_2 and z_1 are the distances of the centers of spots 1 and 2 from the start, and w_1 and w_2 are the spot widths. Resolution is considered to be acceptable when it is higher than or equal to 1 (baseline separation = 1.5). Equation 1.5 indicates that resolution can be enhanced either by increasing the distance between the two spots (increasing selectivity) or by reducing peak width (enhanced efficiency).

D. RETENTION MECHANISM

The role of various hydrophobic and hydrophilic (electrostatic) forces in the chromatographic separation has been vigorously discussed. It is generally accepted that the three main physicochemical processes — adsorption, partition, and ion exchange — play the most important role in the determination of retention capacity and selectivity. Adsorption takes place between the polar substructures of solutes and the various adsorptive centers of the sorbent. Solutes with high adsorption capacity bind more strongly to the sorbent, resulting in enhanced retention, whereas solutes with lower adsorption strength elute more easily. The character of adsorptive forces involved in the solute–sorbent interaction is not well elucidated. It is probable that the solute substructures with permanent dipole moments, with inducible dipole moments, and with capacity to form hydrogen bonds, may equally influence the interaction. Partition governed by molecular hydrophobicity occurs between the solute molecules and the mobile phase. It is supposed that solutes with higher preference for the mobile phase (related to their higher solubility) elute before the compounds with lower solubility in the eluent. Ion exchange refers to the hydrophilic (electrostatic) forces occurring between the dissociable polar parts of the solute molecules and the ionic centers of the sorbent surface.

Separation mechanisms can be divided in two groups according to the relative polarity of the stationary and mobile phases. In the case of adsorptive (direct or normal phase) TLC, the sorbent is polar (generally silica or alumina), whereas the mobile phase is relatively apolar (various organic solvents). Reversed-phase TLC (RP-TLC) is characterized by an apolar stationary phase (most frequently silica sorbent with covalently bonded hydrocarbons on the surface) and a polar mobile phase (water or water and organic solvent miscible with water). Modified silicas with cyano, diol, and amino groups on the surface can be used both in adsorption and reversed-phase separation modes depending on the polarity of the mobile phase. They offer a wide range of solute-sorbent interactions, including ion exchange, adsorption, and hydrophobic interactions.

III. PRACTICAL CONSIDERATIONS

A. SAMPLE PREPARATION

TLC is a suitable chromatographic method for the separation and quantitative determination of solutes that are present in a low quantity in complicated organic and/or inorganic matrices. The presence of a considerable quantity of impurities (e.g., mono-, oligo-, and polymeric carbohydrates; peptides; proteins; and inorganic salts) deteriorates the efficiency of the TLC separation, resulting in modified retention and distorted spot shapes. As TLC plates are disposable, the sample preparation is generally less demanding than in the case of HPLC, and this preparation step can be easily combined with prechromatographic derivatization. Two main methods have been developed for the prepurification of samples: liquid-liquid and liquid-solid (solid-phase) extractions. In liquid-liquid extraction, two immiscible solvents are used. The two solvents have to be selected so that the solute or solutes have the highest possible solubility in one solvent, whereas the compounds of the accompanying matrix preferably have high solubility in the other solvent. In the ideal case, complete separation of the solutes and impurities can be achieved. This ideal condition is practically never achieved in practice. The choice of the solvents depends on the type of solutes and impurities and on the expertise of the chromatographer. The polar phase is generally water or a relatively polar organic solvent (e.g., methanol), whereas the apolar phase is normal- or cycloalkanes, diethyl- or diisopropyl ethers, etc. The principles and practices of solvent extraction have been recently discussed in detail.[59] Generally, sample purification by liquid-liquid extraction is very rudimentary. Fortunately, TLC methods do not require the same sample purity as most HPLC methods. One of the advantages of TLC is that impurities moving with the mobile phase front or remaining at the start do not affect the end result of the analysis. This simple purification method has been used frequently and successfully in many different TLC analyses.

The character of the extraction agent strongly depends on the solutes to be extracted and on the accompanying impurity matrix. Thus, flavonoids and flavonoid glycosides were extracted from the crushed leaves of *Olea europea* L. with 80% methanol, but the beflavonoids were extracted with 70% ethanol.[60] The bioactive components of *Erigeron canadensis* and *Matricaria chamomilla* were extracted with ethyl acetate.[61] Not only the character of the solutes to be extracted, but also the type of the accompanying matrix, influences the selection of the extracting agent. Vanillin and related flavor compounds were extracted from chocolates, vanilla pudding, and ice cream with ethanol, from soft drinks and vanilla-flavored ground coffee with ethyl acetate, and from chocolate milk with chloroform.[62] *Digitalis lanata* and *Digitalis purpurea* were extracted with ethanol followed by chloroform.[63] For the determination of granulocyte-specific glycosphingolipids in preterm labor amniotic fluid, the samples were lyophilized and then extracted with chloroform–methanol–water (4:8:3).[64] For the assessment of the content of vitamin K in bovine liver, samples were extracted with hexane–chloroform (95:5) and 95% ethanol.[65] Oxidized derivatives of pheophorbide a and b were extracted with *N,N*-dimethylformamide from olives, and they were separated on both silica plates impregnated with maize oil in light petroleum and C_{18} layers.[66] Limonoids and limonoid glucosides were extracted from citrus seeds with acetone followed by methanol.[67]

Solid-phase extraction uses a polar (e.g., silica, Florisil) or apolar (modified silicas with covalently bonded alkyl chains on the surface) sorbent filled in a column. Sample is added to the column and the sample components are washed into the column and adsorbed on the sorbent surface. The solute or solutes can be eluted from the column with an appropriate solvent or solvent mixture before or after the impurities. The choice of the sorbent and solvents depends entirely on the chemical character of the solutes and the constituents of the accompanying matrix. A considerably higher sample purity can be achieved with the use of the solid-phase extraction method compared to conventional extraction; however, it is generally more time consuming and this degree of sample purity is not always required in TLC. Solid phase extraction was used for the determination of diflubenzuron residues in water.[68]

In some cases it is possible to apply the samples directly to the layer without any pretreatment. This procedure was used for the determination of organic acid preservatives (sorbic acid, benzoic acid, and dehydroacetic acid) from beverages on silica HPTLC plates. The samples were directly spotted onto the layers without extraction steps. The coefficient of variation was 2 to 5%.[69]

B. SAMPLE APPLICATION

The method used for the application of samples onto TLC plates depends on the purpose of the analysis (analytical or preparative) and on the volume of the sample and on the character of the sample solvent (organic or aqueous). The sample volume is frequently determined by the detection limit of the solute. It is necessary to use high sample volumes when the concentration of the solute to be determined in the sample is very low. Narrow-band sample application results in the highest possible resolution, which is equally important for qualitative and quantitative TLC. The application of samples liable for rapid oxidation has to be carried out in a nitrogen atmosphere. Samples can be applied onto the TLC layers either in spots or in bands. It is generally accepted that the application of bands results in better separation; however, it requires more layer surface and more application time. The small spot diameter also increases the efficiency of TLC separation; 3 mm in traditional TLC and 1 mm in HPTLC is the maximum acceptable diameter. Sample application is a very important step in all TLC separations, especially in quantitative analysis. The influence of the distance between the eluent entry and the position of the starting zone has a considerable influence on the theoretical plate height. The optimum distance is 6 and 10 mm for HPTLC and TLC plates, respectively.[70]

Capillaries and loops were originally used for the application of samples. In these instances the sample volume was not exactly defined, and the spot diameter was highly dependent on the expertise of the chromatographer. Syringes suitable for the exact application of µL and nL volumes provided more precise volumes and smaller spot diameters. In each instance it is very important not to damage the layer with the application device. It is also important to wash the application device thoroughly between the various samples and standards to avoid cross-contamination. For the application of large sample volumes, the use of syringes with pistons is preferable to avoid the possible inaccuracies caused by the high density (chloroform), volatility (diethyl ether), or high surface tension (water) of the solvent. Automated application devices are able to apply various sample volumes in a predefined order onto the plates without direct contact between the sorbent layer and the syringe. In some devices the sample is transferred onto the plate by means of a nitrogen or air stream. Many application systems are in the market such as Applicator for TLC (Merck, Dürren, Germany), Nanomat III, Micro and Nano-Applicator, Automatic TLC Sampler III, Linomat IV (each of them CAMAG, Muttenz, Switzerland), PS 01 and TLC Applicator AS 30 (DESAGA, Heidelberg, Germany), and TLS 100 (BARON, Insel Reichenau, Germany). As it has been previously emphasized, the application of spots as small as possible considerably enhances the efficiency of the TLC separation. This aim can be achieved by applying 1 to 5 µL sample volumes in TLC and 50 to 500 nL in HPTLC. However, when the concentration of solute is very low in the sample, a markedly higher sample volume has to be used. In these instances the use of precoated plates with concentrating (preadsorbent) zones is highly recommended. Concentrating zones consist of a bottom, adjacent layer with low retention capacity (diatomaceous earth). A higher sample volume can be applied onto the concentrating zone without affecting the separation power. First, the solutes in the sample move together with the mobile phase front and the separation begins when the front reaches the adsorption layer of the plate. This easy to perform and elegant method eliminates the problems caused by the application of large sample volumes. A traditional method of the reduction of spot diameter is the predevelopment of the plates with an extremely strong mobile phase up until the upper part of the starting zone. The spot becomes a thin band at the front. This method produces good results; however, it is time consuming and the strong mobile phase has to be completely removed before the TLC separation. The spot diameter can also be decreased by the appropriate choice of sample solvent, apolar ones (preferably *n*-hexane) for adsorption TLC and polar ones (water or methanol) for RP-TLC. The sample volume can also be increased by portionwise application with intermediate drying. It is preferable that the spots can be separated from each other by 15 to 20 mm on TLC and by 5 mm on HPTLC plates. Spots can be applied onto the plates about 10 or 15 mm above the lower edge in TLC and HPTLC, respectively. The usual distance of development is 10 to 15 cm for TLC and 7 cm for HPTLC plates. It is recommended to apply two parallel samples from each standard and sample on the two halves of the same plate to decrease the systematic errors caused by the nonuniformity of sorbent layers (data pair method). After application and drying of the samples, the plates are developed.

C. DEVELOPMENT AND DEVELOPING CHAMBERS

Development can be carried out in linear ascending, linear horizontal, circular, and anticircular modes. Each developing method has advantages and disadvantages; the choice of the appropriate developing system depends on the separation problem to be solved.

1. Linear Ascending Development

The linear ascending developing method is the most frequently used in TLC practice. The mobile phase is added to a tank, and the lower part of the plate is immersed into the mobile phase. Immersion of the initial zones in the mobile phase must be avoided. The mobile phase rises with the help of capillary forces. When the developing distance is reached, the plate is taken out of the tank, the front of the mobile phase is marked for the calculation of retention parameters, and the plate is dried at room temperature or in an electric oven (one-dimensional development). Separation efficiency can be considerably enhanced by using multiple development with various mobile phase systems and drying the plates between each development. In the case of two-dimensional development, the sample is applied at the corner of the plate and developed in one direction. Then the plate is turned 90° and developed again in the same, or in a different, mobile phase. The advantage of the two-dimensional development is higher separation efficiency; however, it is not really suitable for quantitative determination because the standard has to be separated on a different plate, which markedly decreases the reproducibilty and accuracy of the method. Moreover, two-dimensional development is a fairly time consuming process.

It is generally accepted and it has been indicated many times that the gas phase present in the traditional, wide TLC chamber may have a considerable influence on the retention capacity of the stationary phase. The development of TLC plates can be carried out in the presence or in the absence of a gas phase in equilibrium with the mobile phase. Wide chambers having large gas space (normal chambers) allow gas-liquid equilibration. Saturation of the gas phase can be accelerated by lining the chamber with clean, dry filter paper, and by wetting it with the mobile phase. Saturation is a delicate procedure. The temperature of the chamber has to be held as constant as possible to avoid modification of the composition of the gas phase, which can modify the separation capacity and efficiency of the TLC plates. The chambers have to be closed thoroughly and have to be put in places with negligible or zero air movement to ensure the stability of the equilibrated chromatographic system. Normal chambers can also be used without equilibration. Sometimes the separation is better in unsaturated normal chambers than in the saturated ones. This phenomenon was tentatively explained by the supposition that the continous evaporation of the mobile phase into the gas space accelerates the movement of the eluent front, decreasing the time for lateral diffusion of the solutes. The importance of the saturated or unsaturated development is negligible in RP-TLC because the separation is based on the partition of the solutes between the polar mobile and the apolar stationary phases, and the retention capacity of the adsorption centers on the surface of the original sorbent do not have a considerable impact on the separation.

Sandwich chambers have a very low volume and, accordingly, very small vapor space. In these systems, equilibration is not necessary, and development can begin immediately after pouring the mobile phase into the chamber. Sandwich development can be carried out in normal chambers too, placing a glass plate on top of the TLC plate in close proximity to the layer. However, the evaporation of the mobile phase is higher at the edges in this system leading to distorted zone shape. It is also possible to expose the TLC plates to vapors different from the components of the mobile phase (i.e., volatile acids or bases). The twin-trough chamber (CAMAG) contains two separate troughs at the bottom; one of them can contain the solvent designed for the modification of the gas phase and the second one contains the mobile phase.

Many efforts have been devoted to the automation of the process of development. The Automatic Development Chamber (CAMAG) increases the reproducibility of the chromatographic process by controlling plate preconditioning, chamber dimensions (tank or sandwich configuration), development distance, movement of the mobile phase, and completion of the development at any predetermined time or development distance. In the case of the CAMAG Automated Development Chamber (Figure 1.1), the gas washing bottle and the eluent reservoir have to be filled. The separation process begins with the preconditioning of the TLC plate (valve for preconditioning opens and closes after a predetermined time). Counterplate (5) moves to establish development (normal or sandwich) configuration. The eluent valve opens and eluent enters the trough (10). The movement of the eluent front is controlled and stops at a selected migration distance. The remaining eluent flows into the waste bottle. The plate can be dried by heated air circulated by the fan.

The Automated Multiple Development System (CAMAG) has been designed for the automatic application of the stepwise gradient technique. This method is especially suitable for the separation of solutes with a wide range of retention characteristics on one plate. The plate is first developed for a short distance with the mobile phase having the highest elution capacity. In this instance the solutes with the strongest adsorption capacity to the sorbent are separated. Then the mobile phase is entirely removed in vacuum, and a new mobile phase with a lower elution strength is used for a longer development. Compounds with lower adsorption capacity are separated in this step. This procedure can be repeated up to 25 times with mobile phases of decreasing elution order over

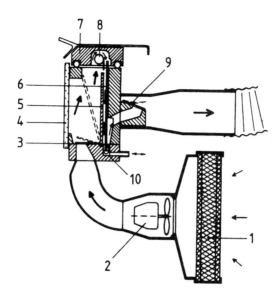

FIGURE 1.1 Schematic of the Automatic Development Chamber (CAMAG). 1, Active carbon filter; 2, blower; 3, air flap; 4, glass window; 5, counterplate; 6, TLC plate; 7, chamber lid; 8, heating coil; 9, air flap; 10, solvent trough. (Reproduced with permission from CAMAG.)

longer distances. The plate can be conditioned between each separation step. The flow diagram of the Automated Multiple Development System is shown in Figure 1.2. This system has been successfully used for the separation and quantitative determination of phosphatidylethanolamine and phosphatidylserine in liposomes. Phospholipids become fluorescent in contact with o-phtalaldehyde.[71] Multiple development was also used for the analysis of various plant extracts.[72]

The Short Bed Continous Development system (SB/CD chamber, Regis Technologies Inc., Morton Grove, IL) is especially suitable for the separation of solutes with high retention capacity. The upper part of the plate projects from the chamber. The mobile phase evaporates upon leaving the chamber and it is renewed continuously.

2. Linear Horizontal Development

Horizontal chambers provide a shorter development time than the traditional ascending developing systems because the effect of gravity does not adversely influence the movement of the mobile phase. The plates are situated horizontally in the chambers, and the transport of mobile phase is ensured either by a glass frit strip (H-separation chamber, DESAGA) or by a capillary split (linear chamber, CAMAG). The horizontal developing chamber of CAMAG consists of PTFE resistent to a wide variety of eluent mixtures (Figure 1.3). The HPTLC plate (1) is placed on the supports with layer downward. When sandwich configuration is required, the plate is covered by a counterplate (2). For normal development the counterplate is removed and the chamber under the plate is filled with the conditioning eluent when it is necessary. The troughs (3) are filled with the eluent, and the development begins by tilting the glass strips (4) to a vertical position. A capillary slit forms between the wall of the trough and the glass strip, and the eluent moves upward and enters the surface of the HPTLC plate. The chamber has to be covered by a glass plate (5) during equilibration and development.

The Mobile-R_f chamber makes possible the use of various mobile phases on the same plates. This method is suitable for the separation of solutes present in low quantities in complicated accompanying matrices because the impurities can be removed by using a different mobile phase. Multipurpose chambers such as the Vario chamber (CAMAG) make possible the selection of the optimal mobile phase composition as well as being suitable for different conditions of layer conditioning. Overpressure layer chromatographic equipment using a forced flow technique offers

Introduction to Techniques and Instrumentation

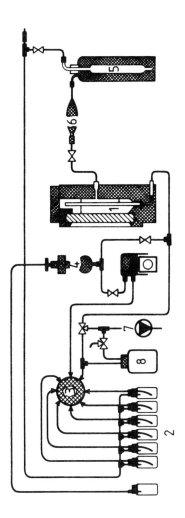

FIGURE 1.2 Flow diagram of the Automated Multiple Development System. 1, Developing chamber; 2, eluent reservoir bottles; 3, 7-port valve; 4, gradient mixer; 5, wash bottle for the preparation of the gas phase; 6, gas phase reservoir; 7, vacuum pump; 8, waste collection bottle. (Reproduced with permission from CAMAG.)

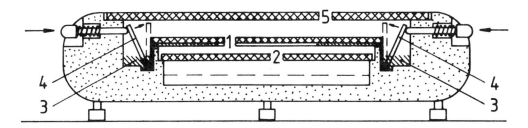

FIGURE 1.3 Schematic representation of a horizontal developing chamber. 1, HPTLC plate; 2, counterplate used only for sandwich development; 3, troughs for the eluent; 4, glass strip to regulate eluent flow; 5, covering glass plate. (Reproduced with permission from CAMAG.)

various possibilities to speed up the separation procedure (Biotech Instruments Ltd., Luton, Bedfordshire, UK; The Munhall Co., Worthington, OH). The gas space is entirely eliminated, and a pump ensures the even speed of the movement of the mobile phase. These methods are theoretically more rapid than any other TLC developing methods based on the use of capillary forces to move the eluent along the plate. However, the preparation of plates and the setup of the instrument frequently decrease or entirely eliminate the gain in time. These methods can be successfully used with highly viscous mobile phases when the normal methods of development require a lot of time or cannot be carried out.

3. Circular Development Methods

These methods transfer the mobile phase to the center of a circular TLC plate and the development occurs out of the center of the plate. The U-chamber (CAMAG) is suitable only for the development of HPTLC plates. Samples can be spotted either onto the dry layer or onto the layer under eluent flow. Constant flow rate of the mobile phase is ensured by a syringe used for its delivery. It is possible to apply a maximum of 12 samples onto the layer. The solutes move outward on the plate as concentric segments. The Selecta Sol™ chamber (Schleicher & Schüll, Dassel, Germany) has been designed for the application of various mobile phases. The mobile phases enter through holes connected to reservoirs of the different mobile phases. Forced flow techniques also have been used to carry out circular development. These instruments exploit centrifugal force to make the mobile phase move through the TLC layers (Chromatotron, Harrison Research Inc., Palo Alto, CA; Rotachrom, Petazon). The use of higher pressures considerably decreases the development time and results in more compact spots, facilitating quantitative analysis (High Pressure Planar Liquid Chromatography, Gerstel).

4. Anticircular Development

In this technique the samples are applied on the outer sphere of the plate, and the mobile phase is transported from a peripheral circle toward the center of the plate. The analysis time is fairly short because the velocity of the mobile phase increases during development. It has been stated that this technique is most suitable for the separation of solutes showing low retention.

D. DETECTION OF SOLUTES

After finishing the chromatographic development, the plate is removed from the chamber and the components of the mobile phase are evaporated as soon as possible in a well-ventilated area with ambient air or hot air. In the case of volatile solutes, removal of the mobile phase has to be carried out in such a manner that each solute will remain without loss on the plate. In some instances, oxidation of the solutes has to be prevented by using nitrogen flow instead of air. Some automated developing systems evaporate the mobile phase under vacuum.

Colored substances can be visually detected on the plates. However, the majority of solutes do not absorb in the visible part of the spectrum, and these solutes have to be detected by other means. Many organic compounds contain one or more conjugated electron systems that absorb light in the ultraviolet (UV) part of the spectrum. These solutes can be easily detected on plates containing one or more fluorescent indicators. These indicators can be excited by UV light (usually 254 nm) and they emit visible radiations. Solutes with absorbing capacity in the UV region absorb the radiation and are detected as dark spots on a green or pale blue background (fluorescence quenching). Some solutes have the capacity to emit fluorescent light when excited with UV light. The detection of these solutes is easier on nonfluorescent layers. Commercial UV lamps emitting about 365 nm can often be successfully used for the detection. As the number of fluorescent compounds is relatively low, this method offers a unique possibility for the sensitive detection of solutes present in complicated nonfluorescent matrices. Impregnating plates containing fluorescent solutes with paraffin oil or with polyethylene glycol increases the stability of fluorescence. An interesting possibility for the documentation and evaluation of TLC results is the image analysis of photochemically derivatized solutes.[73]

1. Derivatization

Solutes that do not absorb UV light and are not colored have to be visualized or detected with an appropriate reagent after development (postchromatographic derivatization). One of the main advantages of TLC is that a huge number of different reagents can be used to facilitate the identification of unknown compounds because the color of the reaction product contains some information about the chemical character of the solute. However, many solutes can be derivatized before development (prechromatographic derivatization) by introducing into the molecule a chromophore or UV absorbing group. Prechromatographic derivatization may improve the detection limit of the solute, and may increase the stability of the original molecule during the development. Thus, dansylation has been used for the separation and quantitative determination of histamine released from rat serosal mast cells.[74]

Aliphatic thiols of biological (cysteine, N-acetylcysteine, homo-cysteine, captopril, glutathione, mercaptopropionylglycine) and other (thioglycolic acid, monothiolglycerol, ammonium thiolacetate) types of interest were separated on silica plates as fluorescent methyl-4-(6-methoxynaphtalene-2-yl)-4-oxo-butenoate derivatives. Derivatization was carried out for 10 minutes at room temperature in an aqueous sodium acetate solution (5%).[75] Trifluoroacetic anhydride sodium iodide reagent was used for selective detection and differentiation of nitrones, nitroxide radicals, and nitrosoamines in mixtures,[76] and 2-trichloromethylbenzimidazole proved to be a selective chromogenic reagent for the detection of some azoles.[77] Diphenyltin chloride as a chromogenic reagent can be used for the detection of flavonoids on silica, polyamide, and cellulose plates.[78] The performance of various derivatization procedures has been frequently compared. It has been established that dansyl chloride was a better derivatizing agent than dabsyl chloride and 7-chloro-4-nitrobenzaxole for the derivatization of some biogenic amines.[79]

Optimal detection procedures require the even distribution of the reagent spray on the surface of the layer in drops as small as possible. Pump systems, as well as various spraying equipment working with compressed air or nitrogen, can be commercially purchased (Aldrich Chemical Co. Inc., Milwaukee, WI; Fisons Scientific Equipment, Loughborough, Leicestershire, UK). Some detection reagents that are frequently used in TLC are commercially available; however, in many instances the preparation of special reagents will remain the task of the chromatographer. Spraying of plates has to be carried out under ventilation. Dipping of plates in a reagent solution results in a more homogeneous distribution of the reagent on the surface of the layer, increasing the reproducibility of the analysis and minimizing the influence of errors inherent in manual spraying. Special immersing chambers have been developed to facilitate this procedure (DC Tauchfix, Lothar Baron, Laborgeräte, Insel Reichenau, Germany). Five seconds is generally enough time to saturate the plate with the detection reagent. An even distribution of the reagent can be achieved by adding it to the mobile phase. This procedure can be used only for reagents that move together with the eluent front; however, the application of this method requires special caution because the reagent may modify the retention strength and selectivity of the TLC system. The heating of plates (5–10 minutes at 100–120°C) is generally necessary for successful derivatization. The use of ventilated electric ovens is highly recommended because the heat distribution of hot plates and IR lamps is generally not even, leading to different degrees of derivatization (consequently different detection sensitivity) on different parts of the plate. Chromatograms can be photographed both in the UV and visible light for documentation (documentation systems are available from Analtech, Newark, DE or Fotodyne, Hartland, WI).

E. QUALITATIVE AND QUANTITATIVE EVALUATION

The reproducibility of traditional TLC has been markedly lower than that of the corresponding HPLC methods. However, with the development and commercialization of various TLC instruments (spotting apparatuses, automated developing systems, and chromatogram scanners), the reproducibility of any TLC method considerably increased and became comparable with that of HPLC. Solutes can be tentatively identified by their R_f values. However, the reliability of the identification can be considerably increased by spotting an authentic standard on the neighboring track and recording the visible, UV, fluorescence, or infrared (IR) spectra of the standard and the solute to be identified and then comparing them. It has been proven that the use of multivariate mathematical statistical methods, such as principal component analysis, facilitates the identification of drugs and their metabolites when four eluent systems are applied.[80]

Semiquantitative determination of the amounts of the separated compounds can be carried out visually. Standard solutions are spotted into the layer and the samples are spotted between. After development, the spot size and intensity of the standard series and those of the unknown solute or solutes can be visually compared. This method is rapid and does not require complicated instrumentation; however, it needs trained personnel, and the reproducibility is rarely better than 10 to 30%. Reliable quantitative evaluation can be carried out either by extraction of the solute from the plate and by the determination of its concentration by any other methods (e.g., spectrometry) or by the determination of the intensity of spots on the plate (*in situ* evaluation). The extraction of the solutes from the plates is time consuming, and reproducibility of the method is lower than that achieved with the modern densitometers.[81] Sensitive chromatogram spectrophotometers (densitometers) have been developed for the determination of the visible, UV, or fluorescence intensity of the various solutes depending on their absorption characteristics (Analtech Inc., Newark, DE; Merck Ltd., Darmstadt, Germany; Shimadzu Europa GmbH, Duisburg, Germany). Scanning is generally carried out at the maximum absorption wavelength of the solute to be determined. As the absorption maximum of a compound may be slightly different on a plate (adsorbed state) than in solution, it is advisable to determine the spectra of the solute with the same scanner and to carry out the scanning at the maximum wavelength determined under these conditions. In most cases, scanning is carried out in reflectance mode. Monochromatic light is directed to the surface of the layer, and the intensity of the reflected diffuse light is measured with an appropriate detector system. Calibration curves are a prerequisite for accurate analytical work. The quantity of analyte has to be in the concentration range of the calibration curve; extrapolation from the calibration curve beyond the lowest and highest standards may lead to severe inaccuracy. Both peak height and peak area can be used for the construction of the calibration curve. It is generally accepted that in the case of narrow and symmetric peaks, both methods give reliable results. In any other cases, the use of peak area is preferred. Isotopes have been extensively applied in metabolism and toxicology studies; therefore their detection and quantitative determination is of high theoretical and practical importance. Isotopes such as ^3H, ^{14}C, ^{32}P, ^{131}I, ^{18}F, ^{11}C, and $^{99m}T_{cm}$ can be detected by use of a radioimaging and analysis system developed for this special purpose (Bioscan, Chester, UK).

F. SENSITIVITY AND REPRODUCIBILITY IN TLC

The sensitivity of traditional TLC analytical procedures was markedly lower than that of the corresponding HPLC and GC methods. However, with the advent of modern HPTLC methods, sensitivity has been considerably enhanced and it is similar or identical to that of other chromatographic methods. Sensitivity depends significantly on the TLC system (stationary and mobile phases, detection method). Using silica HPTLC plates coupled with scanning densitometry, 1 ng of 5-(hydroxymethyl)-2-furfuralcans was detected.[82] However, using similar TLC conditions, the detection limit of buprenorphine, a semisynthetic opioid derived from thebaine, was 16 ng.[83] Carbaryl was detected at the quantity of 1 µg per spot when phenylhydrazine hydrochloride was used as the reagent.[84] The detection limit for phospholipids was 100 ng using silica plates, 10% $CuSO_4$ in 10% H_3PO_4 as a charring agent, and scanning densitometry.[85] A similar detection limit (100 ng) was found in the TLC analysis of flavonoids.[86] Exploiting the higher sensitivity of fluorescence detection, the detection limit of methoxymorpholinodoxorubicin hydrochloride, a new antitumor agent, was found to be lower than 1 ng.[87] An interesting study established that the prewashing of plates with methanol increased the sensitivity by about 40% in the analysis of bioactive cucurbitans extracted with methanol from *Cucurbita andreana*, *C. taxana*, and *C. okeechobeensis*.[88]

The worldwide acceptance of GLP (good laboratory practice) or any other system aiming at the enhancement of the precision of analytical data considerably increased the requirements with which an analytical procedure has to comply. Each analytical method has to be tested for linearity, selectivity, ruggedness, and detection limit, before it is used as an officially acknowledged analytical procedure. TLC supports have to comply with numerous requirements, such as stability of the retention characteristics (separation capacity, retention strength), surface homogeneity and stability, uniformity of specific surface area, and pore volume. A validation process is a prerequisite for any up-to-date TLC analytical method.[89] Validation processes must include not only the separation procedure but also the extraction step as was proved in the case of alkaloid from *Catharantus roseus*.[90] The demand on quality, validation data, and performance of the TLC methods has markedly increased also in the pharmaceutical industry. Modern TLC is a convenient, reliable, and specific analytical method that can be successfully used in contemporary pharmaceutical product control. A modern TLC method uses prewashed stationary phase, automated sample application, well-defined developing chambers or forced-flow techniques, and densitometry for data evaluation. These modifications result in simple, rapid, relatively cheap, and rugged analytical methods that require minimal sample preparation. The cost per analysis is relatively low due to the high sample throughput. It has been found that TLC is especially suitable for the analyses of samples with high percentages of impurities that remain at the start during the development. The relative standard deviation of an instrumental TLC analysis may be as low as 1.5 to 3.0%.[91] The relative standard deviations in the analysis of the antibiotic metacycline were < 2 for TLC-UV, < 3% for TLC-fluorescence detection, and < 1% for HPLC.[92]

G. TLC AS A PILOT METHOD FOR HPLC

TLC and HPLC methods show many similarities, and a wide variety of sorbent–eluent combinations can be used in both methods. It has been established that TLC can been successfully applied to complete the information of HPLC in the pharmaceutical industry[93] and in the analysis of ajmaline stereoisomers.[94] A good agreement was found between the results of TLC and HPLC in the determination of coumarins.[95]

TLC has some advantages as a pilot method for HPLC, such as simultaneous determination of the retention behavior of a considerable number of solutes and detection of solute fractions remaining at the start or show very low mobility. The detection of these solutes is difficult in HPLC, and the removal of the adsorbed fractions from the upper part of the HPLC column is frequently time consuming. Theoretical calculations proved the relationship between the retention behavior of solutes in TLC and in HPLC.[96–98] A study found good correlations between the TLC and HPLC retention behavior of xanthine and adenosine derivatives using reversed-phase systems[99]; however, in some instances the TLC method showed a fairly weak predictive power for the retention behavior of solutes in HPLC. The retention of nonhomologous series of commercial pesticides on an alumina layer and an alumina HPLC column showed poor correlation in *n*-hexane–dioxane eluents.[100] It was assumed that the different surface pH of the sorbents may have accounted for the difference. The retention behavior of antihypoxia drugs differed considerably in RP-TLC and in RP-HPLC, probably due to the different carbon load of the sorbents.[101]

H. DETERMINATION OF MOLECULAR PARAMETERS WITH ADSORPTIVE AND REVERSED-PHASE TLC

Various TLC methods can be successfully used for the determination of physicochemical parameters of bioactive compounds such as hydrophobicity (lipophilicity),[102,103] and specific hydrophobic and hydrophilic surface areas. These molecular parameters are generally well correlated with the biological activity.[104,105] The study of chromatographic retention behavior structure–activity relationships may facilitate not only the rational design of drugs, but also a more profound understanding of the underlying biochemical and biophysical procedures.[106] TLC methods have some advantages over other chemical or physical methods used for the determination of various molecular characteristics. They are rapid and relatively simple, a very low quantity of the solute is required, and the solute need not to be very pure because the impurities separate during the chromatographic process.

Lipophilicity (hydrophobicity) is one of the molecular parameters frequently used in quantitative structure-activity relationship (QSAR) studies. The important role of lipophilicity is due to the fact that the penetration of bioactive compounds through the hydrophobic barrier of the target organs or organisms, their distribution between the hydrophobic and hydrophilic compartments of the cell, as well as their binding to the various cell constituents depends considerably on the lipophilicity. RP-TLC has been extensively used for the determination of molecular lipophilicity. Silica plates impregnated by overnight predevelopment in *n*-hexane–paraffin oil 95:5[107] or in methylsilicone dissolved in diethyl ether[108–111] as well as C_{18} plates with covalently bonded octadecylsilane groups on the silica surface[112,113] have been proved to be equally suitable for the determination of the lipophilicity of a wide variety of compounds such as growth-stimulating amido esters of ethanolamine,[114] alkanol-amines,[115] and some novel aryl-(amino or hydroxy)ethylamino ketones.[116]

REFERENCES

1. Liu, J., Ponder, G. W., and Stewart, J. T., *J. Planar Chromatogr.*, 7, 247, 1994.
2. Görög, S., Balogh, G., Csehi, A., Csizèr, É., Gazdag, M., Halmos, Zs., Hegedùs, B., Herènyi, B., Horváth, P., and Laukó, A., *J. Pharm. Biomed. Anal.*, 11, 1219, 1993.
3. Sherma, J., *Anal. Chem.*, 65, 40R–54R, 1993.
4. McGinnis, S. C. and Sherma, J., *J. Liquid Chromatogr.*, 17, 151, 1994.
5. Abjean, J. P., *Chromatographia*, 36, 359, 1993.
6. Yi, Z., Zhuang, G., and Brown, P. R., *J. Liquid Chromatogr.*, 16, 3133, 1993.
7. Steuckart, C., Berger-Preiss, E., and Levsen, K., *Anal. Chem.*, 66, 2570, 1994.
8. Herold, A., *J. Planar Chromatogr.*, 7, 180, 1994.
9. Sherma, J. and Fried, B., Eds., *Handbook of Thin Layer Chromatography*, Chromatographic Science Series, Vol. 55, Marcel Dekker, New York, 1991.
10. Poole, C. and Poole, S. K., *Anal. Chem.*, 66, 27, 1994.
11. Mohammad, A., Fatima, N., Ahmad, J., and Khan, M. A. M., *J. Chromatogr.*, 642, 445, 1993.
12. Rathore, H. S. and Begum, T., *J. Chromatogr.*, 643, 271, 1993.
13. Sherma, J., *Anal. Chem.*, 64, 134, 1992.
14. Sherma, J., *Anal. Chem.*, 66, 67, 1994.
15. Peischl, R. J., Sabo, M., Dugan, G. E., and Puerschner, G., *J. Planar Chromatogr.*, 7, 211, 1994.
16. Tivert, A.-M. and Backman, A., *J. Planar Chromatogr.*, 6, 216, 1993.
17. Martin, P., Taberner, J., Thorp, R. J., and Wilson, I. D., *J. Planar Chromatogr.*, 5, 99, 1992.
18. Klaus, R., Fischer, W., and Hauck, H. E., *Chromatographia*, 37, 133, 1993.
19. Bartels, R. and Bock, L., *J. Chromatogr. A*, 659, 185, 1994.
20. Leonard, S. A., Wong, S. C., and Nyce, J. W., *J. Chromatogr.*, 645, 189, 1993.
21. Klaus, R., Fischer, W., and Hauck, H. E., *Chromatographia*, 39, 97, 1994.
22. Paesen, J., Quintens, I., Thoithi, G., Roets, E., Reybrouck, G., and Hoogmartens, J., *J. Chromatogr. A*, 677, 377, 1994.
23. Cserháti, T., *J. Planar Chromatogr.*, 6, 70, 1993.
24. Ueda, J., Miyamoto, K., and Kamisaka, S., *J. Chromatogr. A*, 658, 129, 1994.
25. Camara, N. G. and Lederer, M., *J. Pharm. Biomed. Anal.*, 10, 371, 1992.
26. Drescher, J. N., Sherma, J., and Fried, B., *J. Liq. Chromatogr.*, 16, 3557, 1993.
27. Okamoto, M., Ohta, M., Kakamu, H., and Omori, T., *Chromatographia*, 35, 281, 1993.
28. Nawrocki, J., Rigney, M. P., McCormick, A., and Carr, P. W., *J. Chromatogr. A*, 657, 229, 1993.
29. Cserháti, T., *Anal. Chim. Acta*, 292, 17, 1994.
30. Cserháti, T., Forgács, E., and Holló, J., *J. Planar Chromatogr.*, 6, 472, 1993.
31. Norfolk, E., Khan, S. H., Fried, B., and Sherma, J., *J. Liquid Chromatogr.*, 17, 1317, 1994.
32. Janjic, T. J., Zivkovic, V., and Celap, M. B., *Chromatographia*, 38, 447, 1994.
33. Petrovic, S. M., Loncar, E., Perisic-Janjic, N. U., and Popov-Pergal, K., *Chromatographia*, 38, 749, 1994.
34. Fernanda, M., Borged, M., and Pinto, M. M. M., *J. Liquid Chromatogr.*, 17, 1125, 1994.
35. Petrovic, M., Kastelan-Macan, M., Turina, S., and Ivankovic, V., *J. Liquid Chromatogr.*, 16, 2673, 1993.
36. Nikolova-Damyanova, B., Chobanov, D., and Dimov, S., *J. Liq. Chromatogr.*, 16, 3997, 1993.
37. Sharma, S. D., Misra, S., and Gupta, R., *J. Liquid Chromatogr.*, 16, 1833, 1993.
38. Kovács-Hadady, K. and Szilágyi, J., *J. Chromatogr.*, 553, 459, 1991.
39. Kovács-Hadady, K. and Szilágyi, J., *J. Planar Chromatogr.*, 4, 194, 1991.
40. Kovács-Hadady, K., *J. Chromatogr.*, 589, 301, 1992.
41. Kovács-Hadady, K., *J. Planar Chromatogr.*, 4, 456, 1991.

42. Szilágy, J., Kovács-Hadady, K., and Kovács, A., *J. Planar Chromatogr.*, 6, 212, 1993.
43. Kovács-Hadady, K., *J. Pharm. Biomed. Anal.*, 10, 1025, 1992.
44. Kovács-Hadady, K., and Levai, A., *Chromatographia*, 37, 482, 1993.
45. Okamoto, M., Kakamu, H., Oka, H., and Ikai, Y., *Chromatographia*, 36, 293, 1993.
46. Mohammad, A., and Khan, M. A. M., *J. Chromatogr.*, 642, 455, 1993.
47. Iskric, S., Hadzija, O., and Kveder, S., *J. Liquid Chromatogr.*, 17, 1653, 1994.
48. Lepri, L., Coas, V., Desideri, G. P., and Pettini, L., *J. Planar Chromatogr.*, 6, 100, 1993.
49. Lepri, L., Desideri, P. G., and Santianni, D., *Chromatographia*, 36, 297, 1993.
50. LeFevre, J. W., *J. Chromatogr.*, 653, 293, 1993.
51. Xuan, H. T. K. and Lederer, M., *J. Chromatogr.*, 645, 185, 1993.
52. Xuan, H. T. K. and Lederer, M., *J. Chromatogr. A*, 659, 191, 1994.
53. Armstrong, D. W. and Zhou, Y., *J. Liq. Chromatogr.*, 17, 1695, 1994.
54. Wang, Q. S., Xie, W. Q., and Pan, D. P., *Chromatographia*, 35, 149, 1993.
55. Quin-Sun, W., Bing-Wen, Y., and Zhang Z.-C., *J. Planar Chromatogr.*, 7, 229, 1994.
56. Markowski, W. and Soczewinski, E., *Chromatographia*, 36, 330, 1993.
57. Markowski, W., *J. Chromatogr.*, 635, 283, 1993.
58. Cavalli, E., Truong, T. T., Thomassin, M., and Guinchard, C., *Chromatographia*, 35, 102, 1993.
59. Rydberg, J., Musikas, C., and Choppin, G. R., Eds., *Principles and Practices of Solvent Extraction*, Marcel Dekker, New York, 1992.
60. Heimler, D., Pieroni, A., Tattini, M., and Cimato, A., *Chromatographia*, 33, 369, 1992.
61. Matysik, G., Soczewinski, E., and Polak, B., *Chromatographia*, 39, 497, 1994.
62. Belay, M. T. and Poole, C. F., *Chromatographia*, 37, 365, 1993.
63. Matysik, G., *Chromatographia*, 38, 109, 1994.
64. Ludmir, J. and Alvarez, J. G., *J. Liq. Chromatogr.*, 16, 1685, 1993.
65. Madden, U. A. and Stahr, H. M., *J. Liq. Chromatogr.*, 16, 2825, 1993.
66. Minguez-Mosquera, M. I., Gallardo-Guerrero, L., and Gandul-Rojas, B., *J. Chromatogr.*, 633, 295, 1993.
67. Ohta, H., Fong, C. H., Berhow, M., and Hasegawa, S., *J. Chromatogr.*, 639, 295, 1993.
68. Sherma, J. and Rolfe, C., *J. Chromatogr.*, 643, 337, 1993.
69. Khan, S. H., Murawski, M. P., and Sherma, J., *J. Liquid Chromatogr.*, 17, 855, 1994.
70. Cavalli, E. J. and Guinchard, C., *Chromatographia*, 37, 107, 1993.
71. Zellmer, S. and Lasch, J., *Anal. Biochem.*, 218, 229, 1994.
72. Markowski, W. and Matysik, G., *J. Chromatogr.*, 434, 1993.
73. Sanchez, F. G., Diaz, A. N., and Correa, M. R. F., *J. Chromatogr.*, 655, 31, 1993.
74. Singh, N. N., Periera, C., and Patil, U. M., *J. Liquid Chromatogr.*, 16, 1845, 1993.
75. Roveri, P., Cavrini, V., Andrisano, V., and Gatti, R., *J. Liquid Chromatogr.*, 16, 1859, 1993.
76. Kotynski, A. and Kudzin, Z. H., *J. Chromatogr. A*, 663, 127, 1994.
77. Konopski, L. and Pawlowska, E., *J. Chromatogr. A*, 669, 275, 1994.
78. Hiermann, A. and Bucar, F., *J. Chromatogr. A*, 675, 276, 1994.
79. Price, N. P. J. and Gray, D. O., *J. Chromatogr.*, 635, 165, 1993.
80. Romano, G., Caruso, G., Musumarra, G., Pavone, D., and Cruciani, G., *J. Planar Chromatogr.*, 7, 233, 1994.
81. Matysik, G., Glowniak, K., Soczewinski, E., and Garbacka, M., *Chromatographia*, 38, 766, 1994.
82. Schuck, D. F. and Pavlina, T. M., *J. Planar Chromatogr.*, 7, 242, 1994.
83. Chandrashekhar, T. G., Rao, P. S. N., Sneth, D., Vyas, S. K., and Dutt, C., *J. Planar Chromatogr.*, 7, 249, 1994.
84. Patil, V. B. and Shingare, M. S., *J. Chromatogr.*, 653, 181, 1993.
85. Welsh, C. J. and Schmeichel, K., *J. Liq. Chromatogr.*, 16, 1819, 1993.
86. Garcia, S., Heinzen, H., Martinez, R., and Moyna, P., *Chromatographia*, 35, 430, 1993.
87. Farina, A., Quaglia, M. G., Calandra, S., and Gallo, F. R., *J. Pharm. Biomed. Anal.*, 11, 1215, 1993.
88. Halaweish, F. T. and Tallamy, D. W., *J. Liq. Chromatogr.*, 16, 497, 1993.
89. Sun, S. W. and Fabre, H., *J. Liq. Chromatogr.*, 17, 433, 1994.
90. Nagy-Turak, A. and Vegh, Z., *J. Chromatogr. A*, 668, 1, 1994.
91. Renger, B., *J. AOAC Intern.*, 76, 7, 1993.
92. Naidong, W., Hua, S., Verresen, K., Roets, E., and Hoogmartens, J., *J. Pharm. Biomed. Anal.*, 9, 717, 1991.
93. Bliesner, D. M., *J. Planar Chromatogr.*, 7, 197, 1994.
94. Sosa, M. E., Valdes, J. R., and Martinez, J. A., *J. Chromatogr. A*, 662, 251, 1994.
95. Vuorela, P., Rahko, E.-L., Hiltunene, R., and Vuorela, H., *J. Chromatogr. A*, 670, 191, 1994.
96. Rozylo, J. and Janicka, M., *J. Planar Chromatogr.*, 4, 241, 1991.
97. Rozylo, J. and Janicka, M., *J. Liquid Chromatogr.*, 14, 3197, 1991.
98. Kossoy, A. D., Risley, D. S., Kleyle, R. M., and Nurok, D., *Anal. Chem.*, 64, 1345, 1992.
99. Biagi, G. L., Guerra, M. C., Barbaro, A. M., Barbieri, S., Recatini, M., and Borea, P. A., *J. Liq. Chromatogr.*, 13, 913, 1990.
100. Cserháti, T. and Forgács, E., *J. Chromatogr. A*, 668, 495, 1994.
101. Wallerstein, S., Cserháti, T., and Fischer, J., *Chromatographia*, 35, 275, 1993.

102. **Biagi, G. L., Guerra, M. C., Barbaro, A. M., Barbieri, S., Recatini, M., Borea, P. A., and Pietrogrande, C.**, *J. Chromatogr.*, 498, 179, 1990.
103. **Dross, K. P., Mannhold, R., and Rekker, R. F.**, *Quant. Struct. Act. Relat.*, 11, 36, 1992.
104. **Kaliszan, R.**, *Anal. Chem.*, 64, 619, 1992.
105. **Kaliszan, R.**, In *Advances in Chromatography.* Vol. 33, Brown, P. R., and Grushka, E., Eds., Marcel Dekker, New York, 1993, 147–176.
106. **Kaliszan, R.**, *Chemom. Int. Lab. Syst.*, 24, 89, 1994.
107. **Cserháti, T.**, *J. Biochem. Biophys. Meth.*, 27, 133, 1993.
108. **Houngbedji, N. and Waisser, K.**, *J. Chromatogr.*, 509, 400, 1990.
109. **Biagi, G. L., Barbaro, A. M., Sapone, A., and Recatini, M.**, *J. Chromatogr. A*, 662, 341, 1994.
110. **Biagi, G. L., Barbaro, A. M., Sapone, A., and Recatini, M.**, *J. Chromatogr. A*, 669, 246, 1994.
111. **Biagi, G. L., Barbaro, A. M., and Recatini, M.**, *J. Chromatogr. A*, 678, 127, 1994.
112. **Dross, K., Sonntag, C., and Mannhold, R.**, *J. Chromatogr.*, 639, 287, 1993.
113. **Dross, K., Sonntag, C., and Mannhold, R.**, *J. Chromatogr. A*, 673, 113, 1994.
114. **Gocan, S., Irimie, F., and Cimpan, G.**, *J. Chromatogr. A*, 675, 282, 1994.
115. **Bazylak, G.**, *J. Planar Chromatogr.*, 7, 202, 1994.
116. **Hadjipavlou-Litina, D., Rekka, E., Hadjipavlou-Kourounakis, L., and Kourounakis, P. N.**, *Eur. J. Med. Chem.*, 27, 1, 1992.

Chapter 2

THIN-LAYER CHROMATOGRAPHY IN BACTERIOLOGY

Mark Maloney

CONTENTS

I. Introduction 19

II. Lipid Sample Preparation 20
    A. Obtaining Samples for Lipid Extraction 20
    B. Extraction Procedures 20
    C. Column Chromatography 21
    D. Neoglycolipids 21

III. General TLC Methods 21
    A. Choice of TLC Plates 21
    B. Choice of Solvent 22
    C. Preparative TLC in Lipid Purification 22
    D. Identification of Lipids on TLC 23

IV. Use of TLC in the Identification of Bacteria and Bacterial Strains 23
    A. TLC in the Determination of Bacterial Lipid Composition 23
    B. Use of TLC in Assays of Lipase Activity 23

V. Overlay Methods for the Identification of Bacterial Receptors 23
    A. Background 23
    B. Preparation of Chromatograms 25
    C. Peroxidase-based Assays 25
    D. Radiolabel-based Assays 27

VI. Protocols 27

VII. Conclusions 28

References 28

I. INTRODUCTION

Thin-layer chromatography (TLC) has many applications in the study of microorganisms. These studies include the use of TLC in the determination of microbial lipid composition and the study of microbial lipases. Both of these applications are still used to some extent in the identification of bacteria and bacterial strains. TLC has also been an indispensable tool for investigations concerning host lipids that function as receptors for microbial pathogens and their toxins. Though gas liquid chromatography–mass spectrometry (GC-MS) has, in many cases, replaced TLC for the definitive identification of lipids, TLC procedures remain the most suitable methods for lipid receptor investigation. Although other applications of TLC in bacteriology will be addressed, the focus of this chapter will be on TLC assays for the study of lipid receptors of bacteria and bacterial proteins, including purified toxins and "adhesin" proteins that mediate bacterial adhesion. Lipids that function as receptors for viruses have also been well-documented.[1] It should be noted that yeasts such as *Cryptococcus neoformans* and *Candida albicans* possibly use glycolipids or phospholipids as receptors as well.[2] However, results of such studies have been difficult to reproduce, perhaps due to strain differences in yeasts or anomalies in culture conditions.

Radiolabeled cholera toxin was the first protein ligand used in the TLC overlay that demonstrated unequivocally that cholera toxin used GM1 as a receptor.[3] This assay was performed on gangliosides separated on TLC plates that were blocked with 1% polyvinylpyrrolidone in phosphate-buffered saline (PBS) to prevent nonspecific binding of ^{125}I-labeled cholera toxin. Subsequently, polyisobutylmethacrylate (PIBM) was used as a blocking agent and to prevent the loss of silica from thin-layer plates following immersion in the aqueous saline solutions required for the assay.[4] This achievement opened the door for the identification of several other lipid receptors for toxins, as well as fimbrial ("fimbriae," sometimes included in the term "pili," are filamentous appendages often present on bacterial cells) and other adhesin proteins[5]; for the use of antilipid antibodies as probes[6]; and for the use of labeled whole microorganisms and viruses in a modified assay for the identification of receptors.[1]

The majority of applications have, to date, been in the field of bacteriology. Using some variation of Magnani's TLC overlay technique,[3] several glycosphingolipids and phospholipids have been identified as receptors for a large number of bacteria and their toxins, which colonize or target a variety of host tissues.[5,7]

II. LIPID SAMPLE PREPARATION

A. OBTAINING SAMPLES FOR LIPID EXTRACTION

Immediately after harvesting, tissues and cells may be kept on ice for no more than 1 h prior to extraction in order to prevent enzymatic degradation of the lipids. Alternatively, samples may be frozen immediately at –20°C or –70°C for later use. Tissues and cells typically are washed free of blood or tissue culture media with buffered saline solutions such as Hanks' balanced salts solution or PBS, pH 7.4.

B. EXTRACTION PROCEDURES

Tissues or cells are typically extracted in solutions of chloroform–methanol, but isopropanol–hexane (I–H) solutions (2:3, v/v) may be used as a less toxic alternative.[8] However, I–H extractions can give reduced yields of lipid and are not recommended for quantitative analyses. Typically, tissues or cells are extracted with 10 volumes of chloroform–methanol (2:1,v/v). The residue is then re-extracted with five volumes of chloroform–methanol (2:1, 1:1, or 1:2) and the supernatants pooled. Following extraction, gangliosides and nonlipid contaminants such as sugars, amino acids, and salts may be separated from other lipids by aqueous Folch partition.[9,10] Neutral glycolipids with longer chain carbohydrates (four or more sugar residues) will also partition to some extent into the upper phase. Gangliosides and long chain glycolipids can be recovered by using a rotary evaporator to remove organic solvents, resuspending the residue in distilled water with sonication, and dialyzing against distilled water at 5°C. The nonlipid contaminants are dialyzable while the glycolipids form micellar aggregates that are too large to fit through the pores of the dialysis tubing. Alternatively, lipids in aqueous solution may be applied to a Sep-Pak C_{18} cartridge (Millipore Corp., Milford, MA).[11] The nonlipid contaminants are removed by thorough washing, and the lipids that adhere to the C_{18} matrix are then eluted with methanol or solvent mixtures containing methanol.[12] While the majority of gangliosides preferentially partition into the aqueous phase, a percentage of short chain gangliosides such as GM4 will remain in the lower phase. For the separation of all gangliosides from uncharged lipids, it is best to use ion exchange chromatography on DEAE-columns such as DEAE-Sephadex (see below).

C. COLUMN CHROMATOGRAPHY

Alternatives to the Folch partition for removal of nonlipid contaminants include column chromatography using Sephadex G25 or LH20. LH20 chromatography has the advantage over Sephadex G25[10] and the Folch method in that all lipids are obtained in one fraction.[13] The other methods will require that gangliosides and longer chain glycolipids that partition into aqueous phases be recovered separately.

Column chromatography is also used to separate lipids into classes. Aminopropyl columns can be used to separate neutral lipids into cholesterol ester, triacylglycerol, diacylglycerol, monoacylglycerol, cholesterol, and free fatty acid fractions by stepwise elution.[14] However, this technique does not allow for the separation of phospholipids from glycolipids. This is of major importance in microbiological applications because phospholipids and glycolipids migrate similarly on TLC plates and, ideally, should be separated prior to performance of TLC overlay procedures for the investigation of potential bacterial receptors. Glycolipids can be separated from phospholipids by stepwise elution from Florisil columns.[15] However, gangliosides are not soluble in acetone and thus are not amenable to this technique. Saponification with sodium hydroxide will hydrolyze phosphoglycerolipids into components that can easily be separated from glycolipids.[10] However, O-acyl and other labile groups on glycolipids will be lost, and sphingomyelin, due to its ceramide base, and plasmalogens are not removed by this technique.

Both gangliosides and neutral glycolipids can be isolated from all other lipids as acetylated derivatives by stepwise elution from Florisil columns.[16] This technique is suitable for all glycolipids except gangliosides containing labile groups such as 9-O acyl chains, which are relatively uncommon; the 9-O acyl chains are lost upon deacetylation. This potential drawback can be averted by prior removal of the ganglioside fraction using ion exchange chromatography with DEAE-Sephadex.[17] Florisil chromatography of acetylated glycolipids is especially useful when the overall glycolipid composition of tissue is previously unknown[18] or when the glycolipid compositions of different tissues are being compared.[19] However, the technique is time-consuming, and special care must be taken to use anhydrous reagents throughout the procedures.

D. NEOGLYCOLIPIDS

In some studies, it is desirable to have a carbohydrate group artificially bound to a defined synthetic lipid base. This recent development in TLC[20,21] has been applied to microbial adhesion to show, for example, that type 1 (Galactose β1-3 N-acetyl glucosamine or Galβ1-3GlcNAc) and type 2 (Galβ1-4GlcNAc) disaccharides isolated from mucin glycoproteins are potential components of receptors for *Pseudomonas aeruginosa*.[22] These studies could not be performed on whole mucins because of their carbohydrate heterogeneity. Other studies have used the technique to vary the lipid base and fatty acid content of potential glycolipid receptors in order to show the importance and effect of the lipid moiety in presentation of the carbohydrate for binding.[23,24]

III. GENERAL TLC METHODS

A. CHOICE OF TLC PLATES

There are several factors to be considered in choosing TLC plates, including cost, efficiency of resolution of the compounds under study, and the amount of sample available. In general, standard TLC plates using silica gel on glass plates and plastic or aluminum sheets are sufficient for the separation of many compounds of interest, including most lipids. However, total phospholipids isolated from cells or tissues have components that migrate similarly on one-dimensional TLC. Therefore, two-dimensional TLC is recommended for the separation of phospholipid mixtures. Plates that have hard, thin sheets of silica gel tightly bound to plastic, glass, or aluminum, including high-performance TLC (HPTLC) plates, require less sample and often give better separation of compounds. Often glycolipids that vary only in their fatty acid composition can be separated as closely migrating spots or "doublets" on these plates.

With regard to TLC overlay procedures, the choice of TLC plate is of critical importance. Some plates do not require treatment with the plastic polyisobutylmethacrylate (PIBM) prior to their use in this assay. PIBM has been shown to cause artifacts regarding functional glycolipid receptors of bacterial toxins. Specifically, in the presence of PIBM, verotoxin-1 (VT1) binds globotetraosyl ceramide, which is not a functional receptor for VT1 because its Galα1-4Gal residues are not terminal, and shows reduced binding to globotriaosyl ceramide and P1 glycolipid, which are known functional receptors.[25] It appears that PIBM can distort the carbohydrate conformation of glycolipids and alter their receptor properties. For this reason it is highly recommended that plates be used that do not require pretreatment with PIBM. Plastic-backed plates available from Brinkmann Instruments (Westbury, NY) and J.T. Baker Chemical Co. (Phillipsburg, NJ) can be used in this manner. Other factors known to predispose VT1 to artifactually recognize globotetraosyl ceramide include the presence of the antibiotic polymyxin B, which is commonly used in toxin purification protocols and must be removed from toxin preparations.[26] Long-term storage of toxin at 4°C can also produce toxin that recognizes globotetraosyl ceramide, probably due to slight denaturation of the toxin structure.

B. CHOICE OF SOLVENT

In general, more hydrophobic solvents should be used to separate more hydrophobic lipids. It is relatively easy to choose a solvent system that will separate simple lipids such as cholesterol esters, triacylglycerols, fatty acids, cholesterol, diacylglycerols, and monoacylglycerols, while leaving complex lipids at the origin. Conversely, by increasing the amount of polar solvents, such as methanol and water, added to nonpolar solvents such as chloroform, the complex lipids (phospholipids and glycolipids) of interest can be optimally separated by TLC, while any contaminating simple lipids will migrate with the solvent front. Typical chloroform-methanol-water mixtures used as mobile phases for the separation of complex lipids include 65:25:4 for phospholipids and neutral glycolipids, and 60:35:8 for gangliosides. These solvent mixtures are volatile and are easily removed from the silica gel.

Two-dimensional TLC of phospholipids requires that mobile phases that include acids and bases be used to optimize separation of the lipid constituents.[27] Suitable mobile phases include chloroform–methanol–28% ammonium hydroxide, 65:25:4, for development in the first direction. The plate is then allowed to dry, turned so that the developed lane is now at the bottom of the plate and developed in the second direction in chloroform–acetone–methanol–acetic acid–water (30:40:10:10:1). These mobile phases contain solvents (acids and bases) that are not particularly volatile, and care must be taken to remove them by prolonged drying or flushing with nitrogen gas prior to subsequent analysis.

C. PREPARATIVE TLC IN LIPID PURIFICATION

For preparative TLC, the lipid sample is applied as a long band across the bottom of the plate with spots of standard lipids at either end of the plate. The plates are then run in solvent and dried, and the ends of the plates are either cut, if plastic or aluminum, or scored and snapped, if glass, so that they contain the standard lanes and part of the sample. These end plates are sprayed so that the location of the desired sample band can be identified. Alternatively, the entire TLC plate may be sprayed with water or placed in a chamber containing iodine vapors in order to visualize the desired band. The silica gel containing this band is then scraped from the TLC plate, preferably before it has completely dried, and placed in a funnel or column plugged with glass wool or a fritted disc. An appropriate solvent is then applied to elute the lipid from the silica gel.[27] In general, the recovery of lipid from the silica gel is only a fraction of that originally applied, however. This varies with different brands of TLC plates, possibly due to differences in the chemical binder used in the production of the TLC plates. Also, HPTLC and other "hard" plates that are designed to separate small amounts of sample are not suitable for the application of large amounts of lipid. If large amounts of purified lipid are needed, then TLC plates designed specifically for preparative TLC should be used or purification should be on silica-based columns using solvent gradients as required.[10]

D. IDENTIFICATION OF LIPIDS ON TLC*

Omitting overlay procedures that identify lipids on the basis of antibody or toxin-binding specificities, identification of lipids on thin-layer chromatograms typically involves the use of chemical spray reagents that react with specific lipid components to give visible, colored bands and comigration of lipids with standard lipids applied and run on adjacent lanes of the same TLC plate (in the case of two-dimensional TLC, separate standard plates must be run). Common spray reagents are molybdenum blue for phospholipids (blue indicates presence of phosphate); orcinol for glycolipids (purple indicates the presence of sugar); resorcinol for gangliosides (blue indicates the presence of sialic acid); acidic ferric chloride spray for steroids (steroids turn red); phosphomolybdic acid (blue/violet spots on a pale green/yellow background), iodine vapor (brown spots) or even water (white spots) as general lipid sprays; and sulfuric acid–alcohol mixtures as general charring reagents for carbon-rich compounds. Several ready-to-use spray reagents are commercially available from Sigma Chemical Company (St. Louis, MO).

Neutral lipid, sterol, fatty acid, phospholipid and glycolipid standards are all commercially available from various sources, including Matreya (Pleasant Gap, PA), BioCarb Chemicals (Lund, Sweden) and Sigma. Sigma lipid standards often contain some degree of lipid impurities, but are fine for use as TLC standards, while Matreya standards are very pure.

IV. USE OF TLC IN THE IDENTIFICATION OF BACTERIA AND BACTERIAL STRAINS

A. TLC IN THE DETERMINATION OF BACTERIAL LIPID COMPOSITION

TLC has had many applications in bacteriology in the area of the separation and identification of bacterial components. Many of these applications have been replaced by high-performance liquid chromatography (HPLC) and GC-MS methodology in recent times. However, TLC is still used in some clinical and taxonomic applications due, in part, to its ease of use and affordability. Recent studies have used TLC in the comparison of quinone structure in different bacteria,[28] the determination of fatty acids associated with Lipid A in gram-negative bacteria,[29] and the identification of *Mycobacterium* species by their unique glycolipid and mycobactin compositions.[30-33]

B. USE OF TLC IN ASSAYS OF LIPASE ACTIVITY

TLC has applications in monitoring enzyme activity especially with regard to lipases or glycosidases that use glycolipids as substrates. Lipases that have been investigated using TLC assays include the phospholipases of *Mycoplasma* and *Ureaplasma*, which have potential diagnostic applications.[34,35] Investigators have also used TLC to determine the effect of bile salts on *Ruminococcus* and *Bifidobacterium* glycosidases that degrade glycosphingolipids.[36] The glycosidase activities of these and other genera of normal intestinal microbial flora have important implications for the availability of glycolipid receptors for the adhesins of potential pathogens (see below).

V. OVERLAY METHODS FOR THE IDENTIFICATION OF BACTERIAL RECEPTORS

A. BACKGROUND

Soon after the TLC overlay technique was used to identify GM1 as the receptor for cholera toxin,[3] the technique was applied to the identification of lipid receptors for whole bacteria.[37] This modified assay has been used to identify potential glycolipid and phospholipid receptors for a number of bacteria, including many pathogenic species.[7] However, due to the complex nature of bacterial adhesins and bacteria-host tissue interactions, the results of such assays must be interpreted with care (see below). Perhaps the most common glycolipid that functions as a receptor for bacteria is lactosyl ceramide. Many lactosyl ceramide-binding bacteria, including *Bacteroides*, *Fusobacterium*,

---

* See also "Overlay Techniques."

and *Lactobacillus* species, are normal residents of the large intestine, while potential pathogens with this binding specificity include *Clostridium difficile* and *C. botulinum*, *Vibrio cholerae*, and *Shigella* and *Yersinia* species.[7] The normal state of the intestine may require that nonpathogenic bacteria predominate over pathogenic species for available lactosyl ceramide receptor sites. The activity of glycosphingolipid-degrading enzymes of intestinal bacteria mentioned previously may expose the lactosyl ceramide of longer chain glycolipids (or prevent its being masked by longer chain glycolipids) for binding to bacterial adhesins. It is also of interest that Galα1-4Gal sequences of glycosphingolipids were not degraded by any of the bacterial glycosidases studied.[36] Galα1-4Gal containing glycolipids are receptors for potentially pathogenic strains of *Escherichia coli* as well as for the verotoxins (also called Shiga-like toxins) of enterohemorrhagic *E. coli* and Shiga toxin of *Shigella dysenteriae*, respectively (see below). Other lactosyl ceramide-binding bacteria include pathogens that colonize the genitals (*Neisseria gonorrhoeae*), skin (*Propionibacterium*), and respiratory tract (*Hemophilus influenzae* and possibly *Pseudomonas aeruginosa*).[7]

Many potential pulmonary pathogens bind to GalNAcβ1-4Gal sequences found in asialo GM1, asialo GM2, and other glycolipids.[38] These bacteria include *P. aeruginosa*, *H. influenzae*, *Staphylococcus aureus*, *Streptococcus pneumonia*, *Klebsiella pneumoniae*, and various strains of *E. coli*. The results of receptor binding studies using whole *P. aeruginosa*, a major pathogen of pulmonary tissue, illustrate the potential difficulties in interpretation of TLC overlay assays. Initial studies using bovine serum albumin (BSA) as a blocking agent indicated that asialo GM1 and other glycolipids containing GalNAcβ1-4Gal sequences were the only functional glycolipid receptors for *P. aeruginosa*.[39] This correlated well with the presence of asialo GM1 in lung tissue. Subsequent studies indicated that additional binding to lactosyl ceramide and sialic acid containing glycolipids occurred in most strains tested using gelatin as the blocking agent, but that binding to glycolipids containing sialic acid did not occur with BSA as the blocker in the overlay procedures.[40] Although sialic acid has been implicated as a component of the receptor for *P. aeruginosa* on buccal epithelial cells, the role of sialic acid-containing glycolipids as functional receptors is far from proven. It is difficult to determine the *in vitro* artifact from the *in vivo* receptor in such instances. As stated previously, bacteria often possess several adhesion mechanisms with distinct receptors, which may vary with the bacterial strain, and isolating the functional receptors among the interacting components of host membranes is no less complex a task.

Care should be taken regarding the interpretation of bacterial adhesion as a virulence factor. The binding of uropathogenic *E. coli* to glycolipid receptors of urinary tract tissue was assumed to be an initial step in successful urinary tract colonization. However, recent evidence indicates that binding of these glycolipids enhances the clearance of these potential pathogens from the urinary tract.[41] Glycolipid-binding adhesins could still function as virulence factors by maintaining bacterial colonization in other sites, such as the vagina, that could seed bacteria into the urinary tract.[42]

Researchers have isolated several bacterial proteins with binding specificities for glycolipids and phospholipids that reflect, and are likely responsible for, the observed binding of whole bacteria. The proteins are either purified directly from bacterial extracts or cloned and expressed in various systems. These bacterial "adhesins" include several fimbriae proteins as well as nonfimbriae proteins such as exoenzyme S and exoenzyme-like proteins that are relatively loosely attached to the bacterial surface. Fimbriae proteins with specificity for lactosyl ceramide as well as Galα1-4Gal-containing sequences[43-47] and other carbohydrate sequences[48,49] on glycolipids have been identified from various strains of *E. coli*. Exoenzyme S isolated from *P. aeruginosa*[50-52] and a similar adhesin on *Helicobacter pylori*[53] bind phosphatidylethanolamine in addition to the GalNAcβ1-4Galβ1-4Glc sequences of asialo-GM1 and asialo-GM2. A similar binding pattern has been observed for the *Chlamydia* species *C. trachomatis* and *C. pneumoniae*.[54] *H. pylori* has also been reported to bind sulfatide and GM3[55] in TLC overlays.

The TLC overlay system also has been instrumental in studies that indicate that bacterial toxins bind specifically to glycolipid receptors.[7] Excluding cholera toxin, the most extensively studied glycolipid-binding toxins are Shiga toxin and the related verotoxins (Shiga-like toxins), which recognize primarily terminal Galα1-4Gal sequences on globo-series glycosphingolipids.[56,57]

Because glycolipid expression is highly specific for host cell type, glycolipid receptors can be used to target cells and tissues for colonization through the use of adhesins or for destruction through the action of toxins.

Often, bacterial products are not readily available for use as probes for the study of the receptor in host tissue. Because antibodies can be raised against cell surface antigens and antibody assays are familiar to most researchers, antibodies against the lipid receptors can be employed as probes in place of the bacterial products themselves.[3] However, because the lipid moiety, especially the fatty acid composition, of glycolipids can affect receptor function for some bacterial proteins,[23,24,58] antibody binding may not reflect the ligand complexity of toxins and adhesins in this regard. Antiglycolipid or phospholipid antibodies[59] often have a much lower affinity for these receptors than bacterial toxins and adhesins.

B. PREPARATION OF CHROMATOGRAMS

As stated above, the choice of TLC plate is important in regard to preventing artifacts. Plates that do not require the use of PIBM are highly recommended. It is also important that the plates be dried following their development so that they are free of organic solvent. This is especially true if glacial acetic acid or ammonium hydroxide are used in the mobile phase. Plates can be flushed with nitrogen gas if necessary until they appear dry and no odor of solvent remains.

It is recommended that identical plates be spotted and developed in parallel. In this way, one plate can be used to identify lipids by their migration relative to standard lipids and by their reaction with spray reagents such as orcinol for glycolipids and molybdenum blue for phospholipids. Purified standard lipids should be included on each plate to help identify sample lipids by their migration patterns, and a known functional lipid receptor, if available, should be included as a positive control.

Some purification of the lipid sample may be required prior to performing the overlay procedure. This is especially true regarding phospholipids and glycolipids, which tend to migrate with similar R_f values on TLC plates. For example, the phospholipid composition of some tissues can interfere with verotoxin-Gb3 binding assays (see above under "Sample Preparation").

The choice of blocking reagent is important. Without a blocking agent, background staining is usually quite dark, making the results unacceptable. One percent gelatin in 50 mM tris-buffered saline (TBS) is often used, although blocking with typically 1 to 2% or less of albumin, serum from various sources, powdered milk, Tween 20, and trimethyl urea have been used with success. The ideal blocking protocol must be determined empirically for each system. The choice of BSA over gelatin has been shown to diminish the binding of bacteria to some glycolipids in at least one study.[40] However, there is some doubt as to whether the observed binding in the absence of BSA was physiologically relevant or artifactual. Blocking with gelatin should be performed at 37°C for approximately 1 h or longer if required. Excessive blocking may result in background staining of the plate. Again, the ideal blocking scheme must be determined for each system.

C. PEROXIDASE-BASED ASSAYS

Suspensions of bacteria are used at 5×10^7–10^8 cells/ml. Solutions of bacterial lipid-binding proteins are typically used at approximately 1 to 10 µg per TLC chromatogram (plates are typically cut to less than 10×10 cm) in some type of buffered saline such as 50 mM TBS, pH 7.4. The pH of the saline solutions may also have an effect on binding. Perhaps it is best to match the pH to that of the bacterial media or relevant tissue site of colonization *in vivo* in at least one trial, in addition to using the typical pH 7.4 of laboratory saline preparations. Because antiglycolipid or phospholipid antibodies are often of low affinity, concentrations of 200 µg/ml or more may be required. Also, antiglycolipid antibodies are more prone to background staining than toxins that may require that a blocker such as 1% albumin, gelatin, or Tween 20 be included in each step of the incubation and washing procedures.

Following several washing steps in buffered saline, antibody against the lipid-binding protein (bacterial protein or antibody) is added. The appropriate antibody dilution will vary according to the affinity of the antibody and the concentration of the stock solution. Final concentrations of 10 to 200 µg/ml can be used as a guideline. Commercially available secondary antibodies (typically affinity-purified goat or rabbit antibodies raised against primary antibody epitopes) labeled with horseradish peroxidase are typically suitable at a 1:1000 or 1:2000 dilution. Washing steps in buffered saline are required between each incubation with antibody. Incubation times of 1 h at 22 to 24°C are suitable for most primary and secondary antibody solutions. However, longer incubation times for primary antibody solutions, including overnight at 4°C, are sometimes required.

Control plates developed without primary "protein" (such as toxin or primary antibody), without secondary antibodies, and with developing reagent alone (see below) should be performed to rule out artifacts. This is especially important when working with this type of assay for the first time or when novel reagents are being introduced. Ideally, a mutant protein that has lost the ability to bind its receptor or an isotype-matched primary antibody with specificity for an unrelated antigen would be used as a control. Such an ideal control is not always available. Non-specific binding of secondary antibodies to asialo GM1 may occasionally occur in these assays (Maloney, unpublished).

If the second antibody is conjugated to peroxidase, then 4-chloro-1-naphthol, DAB, and DEPDA (listed more or less in order of increasing sensitivity) are suitable substrates for development of the plates. This avoids the need for radiolabeling procedures. Procedures using 4-chloro-1-naphthol are most frequently used and give the least amount of background staining. Color reaction is stopped by washing thoroughly in water. An example of this type of assay is given in Figure 2.1.

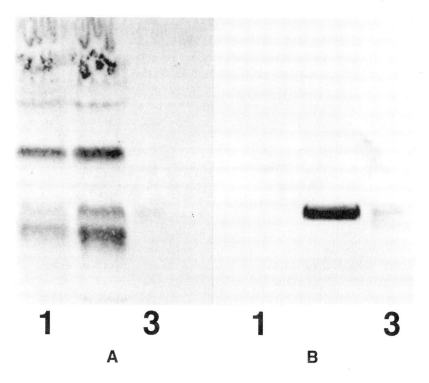

FIGURE 2.1 Peroxidase-based overlay of verotoxin receptor glycolipids. (A) Orcinol-sprayed TLC plate of lipids extracted from verotoxin-resistant U266 cells (lane 1), wild-type U266 cells (lane 2) and globotriaosyl ceramide standard (lane 3). (B) Corresponding peroxidase-based overlay performed using verotoxin-1 (VT-1), anti-verotoxin antibody and peroxidase-labeled secondary antibodies.

Sometimes shadows appear on wet TLC plates due to the hydrophobic nature of the lipids or defects in the silica gel. If held up to light, these translucent "bands" are easily distinguishable from bands due to the development of a colored precipitate following the addition of developing reagent. Control plates developed without primary protein, without second antibody, and with developing reagent alone should be performed to rule out artifacts.

D. RADIOLABEL-BASED ASSAYS

For radiolabel-based assays, bacteria or toxins are often directly labeled prior to performing the binding assay. Labeling of toxins or antibodies typically involves ^{125}I and the use of Bolton-Hunter reagent N-succinimidyl 3-(4-hydroxyphenylpropionate), iodobeads, or iodogen (1,3,4,6-tetrachloro-3a,6a-diphenylglycoluril),[60] and similar labeling techniques have been applied to whole bacteria.[1] Alternatively, whole bacteria have been metabolically labeled with ^{14}C (^{14}C-glucose) or ^{35}S (^{35}S-methionine or sodium sulfate). The overlay is performed using approximately 10^8 (5×10^7–10^9) bacteria/ml (approximately 10^5–10^8 counts per minute) or 200 to 400 ng of toxin/overlay (10^5–10^6 counts per minute) incubated and washed as in the peroxidase-based assay with the omission of second antibody steps and reagent development. Radiolabeled chromatograms are dried and developed by autoradiography performed using X-ray film such as Kodak X-OMAT AR film at –20°C or –70°C. Typical development times range from a few hours to a few days.[1]

VI. PROTOCOLS

The results of a TLC overlay and corresponding orcinol-sprayed plate are shown in Figure 2.1. All steps were performed at room temperature (22–24°C). Identical TLC plates (plastic-backed Polygram silica gel G plates, Brinkmann Instruments) were spotted with globotriaosyl ceramide standard (commercially available from Matreya) or lipids extracted from 5×10^6 cells with 10 volumes of chloroform-methanol (2:1, v/v) followed by reextraction of the residue with five volumes of chloroform–methanol (1:2). Extracts were pooled, dried on a rotary evaporator, resuspended in a minimal volume (<50 µl/10^6 cells) of chloroform–methanol (2:1). Lanes of lipid extract are as follows: (1) U266 myeloma cells (cell line available through American Type Culture Collector (ATCC)) that have been selected to be verotoxin-resistant; (2) U266 wild-type cells; and (3) 0.5 µg of globotriaosyl ceramide standard. Plates were developed in chloroform–methanol–water (65:25:4) and air dried. In (A) the plate was sprayed with orcinol–sulfuric acid reagent[15] and heated at 100°C for 10 min (glycolipids stain purple). In (B), the plate was blocked for 2 h in 1% gelatin (Type B from bovine skin, Sigma Chemical), washed three times with TBS, pH 7.4, and incubated for 1 h with verotoxin-1 (VT-1, 1 µg in 15 ml of TBS). The plate was again washed with TBS, and incubated for 1 h with a monoclonal antibody (PH1) directed against the B subunit of VT-1 at a 1:1000 dilution (stock solution approximately 1.6 mg/ml—a similar monoclonal antibody is available from the American Type Culture Collection. The chromatograms were washed, incubated with peroxidase-labeled antimouse immunoglobulin antisera, 1:2000 dilution in TBS (horseradish peroxidase-labeled goat antimouse IgG, Sigma) for 1 h, washed in TBS and developed with 4-chloro-1-naphthol solution[60] (Sigma). Glycosphingolipids, which function as verotoxin receptors, appear as purple spots on a clean white background (a yellow camera filter is beneficial when using most black and white films to record the results). Only bands comigrating with the globotriaosyl ceramide standard bound VT-1. Note that the VT-1 overlay procedure is not only specific in that it identifies Galα1-4Gal-containing glycolipids such as globotriaosyl ceramide, it is also more sensitive than orcinol for identifying globotriaosyl ceramide bands on thin-layer chromatograms.

The results of peroxidase-based vs. radiolabel-based TLC overlays performed using identically spotted and developed TLC plates are shown in Figure 2.2. Lipids were prepared and plates were spotted as per protocol given above for Figure 2.1 results. Lanes of lipid extracts from cell lines (available through ATCC) are as follows: (1) globotriaosyl ceramide standard (1 µg); (2) ACHN (renal cell carcinoma cells; (3) KG-1 cells; (4) KG-1a cells (myelocytic leukemia); (5) THP-1 (monocytic leukemia); and (6) Ramos (Burkitt's lymphoma). In (A), the overlay was performed using horseradish peroxidase-labeled second antibody and 4-chloro-1-naphthol as developing reagent as given in Figure 2.1B. In (B), overlay was performed using VT-1 labeled with ^{125}I using iodogen technique.[60] Following blocking with gelatin, 200 ng of ^{125}I-labeled toxin (5×10^6 cpm) was applied to the TLC plate in TBS for 1 h. The chromatogram was washed with TBS, air dried, and developed for 24 h using Kodak X-OMAT AR film.

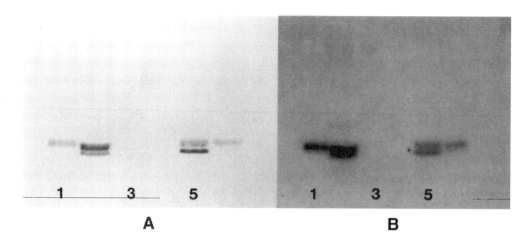

FIGURE 2.2 Peroxidase-based versus radiolabel-based TLC overlays performed using identically spotted and developed TLC plates. Lanes of lipid extracts: (1) globotriaosyl ceramide standard and extracts from the cell lines (2) ACHN; (3) KG-1; (4) KG-1a; (5) THP-1, and (6) Ramos. (A) Peroxidase-based overlay performed as in Figure 2.1. (B) Overlay performed using ^{125}I-labeled VT-1.

VII. CONCLUSIONS

With the advent of TLC overlay procedures, we have seen a dramatic increase in data concerning the receptor function of glycolipids for bacterial pathogens and their toxins. New bacterial species that use glycolipids and/or phospholipids as receptors are identified each year. However, we are just beginning to understand the role of glycolipids in the eukaryotic cells that possess them.[57,61-63] Because of the demonstrated mimicry of host proteins by the verotoxin B-subunit,[64] the bacterial toxin techniques and the research that has accumulated concerning toxin structure and Gb3-binding parameters can be used to guide functional studies of host protein–glycolipid interaction.

In addition to its use in separating and identifying potential bacterial receptors, TLC will continue to be used to investigate bacterial lipid composition and lipase activity. The potential clinical relevance of such studies in diagnosis, for example, has already been mentioned. There are also potential applications in industrial and environmental microbiology. TLC is quick, inexpensive, and requires little equipment. Furthermore, advances such as the introduction of HPTLC plates and enzymatic modification and direct mass spectroscopy of lipids on TLC plates have improved and expanded the analytical capabilities.[4] Certainly with regard to lipid analysis, TLC will continue to play a major role as new functions for glycolipids and phospholipids, including their true functions in eukaryotic cell physiology and cell adhesion, continue to be discovered.

REFERENCES

1. **Karlsson, K.-A. and Stromberg, N.,** Overlay and solid-phase analysis of glycolipid receptors for bacteria and viruses, in *Methods in Enzymology*, Vol. 138, Ginsburg, V., Ed., Academic Press, New York, 1987, 220–233.
2. **Jimenez-Lucho, V., Ginsburg, V., and Krivan, H.,** *Cryptococcus neoformans*, *Candida albicans*, and other fungi bind specifically to the glycosphingolipid lactosylceramide (Galβ1-4Glcβ1-1Cer), a possible adhesin receptor for yeasts, *Infect. Immunol.*, 58, 2085, 1990.
3. **Magnani, J. L., Smith, D. F., and Ginsburg, V.,** Detection of gangliosides that bind cholera toxin: direct binding of ^{125}I-labeled toxin to thin-layer chromatograms, *Anal. Biochem.*, 109, 399, 1980.
4. **Schnaar, R. L. and Needham, L. K.,** Thin-layer chromatography of glycosphingolipids, in *Methods in Enzymology*, Vol. 230, Lennarz, W. J. and Hart, G. W., Eds., Academic Press, New York, 1994, 371–389.
5. **Lingwood, C. A.,** Bacterial cell adhesins/glycolipid receptors, *Curr. Opin. Struct. Biol.*, 2, 693, 1992.

6. **Magnani, J. L., Brockhaus, M., Smith, D. F., Ginsburg, V., Blaszczyk, M., Mitchell, K. F., Steplewski, Z., and Koprowski, H.**, A monosialoganglioside is a monoclonal antibody-defined antigen of colon carcinoma, *Science*, 212, 55, 1981.
7. **Karlsson, K.-A.**, Animal glycosphingolipids as membrane attachment sites for bacteria, *Annu. Rev. Biochem.*, 58, 309, 1989.
8. **Radin, N. S.**, Extraction of tissue lipids with a solvent of low toxicity, in *Methods in Enzymology*, Vol. 72, Lowenstein, J. W., Ed., Academic Press, New York, 1981, 5–7.
9. **Folch, J., Lees, M., and Sloane-Stanley, G. H.**, A simple method for the isolation and purification of total lipids from animal tissues, *J. Biol. Chem.*, 226, 497, 1957.
10. **Hakomori, S.**, Chemistry of glycosphingolipids, in *Sphingolipid Biochemistry, Handbook of Lipid Research*, Vol. 3, Kanfer, J. N. and Hakomori, S., Eds., Plenum Press, New York, 1983, 1–36.
11. **Katz, H. R., Schwarting, G. A., LeBlanc, P. A., Austen, K. F., and Stevens, R. L.**, Identification of the neutral glycosphingolipids of murine mast cells: expression of Forssman glycolipid by the serosal but not the bone marrow-derived subclass, *J. Immunol.*, 134, 2617, 1985.
12. **Schnaar, R. L.**, Isolation of glycolipids, in *Methods in Enzymology*, Vol. 230, Lennarz, W. J. and Hart, G. W., Eds., Academic Press, New York, 1994, 348–370.
13. **Baumann, H., Nudelman, E., Watanabe, K., and Hakomori, S.**, Neutral fucolipids and fucogangliosides of rat hepatoma HTC and H35 cells, rat liver and hepatocytes, *Cancer Res.*, 39, 2637, 1979.
14. **Kaluzny, M. A., Duncan, L. A., Merritt, M. V., and Epps, D. E.**, Rapid separation of lipid classes in high yield and purity using bonded phase columns, *J. Lipid Res.*, 26, 135, 1985.
15. **Christie, W.**, *Lipid Analysis: Isolation, Separation, Identification and Structural Analysis of Lipids*, Pergamon Press, Oxford, 1980.
16. **Saito, T. and Hakomori, S.**, Quantitative isolation of total glycosphingolipids from animal cells, *J. Lipid. Res.*, 12, 257, 1971.
17. **Yu, R. K. and Ledeen, R. W.**, Gangliosides of human, bovine, and rabbit plasma, *J. Lipid. Res.*, 13, 680, 1972.
18. **Maloney, M. D. and Semprevivo, L. H.**, Thin-layer and liquid column chromatographic analyses of the lipids of adult *Onchocerca gibsoni*, *Parasitol. Res.*, 77, 294, 1991.
19. **Maloney, M. D., Semprevivo, L. H., and Coles, G. C.**, A comparison of the glycolipid compositions of cercarial and adult *Schistosoma mansoni* and their associated hosts, *Int. J. Parasitol.*, 20, 1091, 1990.
20. **Childs, R. A., Drickamer, K., Kawasaki, T., Thiel, S., Mizuochi, T., and Feizi, T.**, Neoglycolipids as probes of oligosaccharide recognition by recombinant and natural mannose-binding proteins of the rat and man, *Biochem. J.*, 262, 131, 1989.
21. **Magnusson, G., Ahlfors, S., Dahmén, J., Jansson, K., Nilsson, U., Noori, G., Stenvall, K., and Tjörnebo, A.**, Prespacer glycosides in glycoconjugate chemistry. Dibromoisobutyl glycosides for the synthesis of neoglycolipids, neoglycoproteins, neoglycoparticles, and soluble glycosides, *J. Org. Chem.*, 55, 3932, 1990.
22. **Ramphal, R., Carnoy, C., Fievre, S., Michalski, J., Houdret, N., Lamblin, G., Strecker, G., and Roussel, P.**, *Pseudomonas aeruginosa* recognizes carbohydrate chains containing type 1 (Galβ1-3GlcNAc) or type 2 (Galβ1-4GlcNAc) disaccharide units, *Infect. Immunol.*, 59, 700, 1991.
23. **Fishman, P. H., Pacuszka, T., Hom, B., and Moss, J.**, Modification of ganglioside G_{M1}, *J. Biol. Chem.*, 255, 7657, 1980.
24. **Kiarash, A., Boyd, B., and Lingwood, C. A.**, Glycosphingolipid receptor function is modified by fatty acid content: Verotoxin 1 and Verotoxin 2c preferentially recognize different globotriaosyl ceramide fatty acid homologs, *J. Biol. Chem.*, 269, 11138, 1994.
25. **Yiu, S. and Lingwood, C. A.**, Polyisobutylmethacrylate modifies glycolipid binding specificity of Verotoxin 1 in thin layer chromatogram overlay procedures, *Anal. Biochem.*, 202, 188, 1992.
26. **Head, S., Ramotar, K., and Lingwood, C. A.**, Modification of the glycolipid-binding specificity of verocytotoxin by polymyxin B and other cyclic amphipathic peptides, *Infect. Immunol.*, 58, 1532, 1990.
27. **Fried, B. and Sherma, J.**, *Thin-Layer Chromatography: Techniques and Applications*, Marcel Dekker, New York, 1982, 202.
28. **Hiraishi, A., Shin, Y. K., and Sugiyama, J.**, Rapid profiling of bacterial quinones by two-dimensional thin-layer chromatography, *Lett. Appl. Microbiol.*, 14, 170, 1992.
29. **Hollingsworth, R. I. and Dazzo, F. B.**, The use of phenylcarbamate derivatives for the GC-MS characterization of 3-hydroxyalkanoic acids in the lipopolysaccharides of *Rhizobium*, *J. Microbiol. Meth.*, 7, 295, 1988.
30. **Barclay, R., Furst, V., and Smith, I.**, A simple and rapid method for the detection and identification of mycobacteria using mycobactin, *J. Med. Microbiol.*, 37, 286, 1992.
31. **Laszlo, A., Papa, F., and David, H. L.**, Thin-layer chromatography systems for the identification of *Mycobacterium tuberculosis*, *M. bovis* BCG, *M. kansasii*, *M. gastri* and *M. marinum*, *Res. Microbiol.*, 143, 519, 1992.
32. **Tsukamura, M.**, Differentiation between *Mycobacterium avium* and *Mycobacterium intacellulare* by thin-layer chromatography of lipid fraction after incubation with [^{35}S] methionine, *Microbiol. Immunol.*, 36, 529, 1992.
33. **Hines, M. E. I. and Frazier, K. S.**, Differentiation of mycobacteria on the basis of chemotype profiles by using matrix solid-phase dispersion and thin-layer chromatography, *J. Clin. Microbiol.*, 31, 610, 1993.

34. De Silva, N. S. and Quinn, P. A., Endogenous activity of phospholipases A and C in *Ureaplasma urealyticum*, *J. Clin. Microbiol.*, 23, 354, 1986.
35. De Silva, N. S. and Quinn, P. A., Rapid screening assay for phospholipase C activity in mycoplasmas. *J. Clin. Micro.*, 25, 729, 1987.
36. Falk, P., Hoskins, L. C., and Larson, G., Enhancing effects of bile salts on the degradation of glycosphingolipids by glycosidases from bacteria of the human fecal flora, *Biochim. Biophys. Acta.*, 1084, 139, 1991.
37. Hansson, G. C., Karlsson, K.-A., Larson, G., Stromberg, N., and Thurin, J., Carbohydrate-specific adhesion of bacteria to thin-layer chromatograms: a rationalized approach to the study of host cell glycolipid receptors, *Anal. Biochem.*, 146, 158, 1985.
38. Krivan, H. C., Roberts, D. D., and Ginsburg, V., Many pulmonary pathogenic bacteria bind specifically to the carbohydrate sequence GalNAcβ1-4 Gal found in some glycolipids, *Proc. Natl. Acad. Sci., U.S.A.*, 85, 6157, 1988.
39. Krivan, H. C., Ginsburg, V., and Roberts, D. D., *Pseudomonas aeruginosa* and *Pseudomonas cepacia* isolated from cystic fibrosis patients bind specifically to gangliotetraosylceramide (asialo GM1) and gangliotriaosylceramide (asialo GM2), *Arch. Biochem. Biophys.*, 260, 493, 1988.
40. Baker, N., Hansson, G. C., Leffler, H., Riise, G., and Svanborg-Edën, C., Glycosphingolipid receptors for *Pseudomonas aeruginosa*, *Infect. Immunol.*, 58, 2361, 1990.
41. Andersson, P., Engberg, I., Lidin-Janson, G., Lincoln, K., Hull, R., Hull, S., and Svanborg, C., Persistence of *Escherichia coli* bacteriuria is not determined by bacterial adherence, *Infect. Immunol.*, 59, 2915, 1991.
42. Bloch, C., Stocker, B., and Orndorff, P., A key role for type 1 pili in enterobacterial communicability, *Mol. Microbiol.*, 6, 697, 1992.
43. Lund, B., Lindberg, F., Marklund, B.-I., and Normark, S., The PapG protein is the α-D-galactopyranosyl-('4)-β-D-galactopyranose-binding adhesin of uropathogenic *Escherichia coli*, *Proc. Natl. Acad. Sci., U.S.A.*, 84, 5898, 1987.
44. Lund, B., Marklund, B.-I., Strömberg, N., Lindberg, F., Karlsson, K.-A., and Normark, S., Uropathogenic *Escherichia coli* can express serologically identical pili of different receptor binding specificities, *Mol. Microbiol.*, 2, 255, 1988.
45. Strömberg, N., Marklund, B.-I., Lund, B., Ilver, D., Hamers, A., Gaastra, W., Karlsson, K.-A., and Normand, S., Host-specificity of uropathogenic *Escherichia coli* depends on differences in binding specificity to Galα1-4Gal-containing isoreceptors, *EMBO J.*, 9, 2001, 1990.
46. Karr, J. F., Nowicki, B. J., Truong, L. D., Hull, R. A., Moulds, J. J., and Hull, S. I., Pap-2-encoded fimbriae adhere to the P blood group-related glycosphingolipid stage-specific embryonic antigen 4 in the human kidney, *Infect. Immunol.*, 58, 4055, 1990.
47. Lindstedt, R., Larson, G., Falk, P., Jodal, U., Leffler, H., and Svanborg, C., The receptor repertoire defines the host range for attaching *Escherichia coli* strains that recognize globo-A, *Infect. Immunol.*, 59, 1086, 1991.
48. Rhen, M., Klem, P., and Korhonen, T. K., Identification of two new hemagglutinins of *Escherichia coli*: N-acetyl-D-glucosamine-specific fimbriae and a blood group M-specific N-acetyl-D-glucosamine-specific agglutinin, by cloning the corresponding genes in *Escherichia coli* K-12, *J. Bacteriol.*, 168, 1234, 1986.
49. Korhonen, T. K., Vaisanen-Rhen, V., Rhen, M., Pere, A., Parkkinen, A., and Finne, J., *E. coli* fimbriae recognizing sialyl galactosides, *J. Bacteriol.*, 159, 762, 1984.
50. Baker, N. R., Minor, V., Deal, C., Sahrabadi, M. S., Simpson, D. A., and Woods, D. E., *Pseudomonas aeruginosa* exoenzyme S is an adhesin, *Infect. Immunol.*, 59, 2859, 1991.
51. Lingwood, C. A., Cheng, M., Krivan, H., and Woods, D. E., Glycolipid receptor binding specificity of exoenzyme S from *Pseudomonas aeruginosa*, *Biochem. Biophys. Res. Commun.*, 175, 1076, 1991.
52. Sokol, P. A., Dennis, J. J., MacDougall, P. C., Sexton, M., and Woods, D. E., Cloning and expression of the *Pseudomonas aeruginosa* exoenzyme S toxin gene, *Microbiol. Pathol.*, 8, 243, 1990.
53. Lingwood, C. A., Huesca, M., and Kuksis, A., The glycerolipid receptor for *Helicobacter pylori* (and exoenzyme S) is phosphatidyl ethanolamine, *Infect. Immunol.*, 60, 2470, 1992.
54. Krivan, H., Nilsson, B., Lingwood, C. A., and Ryu, H., *Chlamydia trachomatis* and *Chlamydia pneumoniae* bind specifically to phosphatidylethanolamine in HeLa cells and to GalNacβ1-4Galβ1-4Glc sequences found in asialo-GM$_1$ and asialo-GM$_2$, *Biochem. Biophys. Res. Commun.*, 175, 1082, 1991.
55. Saitoh, T., Natomi, H., Zhao, W., Okuzumi, K., Sugano, K., Iwamori, M., and Nagai, Y., Identification of glycolipid receptors for *Helicobacter pylori* by TLC-immunostaining, *FEBS Lett.*, 282, 385, 1991.
56. Lingwood, C. A., Verotoxins and their glycolipid receptors, in *Advances in Lipid Research*, Vol. 25, Bell, R., Hannun, Y. A., and Merrill, A., Jr., Academic Press, San Diego, 1993, 189–212.
57. Maloney, M. and Lingwood, C., Interaction of verotoxins with glycosphingolipids, *Trends Glycosci. Glycotech.*, 5, 23, 1993.
58. Pellizzari, A., Pang, H., and Lingwood, C. A., Binding of verocytotoxin 1 to its receptor is influenced by differences in receptor fatty acid content, *J. Biochem.*, 31, 1363, 1992.
59. Costello, P. and Green, F., Binding affinity of serum immunoglobin G to cardiolipin and other phospholipids in patients with systemic lupus erythematosus and syphillis, *Infect. Immunol.*, 56, 1738, 1988.
60. Harlow, E. and Lane, D., *Antibodies: A Laboratory Manual*, Cold Spring Harbor Laboratory, New York, 1988, 324–338, 506.

61. **Hakomori, S.-I.,** Bifunctional role of glycosphingolipids, *J. Biol. Chem.*, 265, 18713, 1990.
62. **Bevilacqua, M., Butcher, E., Furie, B., Furie, B., Gallatin, M., Gimbrone, M., Harlan, J., Kishimoto, K., Lasky, L., McEver, R., Paulson, J., Rosen, S., Seed, B., Siegelman, M., Springer, T., Stoolman, L., Tedder, T., Varki, A., Wagner, D., Weissman, I., et al.,** Selectins: a family of adhesin receptors, *Cell*, 67, 233, 1991.
63. **Cummings, R. D. and Smith, D. F.,** The selectin family of carbohydrate-binding proteins: structure and importance of carbohydrate ligands for cell adhesion, *BioEssays*, 14, 849, 1992.
64. **Maloney, M. D. and Lingwood, C. A.,** CD19 has a potential CD77 (globotriaosyl ceramide)-binding site with sequence similarity to verotoxin B-subunits: implications of molecular mimicry for B cell adhesion and enterohemorrhagic *Escherichia coli* pathogenesis, *J. Exp. Med.*, 180, 191, 1994.

Chapter 3

THIN-LAYER CHROMATOGRAPHY IN PLANT SCIENCES

Jacques Pothier

CONTENTS

I. Introduction 34

II. Alkaloids 35
    A. Isolation Methods 35
    B. Detection of Alkaloids in TLC 35
        1. Dragendorff's Reagent 35
        2. Potassium Iodoplatinate Reagent 35
    C. Special Part 35
        1. Introduction 35
        2. Individual Groups of Alkaloids 36
            a. Indole Alkaloids 36
            b. Ergot Alkaloids 36
            c. Quinolizidine Alkaloids 37
            d. Quinoline and Isoquinioline Alkaloids 37
            e. Opium 37
            f. Pyrrolizidine Alkaloids 38
            g. Tropolone Alkaloids 38
            h. Miscellaneous Classes of Alkaloids 38
            i. Tropane Alkaloids 39
            j. Quaternary Alkaloids 39

III. Flavonoids 39
    A. Introduction 39
    B. General Consideration 39
    C. Special Parts 40
        1. Mobile Phases and Eluents 40
            a. Bi-Dimensional TLC 40
        2. Silica Gel 40
        3. Polyamide 40
            a. Aglycones 40
            b. Flavonoids 41
        4. Cellulose 41
        5. Reversed Phase C_{18} 41
        6. Detection 41
            a. Without Chemical Treatment 41
            b. Detection with Reagents 42

IV. Anthocyanins 42
    A. Isolation Methods 42
    B. Special Parts 42

V. Essential Oils 43
    A. Isolation (Extraction of Essential Oils) 43

        B. General Considerations 43
        C. Special Part 43

VI. Cardiac Glycoside Plants 44
        A. Isolation 44
        B. Detection 44
        C. Special Part 45

VII. Saponins 45
        A. Triterpene Saponins 45
        B. Steroid Saponins 45
            1. Isolation 45
            2. Detection 45

References 46

I. INTRODUCTION

Thin-layer chromatography (TLC) has been used since the 1960s to analyze and identify compounds in plants (see Stahl[1] and Randerath[2]).

The special advantages of TLC are sensitivity, speed, and versatility.

Versatility allows for a great number of different sorbents available in commercial form, e.g., silica gel TLC or HPTLC plates, reversed phase C_8 and C_{18} plates, cyano- and diol plates, aluminium oxide and cellulose; in addition to the choice of sorbents, detection in TLC is an advantage over other analytical methods, because it is possible to spray the plates and heat them at 120°C. TLC is *fast* due to the compact nature of the adsorbent, and this is an advantage when working with labile compounds, which is often the case in plant analysis.

Sensitivity allows for separations of less than μg amounts of material. There are other reasons for the popularity of this method in plant analysis, one of which is that in addition to qualitative detection, TLC also provides semiquantitative information on the main active constituents of plant preparations. TLC also provides a chromatographic fingerprint and is suitable for monitoring the identity and purity of plants, and for detecting adulterations and substitutions. TLC is also useful in plant analysis because it is possible to work on crude extracts, which is not the case with other analytical methods.

The literature on the TLC of plant analysis is extensive, with the most important contributions being those of Stahl,[1] Randerath,[2] and Wagner.[3]

Most TLC studies in plant chemistry have used silica gel as the sorbent, but studies on aluminium oxide and reversed phase plates are also available. Many TLC studies are listed in the different pharmacopœias of the world (European, German, French, Swiss, Austrian, and American) as reported by Wagner.[3]

There are numerous compounds in plants. The information in this chapter is limited to the compounds noted in **bold**, since many other substances have been covered in other chapters of this book. It is thus possible to classify the main compounds extracted from plants[4] (see below).

1. Nitrogen compounds: **alkaloids**, amines, amino acids;
2. Sugars and derivatives;
3. Phenolic compounds: Phenols and phenolic acids: **flavonoids**, **anthocyanins**;
4. Terpenoids: **essential oils**, diterpenoids, triterpenoids and steroids, carotenoids, **saponins**, **cardiac glycosides**;
5. Organic acids: lipids and related compounds. Plant acids, fatty acids and lipids, alkanes and related hydrocarbons; sulfur compounds;
6. Macromolecules: Nucleic acids, proteins, polysaccharides.

II. ALKALOIDS

A. ISOLATION METHODS

An important goal of alkaloid investigation is the isolation of alkaloids in their authentic forms. Much of the activity in the field of alkaloid chemistry has been directed toward the development of optimum methods without the formation of artifacts.

Alkaloids occur in plants as salts of organic acids or inorganic acids with complex mixtures of water-soluble compounds, e.g., lipids, gums, resins, and tannins. It is necessary to remove these nonalkaloid compounds during isolation and purification of plant alkaloids. The beginning of the extraction is done with organic solvents immiscible in water. After the displacement of the alkaloid bases from their salts, using treatment with a mineral base, ammonia is sufficiently basic to liberate most of the common alkaloids; ammonia can be removed afterward. In plant materials rich in lipids, a preliminary treatment with hexane or petroleum ether is necessary. In plant materials rich in tannins, sodium hydroxide suffices to cleave the alkaloid tannate salts. Extraction of the alkaloids as salts by means of alcohol–water mixtures is often preferred, with less risk of altering the genuine alkaloids under such conditions. Extraction with organic solvents such as dichloromethane, chloroform, and diethyl ether gives extracts containing gums, lipids, and resins. It is necessary to further purify the crude alkaloid extracts. The isolation and purification methods chosen to investigate alkaloids depend on the chromatographic technique to be used. TLC and OPLC (overpressured layer chromatography) are of great value when crude extracts or mixtures of unknown composition have to be analyzed and little information is available about the components. Many problems concerning alkaloids have been solved using these techniques.

B. DETECTION OF ALKALOIDS IN TLC

Many different methods have been described to characterize alkaloids both nonselective and selective. The nonselective methods, which are used for organic compounds in TLC, such as quenching of UV light on fluorescent plates, iodine spray reagent, or iodine vapors and concentrated sulfuric acid, are often used. Selective and specific alkaloid reagents include those listed above or various modifications of Dragendorff's reagent and potassium iodoplatinate.

1. Dragendorff's Reagent

The first spray reagents described for this purpose were the modifications of Munier and Macheboeuf[5] and Munier.[6] Several modifications of this reagent are described in the work of Baerheim Svendsen, and Verpoorte[7]; various nonalkaloid compounds also react with Dragendorff's reagent,[7] and the sensitivity for such compounds is increased if the plate is sprayed with sulfuric acid or sodium nitrite after spraying with Dragendorff's reagent. The sensitivity of the reagent for alkaloids is in the range 0.01 to 1 µg.

2. Potassium Iodoplatinate Reagent

This reagent exists in various forms. The sensitivity is the same as that for Dragendorff's reagent (0.01–0.1 µg), but iodoplatinate potassium reagent has several advantages. The most interesting advantage is that different colors are obtained with alkaloids, the colors varying from brown through violet to blue. The separation of 81 authentic samples of alkaloids with the colors obtained after spraying with iodoplatinate reagent has been described.[9]

C. SPECIAL PART

1. Introduction

The alkaloids constitute a heterogeneous group of substances that are characterized, with a few exceptions, by the presence of a tertiary or quaternary amino group in the molecule. Common organic solvents are used as mobile phases. Many systems contain chloroform or dichloromethane, the eluting power of which may be decreased by the addition of cyclohexane or increased by acetone, ethanol, or methanol. With untreated silica gel layers, the mobile phase is often made alkaline by the addition of ammonia to the more polar solvents, or diethylamine to the less polar solvents.

Waldi et al.[9] described the analysis of a series of alkaloids by means of eight TLC systems. The alkaloids were identified by their hR_f values in combination with their appearance under UV light at 366 nm and their color reaction obtained by spraying with iodoplatinate reagent. Sunshine et al.[10] described a scheme for 138 drugs, including a series of alkaloids identified by their hR_f values in seven different solvent systems. Noirfalise and Mees[11] presented a scheme for 16 alkaloids with hR_f values obtained with the same mobile phase, on four different stationary phases. Ahrend and Tiess[12] described the identification of 39 alkaloids on silica; the mobile phases used were: methanol, chloroform–acetone (90:10), chloroform–butanol–25% ammonia (70:40:5) and butanol–acetic acid–water (4:1:5 [upper phase]). Moffat et al.[13] presented an identification scheme for basic substances, by determining the discriminating power of some TLC systems, in 37 TLC systems investigated. The best results were obtained with either cyclohexane–toluene–diethylamine (75:15:10), or chloroform–methanol (90:10). These mobile phases are used with silica plates impregnated with 0.1 N sodium hydroxide. Armstrong[14] used silica gel plates with the following eluents: methanol–ammonia (100:1.5), chloroform–diethylamine (90:10), and cyclohexane–diethylamine (90:10). Breiter and Helger[15] described a separation method for alkaloids using ethyl acetate–methanol–ammonia (90:10:1) on silica gel. Pothier et al.[16] described the separation of numerous alkaloids in plant extracts by overpressured layer chromatography (OPLC) on alumina plates with ethyl acetate as a single solvent. The method was applied to the determination of hR_f values of 81 authentic samples of alkaloids.

2. Individual Groups of Alkaloids

The purpose of this section is to describe the chromatographic systems, i.e., sorbents and mobile phases used recently in the analysis of plant alkaloids and also the derivative techniques used to identify various compounds.

a. Indole Alkaloids

Rauwolfia. Most of the *Rauwolfia* alkaloids belong to the β-carboline type of structure. The main alkaloids, serpentine, ajmaline, and reserpine can be separated with the solvent: toluene–ethyl acetate–diethylamine (70:20:10) or heptane–ethylmethylketone–methanol (58:34:8) and detected with Dragendorff's reagent. All *Rauwolfia* alkaloids show a brown color in visible light; ajmaline shows a prominent fluorescence quenching in UV light at 254 nm, and can be made visible by treatment with concentrated nitric acid.[4] The use of cyclohexane–diethyl ether (3:2) and chloroform–methanol (19:1) or (9:1) on silica gel gives a good separation of six indole alkaloids from *Rauwolfia cubana* stem bark.[16]

Catharanthus. Indole alkaloids of *Catharanthus roseus* have been studied recently in plant tissues. Catharanthine, vindoline, leurosine, and vinblastine have been studied on silica gel with ethyl acetate–ethanol–benzene–ammonia (100:5:5:3).[17] Quantitative analysis of serpentine and ajmalicine has been done on silica gel with chloroform–methanol (9:1) with detection by spraying with Dragendorff's reagent followed by spraying with 10% sodium nitrite and scanning after 24 h at 435 nm.[18] Two-dimensional TLC of indole alkaloids on silica gel has been studied with ethyl acetate–methanol (4:1) and dichloromethane–methanol (12:1)[19] as mobile phases.

b. Ergot Alkaloids

The separation of the ergot alkaloids is not easy with the usual solvents and it is often necessary to employ tertiary or quaternary eluents. Detection of compounds is obtained with Van Urk's reagent. Ergometrine, ergotamine, and ergocristine can be separated with toluene–ethyl acetate–diethylamine (70:20:10) or toluene–chloroform–ethanol (28.5:57:14.5).[3]

Two unusual ergopeptines have been separated from the main ergotamine and ergotoxine groups on silica with dichloromethane–isopropanol (92:8), or dichloromethane–methanol (90:10) with detection by UV at 366 nm or Van Urk's reagent.[20] Ergot alkaloids have been separated by TLC with gradient elution using the following solvent systems: chloroform–diethylamine (12 stages,

7 steps) chloroform–acetone–diethylamine (11 stages, 5 steps) with visualization under UV light and also by specific color reactions with Auport's reagent.[21]

c. Quinolizidine Alkaloids

The quinolizidine alkaloids (lupine alkaloids) are present in the family *Fabaceae*. The majority of them are not fluorescent, so it is possible to carry out quantitative analysis principally by derivatization with Dragendorff's reagent or iodoplatinate reagent.[22] In *Chamaecytisus pygmaeus*, lupanine, spartein, cytisin, *N*-methyl cytisin, and chamaecytisin have been separated on silica gel with chloroform–methanol–ammonia (85:14:7), chloroform–methanol (8:2), cyclohexane–diethylamine (7:3) cyclohexane–dichloromethane–diethylamine (4:4:2) and toluene–acetone–ethanol–ammonia (30:40:12:4).[23] Some authors characterized dansylated alkaloid as hydroxynorcytisin from *Laburnum anagyroides*[24] on silica gel with cyclohexane–ethyl acetate (2:3), benzene–diisopropylamine (5:1), chloroform–butanol–water (5:3:1) with visualization under UV light at 366 nm; quantification by GC after elution with methanol in the genus *Virgilia* on silica gel with methanol–chloroform–ammonia (85:14:1) to separate classical lupine alkaloid and virgiboïdine, virgiline, and orobordine.[25] Detection with UV and iodine vapors using TLC with ether–methanol–ammonia (40:4:1) as the mobile phase has also been described to characterize sophocarpine, matrine, and sophoramine on silica gel.[26] OPLC has also been employed on alumina plates with ethyl acetate as a single solvent to separate lupinine, sparteine, lupanine, and 13-hydroxylupanine from the genus *Lupinus*.[27]

There are other methods of derivation, for example: quantification by densitometry at 565 nm after detection by Dragendorff's reagent on silica gel plates with dichloromethane–methanol–10% ammonia (83:15:2)[28]; UV detection by TLC of lupanine and 7-hydroxylupanine on silica gel with chloroform–methanol–ammonia (85:14:1)[29]; detection by iodoplatinate has also been used in TLC on silica gel plates with chloroform–cyclohexane–butylamine (5:4:1), or aluminium oxide with 1.5% methanol in chloroform; also, with an UV detection after heating at 100°C for 3 min and visualization by UV light at 254 nm.[30]

d. Quinoline and Isoquinoline Alkaloids

Cinchona. There are many methods described in the literature regarding *Cinchona* alkaloid analysis.[7] One of the most sensitive quantitative measurements was by fluorescence detection after postchromatographic derivatization with 10% sulphuric acid solution. HPTLC and TLC of 17 *Cinchona* alkaloids on silica with chloroform–acetone–methanol–25% aqueous ammonia (60:20:20:1) have been reported.[31] The European pharmacopœia describes chloroform–diethylamine (90:10) on silica gel to characterize (following treatment with sulphuric acid) the main alkaloids quinine, quinidine, cinchonine, cinchonidine, with toluene–ethyl acetate–diethylamine (70:20:10); it is also possible, in addition to the four main alkaloids, to characterize other *Cinchona* alkaloids, and epiquinine bases can also be detected.

The epibases (threo-compounds) of quinine and quinidine migrate in the lower hR_f range. The dihydrocompounds of quinine, quinidine, cinchonine, and cinchonidine show no fluorescence in UV light at 365 nm. Certain yellow-green fluorescent zones may represent quinonoid structures.[3]

e. Opium

The determination of the opium alkaloids is one of the most important analyses in the alkaloid field; many methods have been described recently with different techniques of derivatization:

TLC of morphine, thebaine, codeine, and papaverine on silica gel with toluene–acetone–ethanol–ammonia (45:45:7:3) as the mobile phase. Visualization was made under UV light at 254 nm.[32] Quantitative determination of the major alkaloids in the seeds and shells of *Papaver somniferum* on silica gel with a mobile phase of cyclohexane–ethylene diamine (4:1). Detection was by densitometry at 230 nm.[33] Preparation of substituted hydrazone derivatives of ketones with a morphine skeleton on silica gel with a mobile phase of either chloroform–acetone–diethylamine

(5:4:1) or chloroform–methanol (9:1) or benzene–methanol (8:2). Detection was by UV light or by spraying with Dragendorff's reagent.[34]

A separation of five opium alkaloids has been described on silica rods with benzene–ethanol (17:1) or (9:1) with quantification by FID[35] as well as a determination of morphine in opium by TLC on silica gel with a mobile phase of ethyl acetate–methanol–ammonia (85:10:5) and detection under UV light at 280 nm.[36] Many pharmacopoeias[3] use toluene–acetone–ethanol–ammonia (40:40:6:2) or toluene–ethyl acetate–diethylamine (70:20:10) mobile phases with silica gel; the quantitative analysis may be carried out by densitometry after visualization by Dragendorff's reagent with $NaNO_2$ reagent or Marquis' reagent; this gives a violet color with morphine and codeine, weak violet with thebaine and noscapine, brown with papaverine.

Chelidonium. The determination of isoquinoline alkaloids in Chelidonium majus is possible using a stepwise gradient with binary and tertiary eluents consisting of toluene–methanol–ethyl acetate-isopropanol with detection under UV light at 366 nm or by spraying with Dragendorff's reagent[37]; eight isoquinoline alkaloids were separated on silica gel with toluene–methanol–diethylamine (60:5:2) saturated with formamide, and detection was made by exposing the plates to iodine vapors and heating at 180°C for 20 min by fluorodensitometry at 350 nm.[38]

f. Pyrrolizidine Alkaloids

Senecio. The TLC of alkaloids from *Senecio vulgaris*, senecionine, senecionine N-oxide, agmatine, spermidine, spermine, ornithine, putrescine, and citrulline, were performed on silica gel with a mobile phase of chloroform–methanol–ammonia 25%–pentane (82:14:2.6:20) or acetone–methanol–ammonia 25% (40:30:20); detection was under UV light at 254 nm.[39]

g. Tropolone Alkaloids

Colchicum. TLC of colchicine can be performed on silica gel with a mobile phase of chloroform–methanol (95:5). Quantification is by densitometry at 243 nm[40] and it is possible to separate colchicine, 3-demethylcolchicine and demecolcine on silica gel with benzene–ethyl acetate–butylamine (50:40:10) or (70:20:10).[41] The determination of tropolone alkaloids by TLC[42] is possible with benzene–ethyl acetate–diethylamine–methane–water (15:12:3:6:1:2); benzene–ethyl acetate–diethylamine (5:4:1 + 8% methanol), with quantification by fluorodensitometry. Autumnaline, isoautumnaline, speciaritchine, speciacolchine, androcymbine, and colchiritchine are separated by chloroform–acetone–diethylamine (80:10:10), ethyl acetate–ethanol (75:25) and benzene–chloroform–diethylamine (80:10:10) on silica gel from *Colchicum ritchii*. Determination of hR_f values of 12 colchicines and 8 colchicinamides has been made by reversed phase TLC on silica gel, impregnated with 5% liquid paraffin.[43] With ethyl acetate–methanol–water (100:13.5:10) colchicine and the minor alkaloids, colchiceine, *N*-acetyldemecolchine, 1-ethyl-2-demethylcolchiceine and carnigenin give a yellow fluorescence in UV light at 365 nm. Colchicine in the solvent system toluene–methanol (86:14) after two developments at 15 cm forms a brown zone with Dragendorff's reagent and a yellow color with 10% ethanolic hydrochloric acid (Wagner).[3]

h. Miscellaneous Classes of Alkaloids

Aconitum. The separation and determination of poisonous alkaloids in the root of *Aconitum kusnezoffi*, aconitine, mesaconitine, and hypoaconitine can be done with the mobile phase ether–chloroform–ammonia (25:10:1) on silica gel[44] or cyclohexane–ethylacetate–ethylenediamine (80:10:10)[45] or hexane–chloroform (60:40); chloroform–methanol (80:20) or (97:3)[46] with detection under UV light at 254 nm; quantification by densitometry at 236 nm; other detection with Dragendorff's reagent, followed by spraying with a 5% $NaNO_2$ solution. Another separation has been done by HPTLC using the same condition and centrifugal TLC on aluminium oxide with a gradient of hexane, ether, and methanol.

Cocaine. Qualitative and quantitative HPTLC determination of cocaine, ecgonine, ecgonine methyl ester, and benzylecgonine were done on silica gel with a mobile phase of water–methanol–sodium acetate buffer (0.2 M aqueous) (28:12:60:10); detection by dipping into Dragendorff's reagent and postchromatographic derivatization after heating and fluorescence enhancing by dipping in light liquid paraffin–hexane (1:2) and by scanning in the fluorescence mode at 313/390 nm with the limit of detection being 10 ng.[47]

i. Tropane Alkaloids

(-) Hyoscyamine (or atropine) is the main alkaloid present in *Atropa belladonna*, *Hyoscyamus niger*, and *Datura stramonium*. The TLC differentiation of these drugs is based on the hyoscyamine/scopolamine ratio, and to a limited extent on the contents of the minor alkaloids. For identification and determination of alkaloid content, the European Pharmacopœia[3] describes a TLC technique on silica gel with a solvent system of acetone–water–ammonia (90:7:3); good results were also obtained with the eluent toluene–ethyl acetate–diethylamine (70:20:10).[3]

j. Quaternary Alkaloids

Quaternary alkaloids are highly polar compounds and their TLC analyses present some special problems; it is not possible to separate these compounds with classical eluents used for chromatography of alkaloids. Overpressured layer chromatography (OPLC) is suitable for this purpose on silica gel with a mobile phase of ethyl acetate–tetrahydrofuran–acetic acid (60:20:20); with this eluent, choline, betaine, carnitine, trigonelline, hydrastinine, berberine, tubocurarine, sanguinarine, and serpentine can be identified.[48] Good results can be obtained by TLC with the following eluents: *n*-propanol–formic acid–water (90:1:9) or toluene–ethyl acetate–diethylamine (70:20:10) on silica and cyclohexane chloroform–glacial acetic acid (45:45:10) on alumina oxide.[3]

III. FLAVONOIDS

A. INTRODUCTION

The main constituents of flavonoid drugs are 2-phenyl-γ benzopyrone. The term *flavonoids* is derived from "flavus" (yellow). Flavonoids are mainly water-soluble compounds that can be extracted with 70% ethanol; they remain in the aqueous layer, following a partition of this extract with hexane. Flavonoids are phenolic and thus change color when treated with base. They are easily detected on chromatograms containing conjugated aromatic systems and show intense absorption bands in UV light and visible regions of the spectrum. Flavonoids are generally present in plants bound as glycosides and any one flavonoid aglycone may occur in a single plant in several glycosidic combinations. When analyzing flavonoids it is better to examine the aglycones present in hydrolyzed plant extracts; flavonoids are present in plants as mixtures of different flavonoid classes. Classification of flavonoid types in plant tissue is based on the study of solubility properties and color reactions. This is followed by a one-dimensional chromatographic analysis of the hydrolyzed plant extract.

B. GENERAL CONSIDERATION

Because there are no general reactions of flavonoids, it is often the characterization of aglycone, which is used in general analytical methods. Therefore, TLC is a good technique for the determination of flavonoids because the choice of reagents is important. Various sorbents and eluents are used to identify flavonoids. The literature on this topic is considerable. Wagner[3] has described five chromatography solvents with the hR_f values of numerous flavonoids. Harbone[49] has provided tables listing suitable sorbents and mobile phases for routine separations of these compounds. Isaksen[50] has provided an extensive review on the TLC of natural pigments including flavonoids and anthocyanins. Optimal chromatographic systems, sorbents, and eluents for flavonoids and anthocyanins are considered in detail in that review. Betti et al.[51] described a chromatographic approach for the analysis of flavonoid compounds by planar chromatographic systems (layer and mobile phases were studied). The choice of the best systems for separating the components of a single class of compounds was also discussed.

C. SPECIAL PARTS
1. Mobile Phases and Eluents
The purpose of this section is to discuss the recent literature on the diversity of techniques used to separate flavonoids.

a. Bi-Dimensional TLC
Many authors have described separation of flavonoids using bi-dimensional systems and different mobile phases. HPTLC of flavone, 5-methoxy-isoflavone is achieved on silica gel with hexane–ether (3:1) (first direction) and chloroform–ethyl acetate (95:5) (second direction).[52] In addition to silica gel, reversed phase plates (C_{18}) are also used with hexane–ethyl acetate (72:27:1) (first direction) and 1M acetic acid in 50% methanol (second direction).[53]

Polyamide has been used as a layer to taxonomic variation in the flavonoid from *Oryza latifolia* with benzene–MEK–methanol–butanol (40:20:15:0.2) (first direction) and water–butanol–acetone–dioxane (8:3:2:1) (second direction).[54]

Cellulose is used as the layer in research on the flavonoids of wood and bark species in the genus *Eucalyptus* with the upper mobile phase of butanol–acetic acid–water (4:1:5) for the first direction and 30% acetic acid for the second.[55]

2. Silica Gel
Silica gel is used in numerous TLC studies of flavonoids. These compounds are polar and therefore many authors use very polar eluents: water, acetic or formic acid, MEK (methyl ethyl ketone), ethyl acetate, or acetone. For example, we can quote different works; the separation of flavonoid glycosides in the taxonomy study of *Ulmus* sp.[56] with ethyl acetate–formic acid–water (6:1:1); the separation of quercetin 3-O-rutinoside, quercetin 3-O-glycoside, kæmpferol 3-O-glycoside[57] with ethyl acetate–formic acid–acetic acid–water (100:11:11:27); the chemotaxonomic study of flavonoids of *Lipocedrus* sp. with chloroform–methanol–formic acid (90:10:1)[58]; the study of flavonol glycosides from *Calendula officinalis* flowers[59] with benzene–methanol–acetic acid (45:8:4). The analysis of flavonoids in the root cortex of *Illicium henryi*[60] with butanol–acetic acid–methanol–water (4:2:1:5). The TLC of various kæmpferol glycosides[61] with ethyl acetate–MEK–acetic acid–water (5:3:1:1). The separation of two phloroglucinol glycosides and flavonol glycosides from *Sedum sediform*[62] with toluene-acetone-formic acid (30:30:6). The determination of flavonoids in *Olea europaea*[63] with ethyl acetate–formic acid–water (6:1:1). The analysis of polyhydroxy and polymethoxy flavones[64]; behavior of 23 flavonoid compounds with different proportions of heptane and slightly 2 acidified propanol.

3. Polyamide
In addition to silica gel, polyamide is often used in the TLC of flavonoids and aglycones. The solvents used with polyamide are less polar than with silica gel, and acids are not used with most of the eluents; methyl ethyl ketone (MEK), toluene ethyl acetate, methanol, and water, for example, have been described in numerous works.

a. Aglycones
The study of variation in the flavonoid aglycones excreted by the leaves of *Alnus glutinosa* has been done using toluene–petroleum ether–MEK–methanol (50:25:11:13).[65] The separation of foliar flavonoid aglycones of the genus *Keckiella* has been done with benzene–MEK–methanol (80:13:7).[66] TLC has been done with toluene–MEK–methanol (12:5:3) on the flavonoid aglycones of *Viguiera* sp.[67]

b. Flavonoids

The flavonoid variation in *Lastenia californica*[68] on polyamide layers with an eluent of water–butanol–acetone–dioxane (70:15:10:5) has been studied. The separation of flavonoids and the identification of *Rosa* cultivars[69] with a mobile phase of methanol–acetic acid–water (18:1:1) has been examined. The study of exudate flavonoids from aerial parts of four *Cleome* sp.[70] with methylene chloride–benzene–methanol (15:1:1), benzene–petrol ether–MEK–methanol (60:60:7:7) or (60:30:7:7), and benzene–MEK–methanol (4:3:3) has been reported. The analysis of flavonoids in the root cortex of *Illicium henryi*[60] with butanol–acetic acid–methanol–water (4:1:2:5) has been done.

The chemotaxonomic study of various flavonoid-containing extracts of *Libocedrus* sp.[58] with ethyl acetate–MEK–formic acid–water (5:3:1:1) has been reported. The study of leaf flavonoids of *Polanisia trachysperma*[71] with petrol ether–toluene–butanol–methanol (12:6:1:1) 5-hydroxy-3,6,7,8,4'-pentamethoxy flavone, 5-hydroxy-3,6,7,8,3',4'-hexamethoxyflavone and 5,3'-dihydroxy-3,6,7,8,4'-pentamethoxyflavone has been reported.

The chemosystematic study of local members of the subtribe *Guaphaliinae* (Asteracee) with benzene–petrol ether–MEK–methanol (60:60:7:7) and toluene–MEK–methanol (60:25:15) calycapterin and 3'-methoxycalycapterin was reported.[72]

The preparative TLC of flavonones, i.e., apigenin 7-O-D-rhamnoglucoside, naringenin 7-O-D-neohesperidoside, and isokuranetin 7-O-D-neohesperidoside has been described with nitromethane–methanol (1:1).[73] The determination of total flavones of *Trollius macropetalus*[74] was done with ethyl acetate–methanol–water (6:3:1) and ethanol–water (4:1).

4. Cellulose

Cellulose is used less frequently than polyamide or silica gel; the utilization of cellulose is based on earlier studies using paper chromatography. Therefore, cellulose is used in addition to polyamide and silica gel to determine some flavonoids as discussed below. The determination of flavones and C-glycosyl flavones from the leaves of some *Arrhenatherum* sp. on cellulose with acetone–water (3:7)[75]; the determination of the flavonoid profiles of *Libacedries* on cellulose with 15% acetic acid.[58] This technique is also used to study flavonoids in wood and bark in the genus *Eucalyptus*.[55] Finally, for the separation of flavonoids of *Rosa* cultivars[69] with the mobile phase methanol–acetic acid–water (18:1:1).

5. Reversed Phase C_{18}

Reversed phase plates are used with polar compounds. This is the case in the separation of polyhydroxy and polymethoxy flavones with methanol–water + 3% sodium chloride (in different proportions). Isoflavonoids from *Streptomycetes*: genistein, 8-chlorogenistein, and 6,8-dichloro-genistein can be separated on RP_{18} silica plates with acetonitrile–water–formic acid (40:60:1). The determination of the flavonoids, flavonoid glycosides, and riboflavonoids in *Olea europaea* was performed[63] on RP_{18} silica plates with methanol–water–acetic acid (50:50:6). The high performance quantitative thin-layer chromatography of flavonoid glycosides and biflavonoids of *Cupressus sempervirens*[76] gave good results with RP_{18} silica plates with a mobile phase of water–methanol–acetic acid (50:50:6).

6. Detection

a. Without Chemical Treatment

UV light at 254 nm. All flavonoids cause fluorescence quenching, which are seen as dark blue zones on a yellow background on the TLC plate.

UV light at 365 nm. Depending on the structural type, flavonoids fluoresce yellow, blue, or green. It is possible to intensify the fluorescence in UV light at 365 nm by the use of various reagents.

b. Detection with Reagents

The reagent used most frequently is NP/PEG. The plate is sprayed with 1% methanolic diphenyl boric acid-β-ethyl amino ester (= diphenylboryl oxyethylamine)(NP) followed by 5% ethanolic polyethylene glycol 4000 (PEG) 10 ml and 8 ml, respectively. Intense fluorescence is produced immediately, or after 15 min in UV light at 365 nm. PEG increases the sensitivity. The fluorescence behavior is based on the structure of the compounds.

For the flavonols:
> Glycosides of quercetin and myricetin give an orange color; glycosides of kæmpferol and isorhamnetin give a yellow-green color.

For the flavones:
> Glycosides of luteolin give an orange color and glycosides of apigenin give a yellow-green color.

In addition to this reagent, it is possible to use general chromatographic sprays such as $AlCl_3$ or antimony chloride in 5% alcohol.

IV. ANTHOCYANINS

The anthocyanins are the most significant group of colored substances in plants; anthocyanins are responsible for the pink, red, mauve, violet, and blue colors of flowers and other plant parts. They are present in plants as glycosides of flavylium salts, principally in the petals, leaves, and fruits of higher plants. For the classification of anthocyanins, see Harbone.[77]

A. ISOLATION METHODS

Anthocyanins are unstable in neutral alkaline solutions, and their color may fade by exposure to light. Anthocyanins must be extracted from plants with solvents containing acetic or hydrochloric acids. Plants are extracted for 15 min with methanol–HCl 25% (9:1); the filtrates are used for chromatography. The anthocyanins yield anthocyanids and various sugars upon acid hydrolysis; in the TLC of anthocyanins, only the anthocyanids are detected.

B. SPECIAL PARTS

In addition to the works of Harbone, there is a chromatographic analysis of anthocyanins with 126 references on TLC.[78] The determination of anthocyanin content in the petals of red poppies during flower development by six stage polyzonal step-wise gradients on silica with mixtures of ethyl acetate, isopropanol, water, and acetic acid and detection under UV light at 365nm has been described.[79]

TLC of anthocyanins (e.g., cyanidine, pelargonine, pelargonidine, and malvinine) on silica with mixtures of ethyl acetate–2-propanol–acetic acid–water containing increasing concentrations of the more polar components has been described. Detection is with 4% aqueous potassium hydroxide solution and 5% ethanolic aluminum chloride. Quantification is by densitometry at 465 nm or 580 nm.[80] TLC of various proanthocyanidins on silica gel with an eluent of ethyl acetate–formic acid–acetic acid–water (140:2:1:59) on cellulose; detection by UV light or by spraying with vanillin–sulphuric acid or anisaldehyde–sulphuric reagent, also by TLC of acetate derivatives on silica gel with toluene-acetone (2:1).[81]

Anthocyanins and anthocyanids are usually separated on microcrystalline cellulose. The determination of the chemical structures of two anthocyanins from *Ipomœa batata* with acetic acid–hydrochloric acid–water (15:3:82), ethyl acetate–acetic acid–water (3:1:1), ethyl acetate–pyridine–water (15:7:5) and phenol–formic acid–water (75:1:25) detection by UV for the anthocyanins and with aniline hydrogene phtalate for reducing sugars has been reported.[82]

V. ESSENTIAL OILS

The essential oils comprise the volatile steam-distillable fraction responsible for the characteristic odor or smell found in many plants. Plant families particularly rich in essential oils include the Asteraceæ *Chamomilla*, Lamiaceæ, e.g., the mints, *Mentha* species, Rosaceæ *Rosa*, Rutaceæ *Citrus* oils, Myrtaceæ *Eucalyptus*, Apiaceæ anise, and cumin. Essential oils are mixtures of many substances; the most prominent ones are terpenes and phenylpropane derivatives. The terpene essential oils can be divided in two classes: the mono- and sesquiterpenes, and the C_{10} and C_{15} isoprenoids. Simple monoterpenes are widespread and tend to occur as components of the majority of essential oils. Some compounds are found in leaf oils, especially α- and β-pinene, limonene, myrcene, and α-phellandrene. Flower and seed oils tend to have more specialized monoterpenes inside.

A. ISOLATION (EXTRACTION OF ESSENTIAL OILS)

To isolate from plant tissues, mono- and sesquiterpenes are extracted with hexane, ether, or acetone. The standard accepted method is steam distillation with a modified distillation apparatus according to Cocking and Middleton (European pharmacopœia). The quantity of plant used is sufficient to yield 0.1 to 0.3 ml of essential oils. The sample weight is 10 to 50 g, depending on the type of plant (European pharmacopœia) and the rate of the distillation is 2 to 3 ml per minute. Distillation is continued for 1 to 4 h; 1 ml of xylene is placed in the distillation flask, so a blank "xylene value" must be determined in a separate distillation in the absence of the plant.

If a quantitative yield of the oil content is not required, a sample of essential oil suitable for investigation by TLC can be obtained by reducing the distillation period to 1 h. With the exception of plants containing, e.g., eugenol, the distillation is performed without xylene. The resulting oil is diluted 1:10 with toluene. Eugenol-containing oils, obtained by distillation in the presence of xylene, can be applied directly to the plate.

The future of extraction of essential oils will be by supercritical fluids because they cause less artifacts.

B. GENERAL CONSIDERATIONS

TLC can be used to advantage in combination with GC because most of the essential oil compounds are volatile. It is the ideal technique for characterization: GC is an essential tool for chemotaxonomic studies of essential oils. For the identification of volatile terpenes in any plant material, it is essential to combine the use of GC with other procedures, especially with TLC. GC is useful for monitoring fractions separated by TLC. Even when GC is available, TLC is useful at all stages for the separation and analysis of terpenes; when dealing with the less volatile sesquiterpenoids, it may even become the method of choice.

C. SPECIAL PART

Silica gel is the most widely used sorbent with solvents such as benzene or toluene, chloroform, methylene chloride, and ethyl acetate as described by Wagner.[3] The eluent toluene–ethyl acetate (93:7) is suitable for the analysis.

The European pharmacopœia describes different solvent systems for individual essential oils:

Benzene (or toluene because this is not as carcinogenic) for *Pimpinella anisum*;
Chloroform for *Melissa officinalis*;
Methylene chloride for *Pimpinella anisum*, *Juniperus*, *Lavandula*, *Rosmarinus*, and *Salvia*;
Toluene–ethyl acetate (90:10) *Eucalyptus*;
Toluene–ethyl acetate (95:5) for *Mentha*; and,
Chloroform–toluene (75:25) for *Thymus vulgaris* and *Chamomilla recutita*.

The stereodifferentiation of borneol and isoborneol on silica gel with the mobile phase toluene–ethyl acetate (93:7)[83] and detection by spraying with anisaldehyde–sulphuric acid reagent has been reported. Isolation of a new β-dihydroagarofuran sesquiterpene from the root bark of *Celastrus rasthornianus* by preparative TLC on silica gel with acetone–benzene (1:4).[84] Final purification is done on RP_{18} plates with methanol–water (4:1); detection under UV light.

The determination of the relation between the antimicrobial activity and the chemical composition of the essential oil of *Sideritis sipylea* by preparative TLC of ether-extracted oil on silica of 0.2 mm thickness with ethyl acetate–hexane (3:17), then quantification by GC after elution with ether[85] has been reported.

The study of biologically active sesquiterpenes of the type isolated from *Polygonum glabrum*; TLC of diesters of 2,3-dihydroxy-isodrimeniol on silica with ethyl acetate–hexane (2:3), and HTPLC on RP_{18} silica with methanol–water (9:1) have all been reported. The four new compounds could only be separated on RP_{18}.[86]

A TLC chromatographic method for establishing the botanical origin of the cinnamins of commerce: HPTLC of cinnamyl alcohol, cinnamyl acetate, transcinnamaldehyde, eugenol, and coumarin on silica with six different solvents systems. Quantification is by densitometry at 270 nm.[87]

VI. CARDIAC GLYCOSIDE PLANTS

These plants contain steroid glycosides which are used in the treatment of cardiac disorders. Drugs such as digitalin and its derivatives are used to regulate heartbeat. The steroids are structurally derived from the tetracyclic 10,13-dimethylcyclopentanoperhydrophenanthrene ring system. They possess a γ-lactone ring for the cardenolies or a δ-lactone ring for the bufadienolide attached in the β-position at C_{17}.

The sugar residues are typically derived from deoxy-, and or C-3-O-methylated sugars, and they are linked glycosidically by the C-3-OH groups of the steroid aglycones. The main plants are *Digitalis*, *Nerium oleander*, *Strophanthus* sp., and *Convallaria majalis* for the cardenolides, *Helleborus* sp. and *Urginea scilla* for the bufadienolides.

A. ISOLATION

Powdered plant is extracted by heating for 15 min under reflux with 20 ml of 5% ethanol, with the addition of 10 ml of lead acetate solution; after filtration the clear solution is treated with a small quantity of acetic acid, then extracted by shaking three times with 15 ml of dichloromethane. The combined lower phases are filtered over anhydrous sodium sulfate and evaporated to dryness. The residue is dissolved in 1 ml of dichloromethane–ethanol (1:1) and the solution obtained is used for chromatography. All cardiac glycoside drugs can be extracted by this method, but when seeds are concerned it is necessary to remove fat with hexane.

B. DETECTION

Detection without chemical treatment
Fluorescence quenching by cardenolides is weak at UV 265 nm; distinct zones of fluorescence quenching are produced by bufadienolides.

Spray reagents. General reagents: Antimony chloride in chloroform and heating at 100°C. Chloramine T, 3% aqueous solution mixed with 25% ethanolic trichloroacetic acid.

Sulfuric acid reagent. The TLC plate is sprayed with this reagent, then heated for 3 to 5 min at 100°C.

Specific reagents. Kedde reagent (alkaline solution of 3,5-dinitrobenzenic acid). Immediately after spraying, cardenolides form a pink or blue-violet color. Bufadienolides do not react. Baljet reagent (alkalin picric solution) form a red-orange color and Raymond-Marthoud (alkaline *m*-dinitrobenzene solution) forms a violet color with cardenolides.

C. SPECIAL PART

Wagner[3] described three solvent systems as follows: ethyl acetate–methanol–water (100:13.5:10), a general solvent system for cardiac glycosides; ethyl acetate–methanol–ethanol–water (81:11:4:8) (the addition of ethanol increases the R_f values of strongly polar compounds like K-strophantoside); MEK–toluene–water–glacial acetic acid (40:5:3:2.5:1) is suitable for the separation of *Scilla* glycosides.

In the recent literature, specific works are of interest. TLC of 14 bufadienolides from the bulb extract *Urginea maritima* on silica gel with chloroform–methanol–water (140:44:7) and ethyl acetate–methanol–water (81:11:8); detection by spraying with anisaldehyde, sulfuric acid.[88]

The determination of chemical differences between species of *Asclepias* sp. on silica gel is done with chloroform–methanol–formamide (first development) and with ethyl acetate–methanol (second development); detection by spraying with 0.4% solution of 2,4,2′,4′-tetranitrodiphenyl in toluene followed by spraying with a 10% solution of KOH in 50% methanol.[89]

The accumulation of lanatosides in *Digitalis lanata* by gradient elution TLC of lanatosides A and C on silica gel with mixtures of ethyl acetate, methanol, and water to form a stepwise gradient with increasing concentrations of methanol was studied. Detection was by chloramine, followed by trichloroacetic acid and heating at 100°C; quantification was by densitometry at 250 nm.[90]

VII. SAPONINS

The formation of persistent foams during plant extraction or during the concentration of plant extracts indicates the presence of saponins. The saponins are mainly triterpene derivatives, with similar amounts of steroids present.

A. TRITERPENE SAPONINS

These saponins possess an oleanane ring system or dammarane and ursane. Many saponins are acidic due to the presence of one or two carboxyl groups in the aglycone; the carbohydrate group contains one to six monosaccharide units, with the most common being glucose, rhamnose, arabinose, xylose, glucuronic acid, and galactorosic acid. All triterpene saponins possess hemolytic activity, which varies with the type of substitution.

B. STEROID SAPONINS

The sapogenins of the steroid saponins are mainly spirostanols; these sapogenins do not have carboxyl groups. Steroid saponins possess less sugar units than the triterpene saponins.

1. Isolation

Usually, plant powder is extracted by heating for 10 min under reflux with 10 ml of 70% ethanol; the clear filtrate is evaporated and 25 to 40 µl of this solution are used for chromatography. Ginseng radix is extracted with 90% ethanol rather than 70% ethanol.

2. Detection

There are several spray reagents, including anisaldehyde–sulphuric acid and vanillin–sulphuric acid, that react with saponins to form blue or blue-violet and yellow colors. Antimony chloride reagent gives red-violet colors in visible light and red-violet-blue and green fluorescence in UV light at 365 nm. Vanillin–phosphoric acid reacts with ginsenosides to give red-violet colors in visible light and red or blue fluorescence in UV light at 365 nm. The main constituents of *Eleutherococcus* root give violet color with visible light and intense yellow, pale blue, and orange fluorescent colors in UV light at 365 nm. Hemolytic saponins are detected as white zones on a red background after drying the plate in warm air.

Separation and quantification of *Ginseng* saponins by two-dimensional TLC on silica gel with chloroform–methanol–water (21:11:4) in the first direction and 1-butanol–ethyl acetate–water

(4:1:1) in the second. Detection is by spraying with vanillin and heating at 120°C for 15 min.[91] Quantification is by densitometry at 600 nm.

Studies on saponin constituents of *Gynostemma longipes* by TLC of the saponins on silica gel with the lower phases of (a) chloroform–methanol–water (65:55:10) and (b) chloroform–methanol–water (30:12:4)–acetic acid (9:1), have been reported. Detection was by densitometry and identification was by FAB-MS and NMR.[92]

REFERENCES

1. **Stahl, E.**, *Dünschicht-Chromatographie*, 2nd ed., Springer-Verlag, Berlin, 1967.
2. **Randerath, K.**, *Dünschicht-Chromatographie*, 2nd ed., Springer-Verlag, Berlin, 1967.
3. **Wagner, H., Bladt, S., and Zgainski, E. M.**, *Plant Drug Analysis*, Springer-Verlag, Berlin, 1965.
4. **Harbone, J. B.**, Phytochemical methods, in *A Guide to Modern Techniques of Plant Analysis*, Chapman and Hall, London, 1973.
5. **Munier, R. and Macheboeuf, M.**, Microchromatography of the separation of alkaloids and of various nitrogen bases, *Bull. Soc. Chim. Biol.*, 31, 1144, 1949.
6. **Munier, R.**, Separation of alkaloids from their N oyl by microchromatography, *Bull. Soc. Chim. Biol.*, 35, 1225, 1963.
7. **Baerheim Svendsen, A. and Verpoorte, R.**, *Chromatography of Alkaloids, Part A : Thin Layer Chromatography*, Elsevier, Amsterdam, 1983, 536.
8. **Pothier, J., Galand, N., and Viel, C.**, Separation of alkaloids in plant extracts by overpressured layer chromatography with ethyl acetate as mobile phase, *J. Planar Chromatogr.–Mod. TLC*, 4, 392, 1991.
9. **Waldi, D., Schnacker, K., and Munter, F.**, Systematic analysis of alkaloids on thin layer plates, *J. Chromatogr.*, 6, 61, 1961.
10. **Sunshine, I., Fike, W. W., and Landesman, H.**, Identification of therapeutically significant organic bases by thin layer chromatography, *J. Forensic Sci.*, 11, 428, 1966.
11. **Noirfasile, A. and Mees, G.**, Thin layer chromatography of some alkaloids and amine bases, *J. Chromatogr.*, 31, 594, 1967.
12. **Ahrend, K. F. and Tiess, D.**, Thin layer chromatographic parameters of 180 toxicologically significant compounds with low volatility in four simple systems with instructions for practical application, *Zbl. Pharm.*, 22, 347, 1973.
13. **Moffat, A. C. and Smalldon, K. W.**, Optimum use of paper, thin layer, and gas-liquid chromatography for the identification of basic drugs, *J. Chromatogr.*, 90, 1, 1974.
14. **Armstrong, R. J.**, Identification by thin-layer chromatography of prescription drugs commonly used in New Zealand, *New Zealand J. Sci.*, 17, 15, 1974.
15. **Breiter, J. and Helger, R.**, A thin layer chromatographic detection tool for drugs and pharmaceuticals, *Med. Labor.*, 30, 149, 1977.
16. **Martinez, J. A., Gomez, C., Santana, T., and Velez, H.**, Alkaloids from *Rauwolfia cubana* stem bark, *Planta Med.*, 55, 283, 1989.
17. **Kretovics, J. E., Szilagyi, L., and Gyurjan, I.**, *In vitro* cultivation and cytological investigation of *Catharanthus roseus*, *Herba Hungarica*, 29, 63, 1990.
18. **Monforte-Gonzalez, M., Ayora-Talavera, T., Maldonado-Menoza, I. E., and Layola-Vargas, U. M.**, Quantitative analysis of serpentine and ajmalicine in plant tissues of *Catharanthus roseus* and hyoscyamine and scopolamine in root tissues of *Datura stramonium* by thin layer chromatography, *Phytochem. Anal.*, 3, 117, 1992.
19. **Balsevich, J., Hogge, L. R., Berry, A. J., Games, D. E., and Mylcheest, I. C.**, Analysis of indole alkaloids from leaves of *Catharanthus roseus* by means of supercritical fluid chromatography and mass spectrometry, *J. Nat. Prod.*, 51, 113, 1988.
20. **Grespi-Perellino, N., Ballabio, M., Gioia, B., and Minghetti, A.**, Two unusual ergopeptines produced by a saprophytic culture of *Claviceps purpurea*, *J. Nat. Prod.*, 50, 1065, 1987.
21. **Cisowski, W. and Lamer-Zaraswska, E.**, Ergot analysis by TLC with gradient elution, *J. Planar Chromatogr.–Mod. TLC*, 3, 47, 1990.
22. **Pothier, J., Petit-Paly, G., Montagu, M., Galand, N., Chenieux, J. C., Rideau, M., and Viel, C.**, Separation of quinolizidine and dihydrofuro[2,3-b]-quinolinium alkaloids by overpressured layer chromatography, *J. Planar Chromatogr.–Mod. TLC*, 3, 356, 1990.
23. **Cubukcu, B., Mericli, A. H., Güner, N., Özhathay, N., Bingöl, S., and Damadyan, B.**, Flavonoids and alkaloids of *Chamæcytsus pygmæus*, *Sci. Pharm.*, 56, 287, 1988.
24. **Hayman, A. R. and Gray, D. A.**, Hydroxynorcytisine, a quinolizidine alkaloid from *Laburnum anagyroides*, *Phytochemistry*, 28, 673, 1989.

25. **Greiswald, R., Veen, G., Vonwyk, B., Witte, L., and Czygan, F. C.,** Distribution taxonomic signifiance of major alkaloids in the genus *Virgilia, Biochem. System. Ecol.*, 17, 231, 1989.
26. **Matsuda, K., Yamada, K., Kimura, M., and Hamada, M.,** Nematicide activity of matrine and its derivatives against pine wood nematodes, *J. Agric. Food Chem.*, 39, 189, 1991.
27. **Pothier, J., Galand, N., and Viel, C.,** Characterization and dosage of alkaloids in plant extracts by overpressured layer chromatography, *Phytotherapy*, 34, 3, 1990.
28. **Tschirch, C. and Kraus, L. J.,** Laburnum alkaloid cytisin, rapid TLC identification, *Dtsch. Apoth. Ztg.*, 131, 1876, 1991.
29. **Greinwald, R., Witte, L., Wray, W., and Czygan, F. C.,** The alkaloid pattern of *Cytisophyllum sentifolium, Biochem. System. Ecol.*, 19, 253, 1991.
30. **Van Wyk, B. E. and Verdoorn, G. H.,** Biochemical variation in the genus *Pærsonia, Biochem. System. Ecol.*, 19, 685, 1991.
31. **Talas, E., Botz, L., Margitfali, J., Sticher, O., and Baiker, A.,** Planar chromatographic separation of *Cinchona* alkaloids formed during enantioselective hydrogenation of ethyl pyruvate, *J. Planar Chromatogr.–Mod. TLC*, 5, 28, 1992.
32. **Kovacs, Z. and Gilanyi, I.,** Importance of the examination of poppy-alkaloids, *Egèszsegtudomàny*, 32, 324, 1988.
33. **Li, Z., Zhao, X., Su, Y., and Chn, Y.,** Quantitative determination of major alkaloids in the seeds and shells of *Papaver somniferum* by thin layer chromatography, *Chin. J. Chromatogr.*, 8, 388, 1990.
34. **Hostafi, S., Szilagyi, L., Markleit, S., and Toth, G.,** Preparation of substituted hydrazone derivatives of ketones with morphine skeleton, *Acta Chimica*, 127, 9, 1990.
35. **Agyangar, M. R., Biswas, S. S., and Tambe, A. S.,** Separation of opium alkaloids by thin layer chromatography combined with flame ionization detection using the peak pyrolysis method, *J. Chromatogr.*, 547, 538, 1991.
36. **Gu, Y., Li, L., and Lin, K.,** Determination of morphine in opium by thin layer chromatography, *Chinese J. Pharm. Anal.*, 13, 236, 1993.
37. **Matysik, G. and Jusiak, L.,** Stepwise gradient in thin layer chromatography of *Chelidonium* alkaloids, *J. Chromatogr.*, 518, 273, 1990.
38. **Niu, C. H. and He, L.,** Determination of isoquinoline alkaloids in *Chelidonium majus* by thin layer chromatography, *Acta Pharm. Sinica*, 27, 69, 1992.
39. **Hartmann, T., Sander, H., Adolph, R., and Toppel, G.,** Metabolic links between the bioynthesis of pyrrolizidine alkaloids and polyamines in root cultures of *Senecio vulgaris, Planta*, 175, 82, 1988.
40. **Sarg, T. M., El-Domiaty, M. M., and Bishr, M. M.,** Thin layer chromatographic scanner, spectrophotometric and high performance liquid chromatographic methods for the determination of colchicine, *Analyst*, 114, 575, 1989.
41. **Stimanek, V., Husek, A., Valka, I., and Sütupinar, N.,** Phytochemical studies of Turkish *Colchicum* and *Merenda* species, *Herba Hungarica*, 29, 64, 1990.
42. **Lukic, V., Gasic, O., Walterowa, D., and Simanek, V.,** Separation and fluorodensitometric determination of some tropolone alkaloids by thin layer chromatography, *Chromatographia*, 28, 516, 1989.
43. **Glavic, D.,** RM values of some colchicines and colchiceinamides determined by reversed-phase thin-layer chromatography, *J. Chromatogr.*, 591, 367, 1992.
44. **Liu, C., Zeng, X., Lin, G., Tang, D., and Mou, J.,** Separation and determination of the poisonous alkaloids in *Aconitum Kusnezoffii* and its processed products, *J. Chin. Trad. Med.*, 83, 1987.
45. **Wang, K. and Tong, Y.,** Determination of the main alkaloids in *Aconitum*, by thin layer chromatography, *Acta Pharm. Sinica*, 25, 387, 1990.
46. **Sayed, H. M., Desai, H. K., Ross, S. A., and Pelletier, S. W.,** New diterpenoid alkaloids from the roots of *Aconitum septentrionale*: isolation by an ion exchange method, *J. Nat. Prod.*, 55, 1595, 1992.
47. **Funk, W. and Dröschel, S. T.,** Qualitative and quantitative HPTLC determination of cocaine, ecgonine, ecgonine methyl ester and benzylecgonine, *Proc. 6th Int. Symp. Instrum. Planar Chromatogr.*, Interlaken, 1991.
48. **Pothier, J., Galand, N., Tivollier, P., and Viel, C.,** Separation of quaternary alkaloids in plant extracts by overpressured layer chromatography, *J. Planar Chromatogr.–Mod. TLC*, 6, 220, 1993.
49. **Harbone, J. B.,** Phenolic compounds, in *Chromatography, 5th ed. – Fundamentals and Applications of Chromatography and Related Differential Migration. Methods*, E. Heftman, Ed., Elsevier, Amsterdam, 1992.
50. **Isaksen, M.,** Natural pigments, in *Handbook of Thin-Layer Chromatography*, J. Sherma and B. Fried, Eds., Marcel Dekker, New York, 1991, 625.
51. **Betti, A., Lodi, G., and Fuzati, N.,** Analysis of flavonoid compounds by planar chromatography: a chemometric approach, *J. Planar Chromatogr.–Mod. TLC*, 6, 232, 1993.
52. **Burton, D. E., Bailey, D. L., and Lillie, C. H.,** Determination of ipriflavone and its impurities by thin-layer chromatography with absorbance and fluorescence detection, *J. Planar Chromatogr.–Mod. TLC*, 6, 223, 1993.
53. **Heimler, D. and Boddi, V.,** Cluster analysis in the comparison of two dimensional chromatograms, *J. Chromatogr.*, 466, 371, 1989.
54. **Boyet, C. H. and Jay, M.,** Taxonomic variation in flavonoids in the *Oryza latifolia* complex, *Biochem. System. Ecol.*, 17, 443, 1989.
55. **Conde, E., Cahadia, E., and Garcia-Vallejo, M. C.,** Optimization of TLC for research on the flavonoids in wood and bark of species of the genus *Eucalyptus, Chromatographia*, 33, 418, 1992.

56. **Heimler, D., Mittempergher, L., Buzzini, P., and Boddi, V.,** Quantitative HPTLC separation of the flavonoid glycosides in the taxonomy of Elm (*Ulmus* sp.), *Chromatographia*, 29, 16, 1990.
57. **Moulton, R. J. and Whittle, S. J.,** The major flavonoids in the leaves of *Milium effusum*, *Biochem. System. Ecol.*, 17, 197, 1989.
58. **Markham, K. R., Franke, A., Molloy, B. P. J., and Webby, R. F.,** Flavonoid profiles of New Zealand *Lipocedrus* and related genera, *Phytochemistry*, 29, 501, 1990.
59. **Ollivier, E. V., Elias, R., Faure, F., Balansard, A., and Bondon, G.,** Flavonol glycosides from *Calendula officinalis* flowers, *Planta Med.*, 55, 73, 1989.
60. **Xie, D., Wang, S., and Cheng, Z.,** Analysis of flavonoids in the root-cortex of *Illicium henryi*, *J. Chin. Herb. Med.*, 21, 447, 1990.
61. **Bashir, A., Hamburger, M., Gupta, M. P., Solis, P. N., and Hostettmann, K.,** Flavonol glycosides from *Monnina sylvatica*, *Phytochemistry*, 30, 3781, 1991.
62. **Sakar, M. K., Petereit, F., and Nahrstedet, A.,** Two phloroglucinol glucosides, flavan gallates and flavonol glycosides from *Sedum sediforme* flowers, *Phytochemistry*, 33, 171, 1993.
63. **Heimler, D., Pieroni, A., Tattini, M., and Cimato, A.,** Determination of flavonoids, flavonoids glycosides and bioflavonoids in *Olea europæa*, *Chromatographia*, 33, 369, 1992.
64. **Hadj-Mahammed, M. and Meklati, B. Y.,** Polyhydroxy- and polymethoxy-flavones: behavior in high performance thin layer chromatography, *J. Planar Chromatogr.–Mod. TLC*, 6, 242, 1993.
65. **Gonnet, J. F. and Daniere, C.,** Individual variation in the flavonoid aglycones excreted on the leaves of *Alnus glutinosa*, *Biochem. System. Ecol.*, 17, 239, 1989.
66. **Mistretta, O. and Scogin, R.,** Foliar flavonoid aglycones of the genus *Keckiella*, *Biochem. System. Ecol.*, 17, 455, 1989.
67. **Schilling, E. E.,** External flavonoid aglycones of Viguiera, *Biochem. System. Ecol.*, 17, 835, 1989.
68. **Desrocheos, A. M. and Bohm, B. A.,** Flavonoid variation in *Lastenica californica* (Asteraeæ), *Biochem. System. Ecol.*, 21, 449, 1993.
69. **Van Sumere, C., Fache, P., Van de Custelle, K., De Cooman, L., Everaert, E., De Loose, R., and Hutsebaut, W.,** Improved extraction and reversed phase HPLC separation of flavonoids and the identification of *Rosa* cultivars, *Phytochemical Analysis*, 4, 279, 1993.
70. **Sharaf, M., Mansour, M. A., and Saleh, N. A. M.,** Exudate flavonoids from aerial parts of four *Cleome* species, *Biochem. System. Ecol.*, 20, 442, 1992.
71. **Wollenberger, E., Stern, S., Roitman, J. N., and Yatskievych, G.,** External leaf flavonoids of *Polanisia trachysperma*, *Phytochemistry*, 28, 303, 1989.
72. **Salek, N. A. M., Mansour, R. M. A., El Kareemy, Z. A. R., and Fayes, A. A.,** The chemosystematics of local members of the subtribe *Guaphaliinae*, *Biochem. System. Ecol.*, 16, 615, 1988.
73. **Bartolome, R. E.,** Crystallization and thin layer chromatographic separation of flavonones from traces of other types of flavonoid compound, *J. Planar Chromatogr.–Mod. TLC*, 7, 70, 1994.
74. **Liu, L., Wang, X., Fu, Q., and Wang, L.,** Bacteriostatic activity of longpetal globeflower, *Trollius macropetalus*, and the determination of its total flavones, *J. Chin. Herb. Med.*, 23, 461, 1992.
75. **Jay, M. and Ismaili, A.,** Flavones and C-glycosyl-flavone from the leaves of some *Arrhenatherum species*, *Phytochemistry*, 28, 3025, 1989.
76. **Heimler, D. and Pieroni, A.,** High performance quantitative thin-layer chromatography of flavonoid glycosides and bioflavonoids of *Cupressus sempervirens* in relation to cypress canker, *Chromatographia*, 31, 247, 1991.
77. **Harbone, J. B.,** *Phytochemical Methods. A Guide to Modern Techniques of Plant Analysis*, 2nd ed., Chapman and Hall, London, 1984.
78. **Lee, H. S. and Hong, V.,** Chromatographic analysis of anthocyanins, *J. Chromatogr.*, 624, 221, 1992.
79. **Matysik, G. and Benesz, M.,** Thin layer chromatography and densitometry of anthocyanins in the petals of red poppy during development of the flower, *Chromatographia*, 32, 19, 1991.
80. **Matysik, G.,** Thin layer chromatography of anthocyanins with stepwise gradient elution, *J. Planar Chromatogr.–Mod. TLC*, 5, 146, 1992.
81. **Balde, A. M., Pieters, L. A., Gergely, A., Kolodziej, H., Claeys, M., and Vlietinck, A. J.,** A-type proanthocyanidins from stem-bark of *Pavetta owariensis*, *Phytochemistry*, 30, 337, 1991.
82. **Odake, K., Terahara, N., Saito, N., Toki, K., and Honda, T.,** Chemical structures of two anthocyanins from purple sweet potato, *Ipomœa batata*, *Phytochemistry*, 31, 2127, 1992.
83. **Kreis, P., Juchelha, D., Motz, C., and Mosandl, A.,** Chiral compounds of essential oils, IX., Stereodifferentiation of borneol, isoborneol and bornylacetate, *Dtsch. Ap. Ztg*, 131, 1984, 1991.
84. **Tu, Y. Q.,** A sesquiterpene polyol ester from *Celastus rosthorianus*, *Phytochemistry*, 31, 2155, 1992.
85. **Gersis, V., Spilitius, V., Argyriadou, N., and Poulos, C.,** Relation between the antimicrobial activity and the chemical composition of the essential oil of *Sideritis sipylea*, *Flavour and Fragance*, 6, 93, 1991.
86. **Jacobsson, U. and Muddathir, A. K.,** Four biologically active sesquiterpenes of the type isolated from *Polygonum glabrum*, *Phytochemistry*, 31, 4207, 1992.
87. **Cikalo, M. J., Poole, S. K., and Poole, C. F.,** A thin layer chromatographic method for establishing the botanical origin of the cinnamins of commerce, *J. Planar Chromatogr.–Mod. TLC*, 5, 135, 1992.

88. **Krenn, L., Ferth, R., Robien, W., and Kopp, B.,** Bufadienolides from *Urginea maritima sensu strictu*, *Planta Med.*, 57, 560, 1991.
89. **Sady, M. B. and Seiber, J. N.,** Chemical differences between species of *Asclepias* from the intermountain region of North America, *Phytochemistry,* 30, 3001, 1991.
90. **Matysik, G., Kowalski, J., Strzelecka, H., and Soczewinski, E.,** Investigation of the accumulation of lanatoside in *Digitalis lanata* by gradient elution TLC and densitometry, *J. Planar Chromatogr.–Mod. TLC*, 7, 129, 1994.
91. **Di, Y., Wu, W., and Sun, Y.,** Separation and quantification of Ginseng saponins by two-dimensional TLC densitometry, *Chin. J. Chromatogr.*, 12, 173, 1994.
92. **Sun, W., Sha, Z., and Yang, J.,** Studies on saponin constituents of *Gynostemma longipes*, *Chin. J. Herb*, 24, 619, 1993.

Chapter 4

THIN-LAYER CHROMATOGRAPHY IN PARASITOLOGY

Bernard Fried and M.A. Haseeb

CONTENTS

I. Introduction 52

II. Obtaining Parasite Material for TLC..56
    A. Parasites of Humans and Other Warm-Blooded Animals................................56
    B. Parasites of Cold-Blooded Animals 57
    C. Protozoan Parasites 58
    D. Storage, Preservation, and Freezing of Parasite Material................................58

III. Sample Preparation 58
    A. General Considerations..58
    B. Dissolving and Simple Extraction Procedures..58
    C. Complex Extraction Procedures — Partition of Extracts................................59
    D. Column Chromatography 59

IV. Chromatographic Systems and Detection of Compounds 60
    A. General Considerations..60
    B. Neutral Lipids 60
    C. Phospholipids..60
    D. Glycolipids (Glycosphingolipids)...61
    E. Amino Acids 61
    F. Carbohydrates 61
    G. Natural Pigments..61

V. Quantification 62

VI. Validation of Results 63

VII. Selected Protocols 64
    A. General Considerations..64
    B. Neutral Lipids in *Echinostoma caproni* Adults 64
    C. Phospholipids in *Biomphalaria glabrata* Snails
       Infected with the Intramolluscan Stages of *Echinostoma caproni* 64
    D. Amino Acids in *Echinostoma trivolvis* Adults 65
    E. Sugars in the Hemolymph of *Biomphalaria glabrata*
       Infected with the Intramolluscan Stages of *Echinostoma caproni* 66
    F. Carotenoid Pigments in *Echinostoma trivolvis* Rediae 66

References 67

I. INTRODUCTION

Thin-layer chromatography (TLC) has been used since the 1960s to analyze and identify lipophilic and hydrophilic compounds in animal parasites and in hosts they infect. Early TLC studies, mainly on lipids in *Schistosoma*, *Hymenolepis*, and *Ascaris*, were done on homemade silica gel plates,[1] but later work on *Leucochloridiomorpha constantiae* and *Echinostoma trivolvis* used commercially available silica gel plates and sheets.[2,3] Although TLC has been used in parasitology mainly as an analytical tool, other studies employed preparative layer chromatography (PLC) for isolation of compounds that were subsequently analyzed by other non-TLC procedures.[4]

Most TLC studies in parasitology have used silica gel as the sorbent, but studies on cellulose[5] and reversed phase plates[6] are also available. Most of the internationally refereed journals in parasitology include articles on TLC, beginning with the early work of Smith et al.[7] up to the more recent work of Fried et al.[8] A change in the emphasis of TLC studies on parasites during the past 10 years has been to more frequent *in situ* densitometric analysis to quantify the analytes.[6] The use of TLC-densitometry is replacing work in which the compounds of interest are eluted from plates and then quantified by non-TLC techniques, usually spectrophotometry.

This section describes the major uses of TLC with protozoan and helminth parasites and provides examples that are selective rather than exhaustive. Table 4.1 provides a selected list of helminth parasites studied by TLC, the purpose and results of the study, the chromatographic systems, i.e., sorbents and mobile phases, and the methods of detection used.

Prior to our early work with TLC, one of us (B.F.) used lipid histochemical procedures to examine release of lipophilic materials in parasitic worms.[9] These procedures were not able to determine specific neutral lipid fractions released by the worms. However, subsequent use of lipid TLC procedures allowed determination of specific neutral lipid fractions released by the metacercariae of *Leucochloridiomorpha constantiae*.[2] Because TLC is simple, inexpensive, and useful for small-sample analysis, it is anticipated that parasitologists will use this separation procedure more frequently in the future.

TLC has been used both as an analytical and preparative technique to solve problems in parasitology. The following representative studies have used TLC to examine the role of lipids in parasites. Analytical TLC studies on parasitic flatworms have detected mainly triacylglycerols, free sterols, and phospholipids as the major lipids in larval and adult worms.[2,10–12] These lipids function mainly as storage depot fats (triacylglycerols) and as structural components of membranes (sterols and phospholipids).

Considerable variation exists in skin-surface lipids in mammals[13] and birds,[14] and these differences have been studied by analytical and preparative TLC. Most variations in skin lipids occur in different components of the neutral lipids. Skin lipids have unique constituents such as squalene and wax esters that are not detected in lipids from other sites. Certain neutral lipids in the skin of vertebrate hosts serve as chemoattractants to facilitate entry of avian and mammalian schistosome cercariae into their hosts.[14,15]

Chemotaxonomy based on lipid profiles has been studied in fungi, yeast, and bacteria, but little information on this topic exists for protozoan or helminth parasites. Recently, neutral and polar lipids were identified in six strains of *Blastocystis hominis* by one- and two-dimensional TLC.[16] Some strain differences were apparent based on the lipid profiles.

Larval and adult digeneans maintained *in vitro* release lipids into the culture media,[2,17] some of which serve as pheromones or carriers of pheromones. Both analytical TLC and PLC have been used to identify presumptive pheromones in helminths.[18]

Hen's egg yolk is a good supplement to provide lipids to the defined medium NCTC 135 used for *in vitro* cultivation of larval digeneans.[19] TLC has helped to determine the role of neutral lipid supplements in media that have enhanced the growth and development of trematodes *in vitro*.[20]

PLC has been used to isolate lipid classes of helminths for subsequent analyses by other analytical techniques. For example, the sterols of several helminths were isolated by PLC and then analyzed by other techniques.[4]

TABLE 4.1
Analysis of Animal Parasite Extracts or their Products by One-dimensional Thin-layer Chromatography on Silica Gel Plates or Sheets Unless Stated Otherwise

Parasite	Objective	Chromatographic system and the detection reagent(s)	Results	Reference
Amblosoma suwaense free metacercariae	Identification of chemoattractants	Petroleum ether–diethyl ether-acetic acid (80:20:1); detection by PMA and heat	Sterol esters as chemoattractants	73
Angiostrongylus cantonensis adults	Analysis of neutral lipids, polar lipids, and glycolipids in whole worm extracts	Neutral lipids in petroleum ether–diethyl ether-acetic acid (80:20:1); glycolipids in chloroform–methanol–water (24:7:1); phospholipids in chloroform–methanol–petroleum ether–acetic acid (40:20:30:10); borate–silica gel for glycolipid and phospholipid separations; detection by I_2 vapors	All major lipid classes from young adults in the brain and gravid adults in the lung of rats; 60% phospholipids, 30% neutral lipids and 10% glycolipids	74
Ascaris lumbricoides adults	Polar lipids in worm extracts	Two-dimensional TLC and differential spray techniques for visualization, e.g., 50% H_2SO_4, α-naphthol, $AgNO_3$, ninhydrin, PMA and I_2 vapors	Major lipids were phosphatidylserine, globosides, lysophosphatidylcholine, phosphatidylethanolamine, cerebrosides, sphingomyelin, ascaroside, cardiolipin, cholesterol, and cholesterol esters	75
Ascaris suum spermatozoa	Analysis of neutral lipids in spermatozoa	Isopropyl ether–acetic acid (96:4) followed by *n*-pentane–diethyl ether–acetic acid (190:10:1) in the same direction; detection by I_2 vapors	Major neutral lipids of spermatozoa migrated as a free fatty acid fraction	76
Callibothrium verticillatum adults	Analysis of neutral and polar lipids	Neutral lipids in pentane–ethyl ether–acetic acid (270:30:3); polar lipids in chloroform–methanol–acetic acid–water (50:30:4:2); both detected by H_2SO_4-$K_2Cr_2O_7$ and heat	Triacylglycerol was the most abundant neutral lipid and phosphatidylcholine the most abundant phospholipid as determined by gravimetric methods	77
Callibothrium verticillatum adults	Analysis of neutral lipids	Pentane–ethyl ether–acetic acid (270:30:3); detection by H_2SO_4-$K_2Cr_2O_7$ and heat	Major neutral lipid was triacylglycerol (88%)	77
Cotylurus strigeoides metacercariae	Analysis of neutral lipids	Petroleum ether-diethyl ether-acetic acid (80:20:1); Skipski's dual solvent system; detection by PMA and heat	Various neutral lipids in metacercariae and their excretory-secretory products	78
Echinococcus granulosus larvae	Analysis of neutral lipids in hydatid cyst fluid	Petroleum ether–diethyl ether–acetic acid (80:20:1); detection by H_2SO_4 and heat	Major lipids: triacylglycerols and diacylglycerols	79
Echinococcus granulosus larvae	Analysis of glycolipids of hydatid cysts	HPTLC; chloroform–methanol–water (75:25:4); detection of glycolipids with orcinol-H_2SO_4 spray	Resolution of the major neutral and acidic glycolipids	80

TABLE 4.1 (continued)
Analysis of Animal Parasite Extracts or their Products by One-dimensional Thin-layer Chromatography on Silica Gel Plates or Sheets Unless Stated Otherwise

Parasite	Objective	Chromatographic system and the detection reagent(s)	Results	Reference
Echinococcus multilocularis larvae	Analysis of glycosphingolipids of hydatid cysts	HPTLC; chloroform–methanol–water (60:35:8); glycosphingolipids detected by spraying with water	Resolution of monohexosylceramides	81
Echinostoma trivolvis adults	Analysis of sterols in adult worms	PTLC; Skipski's dual solvent system; detection by I_2 vapor	Cholesterol as the major free sterol	4
Echinostoma trivolvis adults	Determine system associated with neutral lipid release	Petroleum ether–diethyl ether–acetic acid (80:20:1); detection by PMA and heat	Free sterols released from the excretory system	82
Echinostoma trivolvis adults	Analysis of phospholipids in worms and their secretions	Chloroform–methanol–water (65:24:4); detection by 50% H_2SO_4 and heat	Phosphatidylcholine, phosphatidylethanolamine, and phosphatidylserine as major phospholipids in extracts and secretions	3
Echinostoma trivolvis excysted metacercariae	Identification of glycerides in the incubation medium	Skipski's dual solvent system; detection by PMA and heat	Diacylglycerols in the excretory-secretory products	20
Echinostoma trivolvis excysted metacercariae	Identification of neutral lipids in metacercariae	Petroleum ether–diethyl ether–acetic acid (80:20:1); Skipski's dual solvent system; detection by PMA and heat	Free fatty acids were the major fraction	20
Fasciola hepatica adults	Analysis of excretory-secretory products	Petroleum ether–diethyl ether–acetic acid (80:20:1); Skipski's dual solvent system; chloroform–methanol–water (65:24:4); detection by H_2SO_4 and heat	Complex mixture of neutral and polar lipids released into the medium	83
Fasciola hepatica adults	Analysis of phospholipids in the posterior region of the body without eggs	Chloroform–methanol–acetic acid–water (50:25:7:3); detection by I_2 vapors	Major phospholipids detected were phosphatidylcholine, phosphatidylethanolamine, and phosphatidylserine	84
Foleyella againae adults	Lipid composition of filarid adults	Neutral lipids by petroleum ether–diethyl ether–acetic acid (80:20:1); phospholipids by chloroform–methanol–water (65:25:4); detection by 10% H_2SO_4 and heat; specific detection reagents also used for the various lipid classes; quantification by Iatroscan TH-10 TLC Analyzer (FID-Rods-Chromarod sII)	Total lipids ≈ 7% of fresh adult weight; 56% neutral lipids and 44% polar lipids; major lipid classes are sterol esters, cholesterol, phosphatidylcholine, and phosphatidylethanolamine	64
Gastrodiscoides hominis adults	Analysis of phospholipids	Chloroform–methanol–water (65:25:4); detection by I_2 vapors	Phosphatidylcholine was the major polar lipid (43%)	71
Hymenolepis diminuta adults	Analysis of neutral lipids	Petroleum ether–diethyl ether–acetic acid (80:20:1); detection by H_2SO_4 and heat	Triacylglycerol was the major neutral lipid	85

TABLE 4.1 (continued)
Analysis of Animal Parasite Extracts or their Products by One-dimensional Thin-layer Chromatography on Silica Gel Plates or Sheets Unless Stated Otherwise

Parasite	Objective	Chromatographic system and the detection reagent(s)	Results	Reference
Hymenolepis diminuta adults	Analysis of lipids	Dual solvent system: ethyl ether–benzene–ethanol–acetic acid (40:50:2:0.2); hexane–diethyl ether (94:6); detection by H_2SO_4-$K_2Cr_2O_7$ (1:1) and heat	Major lipids included phospholipids, glycerides, and free sterols	1
Hymenolepis diminuta adults	Qualitative and quantitative analysis of the lipid profile	Dual solvent system: ethyl ether–benzene–ethanol–acetic acid (40:50:2:0.2); hexane–diethyl ether (94:6) for neutral lipids; phospholipids and glycolipids by both one- and two-dimensional TLC using various solvent systems and detection reagents	Detailed quantitative analyses of the various lipid fractions	86
Hymenolepis diminuta adults	Analysis of isoprenoid lipids	Two-dimensional TLC; solvent A, 25% diisopropyl ether in petroleum ether; solvent B, chloroform; detection by I_2 vapors	Five major and three minor nonsaponifiable lipids; most abundant were cholesterol and farnesol	87
Leucochloridiomorpha constantiae metacercariae	Identification of chemoattractants	Petroleum ether–diethyl ether–acetic acid (80:20:1); detection by PMA and heat	Sterols were the chemoattractants	2
Moniezia expansa adults	Analysis of sterols	Sterols separated in benzene–ethyl acetate (9:1); detection by H_2SO_4 and heat	Cholesterol was the major free sterol	88
Paramphistomum epicilitum and other amphistome adults	Characterization of sterols	Petroleum ether–diethyl ether–methanol–acetic acid (50:50:1:1); detection by 20% perchloric acid and heat	Purity of sterols determined by TLC; structure of sterols determined by non-TLC techniques	89
Schistosoma haematobium adults	Analysis of neutral lipids in adults and their excretory-secretory products	Petroleum ether–diethyl ether–acetic acid (80:20:1); PMA; densitometry	Free sterols, triacylglycerols and sterol esters were major lipids in adults; free sterols most abundant in excretory-secretory products; neutral lipids quantified	69
Schistosoma japonicum adults	Analysis of neutral lipids in adults and their excretory-secretory products	Petroleum ether–diethyl ether–acetic acid (80:20:1); PMA; cholesterol and cholesteryl esters detected with ferric chloride (Lowry, 1968)	Free sterols, triacylglycerols, and sterol esters were major lipids in adults; free sterols most abundant in excretory-secretory products; neutral lipids quantified	90
Schistosoma mansoni adults	Analysis of lipids in adults	Petroleum ether–diethyl ether–acetic acid (80:20:1); Skipski's dual solvent system; detection by H_2SO_4 and heat	Phospholipids, free sterols, and triacylglycerols were the major fractions	91

TABLE 4.1 (continued)
Analysis of Animal Parasite Extracts or their Products by One-dimensional Thin-layer Chromatography on Silica Gel Plates or Sheets Unless Stated Otherwise

Parasite	Objective	Chromatographic system and the detection reagent(s)	Results	Reference
Schistosoma mansoni adults	Analysis of neutral lipids in males and females	Skipski's dual solvent system; detection by PMA and heat	Major neutral lipids in whole males and females were free sterols and triacylglycerols; males released free sterols into medium	92
Schistosoma mansoni adults	Uptake and catabolism of radiolabeled cholesterol	Silica gel sheets; toluene-ethyl ether (9:1); autoradiography on X-OMAT film (Kodak) or detection of radioactivity in isolated bands by scintillation counting	Adults incorporated and catabolized cholesterol; schistosomula did not	93
Schistosoma mansoni cercariae	Analysis of lipids in cercariae	Petroleum ether–diethyl ether–acetic acid (80:20:1); Skipski's dual solvent system; detection by H_2SO_4 and heat	Complex mixture of neutral lipids and phospholipids	7
Taenia taeniaeformis larvae	Analysis of lipid droplets in the tegument	Hexane–diethyl ether (95:5); detection by H_2SO_4 and heat	Lipid droplets consisted of triacylglycerols, sterols, sterol esters, and fatty acids	94

PMA = 5% ethanolic phosphomolybdic acid; **Skipski's dual solvent system** = 1: isopropyl ether–acetic acid (96:4); 2: petroleum ether–diethyl ether–acetic acid (90:10:1) (Skipski et al. (1965)).

TLC has been used to study the pathobiochemical effects of parasites on their hosts (see Table 4.2). Thus, *Leucochloridiomorpha constantiae* grown on the chick chorioallantois altered the lipid components of that site,[21] and *Echinostoma caproni* larvae reduced triacylglycerol levels in the digestive gland–gonad complex (DGG) of *Biomphalaria glabrata*.[22] The intestinal mucosa of mice infected with *Echinostoma caproni* adults had an elevated free fatty acid fraction.[23]

In other studies, TLC has been used to analyze hydrophilic compounds and natural pigments. Using silica gel and cellulose sorbents and various solvent systems, free-pool amino acids in an echinostomatid trematode have been analyzed.[24] This work examined the major free amino acids in adult worms and those released into the nonnutrient medium that served as presumptive hydrophilic pheromones in behavioral studies. Amino acid profiles of tissues and hemolymph of mosquitoes have been examined by qualitative TLC to determine the effect of infection with the rodent malarial parasite, *Plasmodium berghei*[25]; subsequently, amino acids were quantified by use of an automated analyzer. Little information is available on carbohydrate analysis by TLC. Recent work examined glucose and trehalose in the hemolymph and DGG of *Biomphalaria glabrata* to determine the effects of infection by larvae of *Echinostoma caproni*.[26] Carotenoids and xanthophylls in snail tissues infected by larval trematodes have been examined by TLC.[27,28] *Helisoma trivolvis* snails infected with larval stages of *Echinostoma trivolvis* had a complicated carotenoid pattern as compared to uninfected control snails.[28] It was later determined that the echinostome rediae isolated from snail tissues had lutein in addition to β-carotene[29]; the findings were confirmed by spectrophotometry *in situ* on the layer.

II. OBTAINING PARASITE MATERIAL FOR TLC

A. PARASITES OF HUMANS AND OTHER WARM-BLOODED ANIMALS

Parasites can be obtained from humans and other warm-blooded animals following anthelmintic treatment. However, it is important to remember that anthelmintics could affect the subsequent

TABLE 4.2
Assessment of Parasite-Induced Alterations in the Host by Thin-Layer Chromatography

Parasite (Host)	Objective	Chromatographic system and the detection reagent(s)	Results	Reference
Austrobilharzia variglandis (domestic chicken)	Identification of host skin lipids that mediate cercarial penetration	Petroleum ether–diethyl ether–acetic acid (80:20:1); Skipski's dual solvent system; PTLC to remove lipids; detection by PMA or H_2SO_4 and heat	Free fatty acids and free sterols permitted host skin penetration by this avian schistosome cercariae	14
Echinostoma caproni larval stages in snail	To determine if larvae alter the snail lipid composition	Petroleum ether–diethyl ether–acetic acid (80:20:1); detection by PMA and heat	Infection reduced triacylglycerol content of snails	22
Leucochloridiomorpha constantiae larval and adult stages	Effects of parasitism on various host sites	Petroleum ether–diethyl ether–acetic acid (80:20:1); detection by PMA and heat	Infection altered neutral lipid profiles in various host sites	21
Microphallus similis larval stages in snail	Effects of larval parasitism on digestive gland of snails	Neutral lipids: diethyl ether–benzene–ethanol–acetic acid (40:50:2:0.2) followed by hexane–diethyl ether (96:4) in the same direction; detection by I_2 vapors. Phospholipids: chloroform–methanol–water (65:24:4); detection by I_2 vapors	Parasitism by sporocysts decreased triacylglycerols and free fatty acids but increased phospholipids in the digestive gland	61
Schistosoma mansoni larval stages in snail	To determine if larvae alter the lipid composition of the snail	Two-dimensional TLC with petroleum ether–diethyl ether–acetic acid (80:20:1); detection by H_2SO_4 and heat; quantification by densitometry	Infection increased total lipids and triacylglycerols in snails	95

PMA = 5% ethanolic phosphomolybdic acid; **Skipski's dual solvent system** = 1: isopropyl ether–acetic acid (96:4); 2: petroleum ether–diethyl ether–acetic acid (90:10:1) (Skipski et al. (1965)).

biochemical analyses. It is probably best to obtain parasites from warm-blooded hosts as soon as possible after the host is necropsied. The parasites should then be washed rapidly in a cold, isotonic solution and prepared for TLC as soon as possible after removal from the host.

Parasites of warm-blooded animals may be obtained from hosts collected in the wild, from recent road-kills (Note: one must be cautious of the deterioration suffered by parasites and of acquiring unusual infection from wild animals), and from experimentally infected hosts maintained in the laboratory. Some adult parasites, e.g., *Hymenolepis diminuta* and *Trichinella spiralis*, may be obtained by purchasing infected rats. Moreover, the life cycles of these parasites can be maintained in the laboratory after the larval stages have been purchased from the suppliers. Larvae of *H. diminuta*, *T. spiralis*, and *Nippostrongylus brasiliensis* can be purchased from Carolina Biological Supply Co. (Burlington, NC) or Ward's Biological Co. (Rochester, NY).

B. PARASITES OF COLD-BLOODED ANIMALS

Fish, frogs, salamanders, snakes, turtles, and numerous invertebrates, including snails, clams, cockroaches, and other insects, are excellent sources of parasites. A number of guides for maintaining, handling, and processing cold-blooded animals are available.[30-32]

Blood or hemolymph (a term used for invertebrate circulatory fluid) of cold-blooded animals can be analyzed by TLC to determine the effects of parasitism. Hemolymph from invertebrates can be obtained by a variety of methods, including a technique of collecting hemolymph from the heart of *Biomphalaria glabrata* snails.[26,33] Such methods can be adapted for other invertebrates or snails experimentally infected with trematodes. Relatively few TLC analyses of invertebrate hemolymph are available. A review of studies on TLC analysis of lipids in the hemolymph of gastropods is available.[34] Hemolymph of anopheline mosquitoes has been analyzed by TLC to determine the effect of *Plasmodium berghei* infection.[35]

C. PROTOZOAN PARASITES

Although parasitic protozoans can be obtained as described in Sections II.A and II.B above, some species and strains are also available in culture. These can be obtained from Carolina, Ward's, and the American Type Culture Collection (ATCC). A list of protozoa obtainable from ATCC has been published.[36] Knowledge of axenic culture is necessary to propagate protozoa for TLC analysis, and these techniques are available both in primary literature and books.[37] Contamination of a pure culture of protozoa with spurious organisms may invalidate the results of TLC analysis. TLC analysis of parasitic protozoa can be useful in chemotaxonomy, as described recently for *Blastocystis hominis*.[38]

D. STORAGE, PRESERVATION, AND FREEZING OF PARASITE MATERIAL

Parasites removed from organs or tissues of infected hosts should be processed for TLC as soon as possible. If this is not practical, parasite samples should be stored overnight at 2 to 4°C or for longer periods at –20°C. Preservatives or fixatives should not be used because they can alter the sample constituents and may produce anomalous results of TLC analysis.

III. SAMPLE PREPARATION

A. GENERAL CONSIDERATIONS

Sample preparation of animal parasites for TLC analysis is similar to that of other organisms. Considerable literature on sample preparation exists and has been reviewed recently.[39,40] This section is concerned mainly with sample preparation techniques that have been used in animal parasitology.

It should be remembered that samples for TLC must be dissolved in a suitable solvent prior to the application of aliquots to a TLC plate or sheet. Typically, volumes of 1 to 10 µl are applied to the origin of the layer, with even less sample being applied if high performance thin-layer chromatography (HPTLC) is used. HPTLC has a number of advantages over conventional TLC.[40] If the compound of interest is present in relatively low concentrations in a complex host or parasite sample, extraction, isolation, and concentration steps normally precede TLC. These steps may be time consuming, but they are important to the success of the TLC procedure because impurities may yield erroneous results.

B. DISSOLVING AND SIMPLE EXTRACTION PROCEDURES

When the compounds of interest are major components of the sample, dissolving the sample in a suitable amount of an appropriate solvent, followed by direct application of aliquots may be adequate for TLC. This procedure has been used to obtain samples of the acanthocephalan *Moniliformes dubius*.[41] Starting with 1 gram or less of worm tissue, 10 ml of 70% ethanol was added and the sample was allowed to sit overnight. The supernatant was decanted and used for analysis of free pool compounds such as simple sugars, organic acids, and amino acids. The worm tissue was then homogenized in 10 ml of 70% ethanol and centrifuged. The supernatant was used for analysis of proteins, nucleic acids, lipids, and polysaccharides. Another example of this technique is that of the sample preparation for analysis of sugars in the hemolymph of *Biomphalaria glabrata* snails infected with larval *Echinostoma caproni*.[26]

About 150 to 200 µl of snail hemolymph was pooled in a microcentrufuge tube and centrifuged at $8000 \times g$ for 2 min. About 100 µl of the supernatant was transferred to another centrifuge tube, 500 µl of 70% ethanol was added, and the solution was centrifuged at $8000 \times g$ for 5 min. The supernatant was transferred to another vial, and the pellet was washed twice with 100 µl of 70% ethanol. The washes were combined with the supernatant in a new vial. The vial was maintained in a water bath at 45°C, and the sample was evaporated to dryness using a stream of air. Prior to TLC, the residue was reconstituted with 200 µl of 70% ethanol (see Section VII.E).

A similar approach was used to prepare samples of *Echinostoma trivolvis* adults for amino acid analysis.[24] Approximately 100 mg of worm tissue was obtained from experimentally infected domestic chicks. Amino acids were extracted in 70% ethanol for 24 h. The supernatant was then removed to another vial, and the worm sample was homogenized in 1 ml of 70% for 2 min. The material was allowed to stand for 10 min at 4°C. The pooled ethanolic extract was evaporated in air and the residue reconstituted in 100 µl of 70% ethanol. The sample was spotted on a TLC sheet as described in Section VII.F.

Dissolving and simple extraction procedures have also been used for TLC analysis of carotenoids and xanthophylls in *Echinostoma trivolvis* rediae.[29] Rediae were removed from the DGG of naturally infected *Helisoma trivolvis* snails, and about 100 mg of sample was extracted in cold acetone in a test tube. The material was allowed to stand for 5 to 10 min, and the supernatant was removed with a Pasteur pipet. The supernatant was passed through a glass wool filter, and the filtrate was concentrated under a stream of nitrogen to about 100 µl. The sample was not allowed to dry or was not exposed to air for any length of time because of the susceptibility of the pigments to decomposition from oxidation, light, and heat. If the sample is to be stored, it should be purged with nitrogen and kept at –20°C until used for TLC.

C. COMPLEX EXTRACTION PROCEDURES — PARTITION OF EXTRACTS

Extracts may not be pure enough for use as described in Section III.B and may require partitioning with an immiscible solvent as an additional cleanup step. The principle of differential partitioning is to leave impurities behind in one solvent, while extracting the compound of interest into another. Liquid-liquid partitioning systems have been reviewed.[42] The most widely used complex extraction procedure in animal parasitology is that using chloroform–methanol (2:1, v/v) extraction followed by washing with 0.88% KCl, a technique first described by Folch et al.[43]

The following description of a technique is based on two recent studies.[23,44] Place 100 mg of tissue or 100 µl of blood or hemolymph in a glass tube and add 1 ml of chloroform–methanol (2:1). Homogenize the tissue or fluid sample with a glass stirring rod. A protein precipitate will form; allow it to settle (about 5 min). Remove the supernatant fluid with a disposable pipet; pass the fluid through a glass wool filter and collect it in another tube. Repeat the extraction procedure twice, each time adding 1 ml of chloroform–methanol (2:1) to the residue. Collect this supernatant by passing it through a glass wool filter into the tube containing the original supernatant. About 3 ml of a clear supernatant should be obtained. This supernatant will contain most of the lipids of the original sample along with nonlipid impurities. To separate the nonlipid impurities, add about 0.7 ml of 0.88% KCl (the Folch wash) to the tube. Stir the mixture by hand or with a Vortex mixer. This procedure will produce a biphasic mixture consisting of a top hydrophilic layer and a bottom lipophilic layer. Remove the top layer with a disposable pipet. This layer contains mainly amino acids, sugars, and organic acids and can be used for TLC analysis or discarded if interest is only in the lipophilic constituents. The bottom layer contains mainly lipids and should be transferred to a small vial for the TLC analysis of lipids and natural pigments.

D. COLUMN CHROMATOGRAPHY

The use of column chromatography is helpful for sample cleanup and fractionation prior to TLC. Columns are usually packed with silicic acid or Florisil® (Floridin) as the adsorbent. After a sample is applied to the column, elution with chloroform will separate neutral lipids from phospholipids, and the latter compounds may be recovered by elution with methanol. Glycolipids of *Onchocerca gibsoni* were separated from lipid extracts by column chromatography.[45] Because the phospholipids interfered with the analysis of glycolipids, it was necessary to isolate the glycolipids from other lipid classes. The glycolipids were acetylated prior to isolation by Florisil column chromatography. The purified glycolipids were then deacylated and analyzed by HPTLC.

IV. CHROMATOGRAPHIC SYSTEMS AND DETECTION OF COMPOUNDS

A. GENERAL CONSIDERATIONS

This section describes the chromatographic systems, i.e., sorbents and solvents (mobile phases), used frequently in analysis of parasites or their products. Described also in this section are the detection (visualization) techniques used for various compounds.

B. NEUTRAL LIPIDS

The most widely used chromatographic system for neutral lipid analysis uses silica gel as the sorbent and a mobile phase consisting of petroleum ether–diethyl ether–acetic acid (80:20:1).[46] Some variants of this system have used solvents in 90:10:1 or 70:30:1 ratios. The latter ratios produce some differences in the mobilities of the neutral lipids, compared to what is obtained in the 80:20:1 system. The Mangold system (80:20:1) provides good separation of the more commonly occurring neutral lipids in extracts of parasite tissues. Neutral lipids usually separated in this system are free sterols, free fatty acids, triacylglycerols, and sterol esters. This system has been employed in the analytical TLC of parasites.[2,47] The Mangold system has also been used to preparatively isolate the sterol fraction of *Biomphalaria glabrata* snails experimentally infected with larval *Echinostoma caproni*.[48] The sterol fraction was subsequently analyzed by gas-liquid chromatography. The most widely used detection methods following neutral lipid separation in the Mangold system consist of spraying or dipping of plates in 5% phosphomolybdic acid (PMA) in absolute ethanol or spraying the plates with 50% aqueous H_2SO_4 and heating for 10 min at 110°C.

Because the Mangold system does not separate all commonly occurring neutral lipids in animal tissues, a dual solvent system (the Skipski system) is often used with silica gel as the sorbent.[49] In this system, glycerides and free fatty acids are unequivocally separated from free sterols by double development in the same direction with two different mobile phases. The plates are first developed in isopropyl ether–acetic acid (65:4), dried, and then developed in the same direction in petroleum ether–diethyl ether–acetic acid (90:10:1). This dual solvent system is useful for both analytical TLC and PTLC. Lipid fractions of *Echinostoma trivolvis* were preparatively isolated using this system for subsequent studies on chemical communication.[50] Detection procedures used with this system include PMA and aqueous H_2SO_4 as described above. Iodine vapors are typically used in nondestructive vizualization after PLC with either of the two solvent systems described above.

C. PHOSPHOLIPIDS

Phospholipids are more polar than neutral lipids and remain at the origin when nonpolar solvents are used. To separate phospholipids, a more polar solvent system (the Wagner system) consisting of chloroform–methanol–water (65:25:4) is used.[51] This system separates the more commonly occurring phospholipids in animal tissues, e.g., phosphatidylcholine, phosphatidylethanolamine, and phosphatidylserine, and moves the neutral lipids as a single band at or near the solvent front. In practice, it is desirable to separate both neutral lipids and phospholipids on the same silica gel plate, as has been done on extracts of *Echinostoma trivolvis* adults.[3] To achieve such a separation, phospholipids were first separated in chloroform–methanol–water (65:25:4), and after the plate was air-dried, it was developed in the same direction in a second mobile phase consisting of hexane–diethyl ether (4:1). Detection of lipids was accomplished by spraying the plate with 50% aqueous H_2SO_4 and heating. This procedure yields good separation of the commonly occurring parasite neutral lipids and phospholipids on the same plate.[3]

Two-dimensional systems are occasionally used to separate phospholipids. A silica gel plate containing an aliquot of parasite extract was first developed in the Wagner system (see above). After the plate was dried, it was turned 90° and developed in the second direction in either *n*-butanol–acetic acid–water (60:20:20) or chloroform–acetone–methanol–acetic acid–water (5:2:1:1:0.5).[52] The two-dimensional procedure will increase the resolution of spots, but it often results in large spots with tails.[3] Two-dimensional TLC was used to separate and identify phospholipids from extracts of the digestive gland–gonad (DGG) complex of *Biomphalaria glabrata* snails infected with larval *Schistosoma mansoni*.[53] Phospholipids were separated on silica gel plates developed with chloroform–methanol–water–40% NH_4OH (86:28:1:3) in the first direction and chloroform–methanol–water–acetic acid (76:9.5:2.5:12) in the second. The phospholipids were identified by spraying with molybdenum blue reagent and heating the plate; more specific detection reagents were used to identify individual phospholipids.

D. GLYCOLIPIDS (GLYCOSPHINGOLIPIDS)

Analysis of glycolipids in parasite material has been undertaken in the 1990s. A recent study analyzed glycolipids in chloroform–methanol (2:1) extracts of *Schistosoma mansoni* adults. The extract was centrifuged and the residue reextracted in chloroform–methanol (2:1). The supernatants were pooled and dried on a rotary evaporator. Glycolipids were separated as acetylated derivatives from other constituents by Florisil® column chromatography. The glycolipid fraction was then deacylated, and samples were applied on HPTLC plates and developed in chloroform–methanol–water (65:25:4). The glycolipids were visualized by spraying with orcinol reagent.[54]

Glycosphingolipids of *Leishmania amazonensis* amastigotes have also been examined by TLC. Excellent separations were achieved when extracts and standard glycosphingolipids were separated on HPTLC plates using a chloroform–methanol–0.02% $CaCl_2$ (60:40:9) mobile phase. The glycolipids were detected by spraying with orcinol reagent.[55]

E. AMINO ACIDS

Free-pool amino acids in adults of *Echinostoma trivolvis*[5] and the hemolymph of *Anopheles stephensi* infected with *Plasmodium berghei* have been analyzed by TLC.[56] The latter authors used microcrystalline cellulose layers as the sorbent and a solvent system composed of n-propanol–methyl ethyl ketone–water–formic acid (10:6:3:1). Amino acids were detected by dipping the plates in a ninhydrin reagent (0.25% ninhydrin, 0.5% glacial acetic acid, and 0.5% pyridine in acetone). Color was allowed to develop by heating the plate for 5 min at 70°C. *E. trivolvis* amino acids were also analyzed using cellulose sheets (Baker-Flex). The sheets were developed for a distance of 12 cm from the origin in a mobile phase composed of n-butanol–acetic acid–water (60:15:25). Prior to development, ninhydrin was incorporated into the solvent system at a concentration of 0.5 g/100 ml. After development, the sheets were air-dried, and the amino acids were visualized by heating the sheets for 2 min at 90°C. Alanine, proline, serine, and methionine were identified[5] (see Section VII.F).

F. CARBOHYDRATES

Carbohydrates in the hemolymph of *Anopheles stephensi* infected with *Plasmodium berghei*[57] and in *Biomphalaria glabrata* infected with larval *Echinostoma caproni*[26] have been examined by TLC. In the former study, glucose and trehalose were detected in the hemolymph of infected mosquitoes using one-dimensional TLC on 5 × 7.5 cm microcrystalline cellulose sheets. Samples and standards were developed for 7 cm from the origin in a mobile phase consisting of n-propanol–methyl ethyl ketone–water–formic acid (10:6:3:1). The chromatogram was dried at room temperature and developed in the same direction in a mobile phase containing ethyl acetate–pyridine–water (12:5:4). Of the many sprays that were used to detect the sugars, silver oxide was the most sensitive.[57] The other study used preadsorbent HPTLC plates (Merck) impregnated with sodium bisulfite and citrate buffer (pH 4.8). The plates were developed with acetonitrile–water (85:15), and sugars were detected with the α-naphthol–sulfuric acid reagent.[26] The major sugars detected in both the hemolymph and DGG of snails were glucose and trehalose. Infection with larval echinostomes caused a marked reduction of these sugars in snails.

G. NATURAL PIGMENTS

The redial stages of trematodes often contain yellow-orange carotenoid pigments. It has been shown that the DGG complex of *Helisoma trivolvis* snails infected with larval *Echinostoma trivolvis* had high levels of carotenoids.[28] A recent TLC study identified the carotenoids in rediae of *E. trivolvis* as β-carotene and lutein.[29] Chloroform–methanol (2:1) extracts of 1000 rediae along with authentic β-carotene and lutein standards were chromatographed on 10 × 20 cm HPTLC preadsorbent silica gel plates. Plates were developed in the Mangold system (Section IV.B), and the pigments were viewed after the plates were air-dried. These pigments have intrinsic colors, and, therefore, a detection reagent was not needed. These findings were confirmed using a second chromatographic system consisting of a C_{18} chemically bonded reversed-phase layer and a mobile phase composed of petroleum ether–acetonitrile–methanol (2:4:4); β-carotene and lutein were thus identified (see Section VII.D).

V. QUANTIFICATION

A variety of methods have been described for quantification of compounds of interest following separation by TLC. These methods can be divided into two categories: (1) scraping and elution of compounds from the TLC plate and quantification by other means; and (2) direct *in situ* quantification, usually by densitometric methods. Both methods have been used in parasitology, and specific examples are given below.

In the first method, the compounds of interest were scraped and eluted from TLC plates and then analyzed by spectrophotometric, gravimetric, or chromatographic techniques. Scraping and elution procedures are used in PLC, and this topic has been reviewed.[58] Prior to scraping and elution of compounds from TLC plates, the bands must be visualized by a nondestructive detection agent such as iodine vapors, water spray, or fluorescent reagents, e.g., fluorescein isothiocyanate, or rhodamine B. Care must be taken to assure that the detection reagent does not react adversely with the compounds of interest.[40]

Lipids in the acanthocephalan *Paratenuisentis ambiguus* have recently been separated and quantified by scraping and elution.[59] Lipids in worm extracts were separated by various TLC procedures, and the total neutral lipid fraction as well as individual neutral lipid classes were quantified gravimetrically after recovery from the TLC plates. Aliquots of total lipids as well as acyl lipids were converted to methyl esters by methanolysis. The fatty acid methyl esters and steryl acetates were purified by TLC on silica gel H with hexane–diethyl ether (4:1) as mobile phase prior to isolation and quantification by gas-liquid chromatography. In this study, sufficient quantities of worm material were available for gravimetric analysis. Gravimetric analysis of samples less than 100 mg wet weight may not be reliable, since small amounts of sorbent, binder, or other impurities may be eluted from the plates and weighed inadvertently.

Phospholipids separated by TLC[51] (see Section IV.C) and then eluted from the plate may be quantified by the spectrometric technique for phosphorus.[60] Phospholipid composition of the digestive gland of the marine snail *Littorina saxatilis rudis* infected with *Microphallus similis* sporocysts was determined by this method.[61]

Flame ionization detection (FID) systems in combination with TLC have also been used for lipid quantification.[62] Iatroscan rods or Chromrods (Iatron Laboratories Inc., Tokyo, Japan) have been the most frequently used separation media.[63] Although used infrequently in parasitology, the Iatroscan TH-10 TLC Analyzer (Iatron Laboratories Inc.) has been used to quantify lipid classes in the filarid nematode *Foleyella agamae*.[64]

The second above-cited quantification method, *in situ* densitometry, is of increasing current interest in TLC and has a large body of literature.[65] A number of recent studies in parasitology have employed this method of quantification.[26,29,44] Many compounds that can be detected in visible or ultraviolet light can be subjected to densitometric analysis. Suitable standards are required and should match the compounds of interest. TLC methods for *in situ* densitometry are similar to those of conventional TLC; however, greater care is needed in sample preparation and application, and choice of development system, detection reagents, and standards. The range of weights of standards used should bracket as closely as possible the weights of the compounds of interest in the applied samples. Densitometry is usually accomplished in the transmittance or reflection mode with a particular commercial densitometer. Considerable choice exists in the selection of densitometers, from simple models to automated instruments coupled to computer systems.[66] Detailed descriptions of quantification by *in situ* densitometry are available elsewhere.[40]

Phospholipids in *Biomphalaria glabrata* infected with larval *Echinostoma caproni* were quantified by *in situ* densitometry following TLC.[67] Pooled samples of choloroform–methanol (2:1) extracts of the DGG of infected vs. uninfected *B. glabrata* were prepared (Section III.C). TLC was performed on 20 × 10 cm laned silica gel HPTLC plates with preadsorbent spotting area (Whatman) in a Camag twin-trough HPTLC chamber using chloroform–methanol–isopropanol–0.25% KCl–ethyl acetate (30:9:25:6:18) as the mobile phase. For quantification of phosphatidylethanolamine (PE), phopsphatidylserine (PS), and phosphatidylcholine (PC), various aliquots of samples and authentic phospholipid standards were applied to the plates in the range of 600 to 1600 ng.

Phospholipids were detected by spraying the plate with a 10% cupric sulfate–8% phosphoric acid solution and subsequent heating at 160°C for 10 min. Standard and sample zone areas were measured by reflectance scanning at 400 nm using a Shimadzu CS-930 (Columbia, MD) densitometer in the single-lane–single-beam mode. Four aliquots of PE, PS, and PC standards were scanned, and a calibration curve was constructed. The weight of each phospholipid in the sample spots that had areas best bracketed by the standards were interpolated from the calibration curve, and the percent phospholipid in the snail was calculated based on the sample weight.

VI. VALIDATION OF RESULTS

A number of techniques have been used to determine the validity of TLC findings on parasites and these are discussed below. To determine the reliability of TLC, replicate experiments (at least two or more replicates) should be carried out on different representative samples. For example, several replicates of extracts of *B. glabrata* infected with *E. caproni* were used to show that this infection depletes triacylglycerols from the snails.[22]

Use of pooled samples is important in TLC studies on parasites or hosts infected with parasites to eliminate intrinsic variations that are often seen in analyses on individual samples. For example, DGG or hemolymph were pooled from five infected or five uninfected snails for each analysis to determine the effect of *Echinostoma caproni* infection on carbohydrates in *B. glabrata* snails.[26] Consistent results were obtained on the sugar content of infected vs. uninfected snail populations using pooled samples.

The presence of certain components in a sample following TLC analysis may indicate contamination or degradation of the sample. For instance, significant amounts of methyl esters rarely occur in animal tissues,[68] and their presence may indicate a preparation artifact. The neutral lipid mixed standard 18-4A (Nu Chek, Elysian, MN) contains methyl oleate. Use of this standard along with the sample will reveal if methyl esters are present in the sample. Significant amounts of methyl esters would also suggest an anomaly in TLC and/or sample preparation. Readers are referred to published chromatograms that show typical neutral lipid profiles in worm samples; note the absence of a methyl oleate zone in these chromatograms.[2,69]

Use of different chromatographic systems is helpful in validating the results of a TLC analysis. For example, natural pigments in the rediae of *Echinostoma trivolvis* were found to be β-carotene and lutein using silica gel sorbent and the Mangold solvent system. This finding was confirmed by using a C_{18} reversed phase plate developed in petroleum ether–acetonitrile–methanol (2:4:4).[29] Confirmation of the results of TLC analysis using two different modes of chromatography, e.g., normal phase and reversed phase, as demonstrated in this study[29] is a good way to validate the TLC results.

Use of various detection reagents is also helpful in validating the results of TLC. Analysis of phospholipids in *E. trivolvis* is an example of such validation.[3] Phospholipids were first visualized by nonspecific detection reagents, and individual phospholipids were then identified by specific detection tests.[3]

Perhaps the best way to validate the findings of TLC is to isolate the fraction of interest by PLC and confirm its identity using another analytical technique, e.g., gas-liquid chromatography

(GLC) or high-performance liquid chromatography. Numerous TLC studies on *E. trivolvis* adults showed that the most abundant free sterol in this organism is cholesterol.[4,50] Capillary GLC analysis of the preparatively isolated sterol fraction of *E. trivolvis* adults confirmed the presence of cholesterol as the major free sterol; numerous other minor free sterols were also identified.[70]

VII. SELECTED PROTOCOLS

A. GENERAL CONSIDERATIONS

The following protocols are based on TLC experience by one of us (B.F.) with echinostome parasites. Echinostomes parasitize a variety of avian and mammalian hosts and can be maintained in the laboratory in rodent or avian hosts and pulmonate snails. It is hoped that the following protocols can be adapted for studies on other organisms.

B. NEUTRAL LIPIDS IN *ECHINOSTOMA CAPRONI* ADULTS*

This study examined the neutral lipid profile in adults of *E. caproni*.

Sample preparation. Adult worms were recovered from the intestines of experimentally infected hosts, rinsed in saline, and pooled to obtain an approximate 50 mg wet weight sample. Lipids from the worms were extracted in chloroform–methanol (2:1) as described under sample preparation (Section III.C). The sample was dried under a stream of nitrogen and reconstituted in 200 µl of chloroform–methanol (2:1) prior to TLC.

Chromatography. Silica gel plates (20 × 20 cm, LK6D, Whatman, Clifton, NJ) with channels and preadsorbent spotting area were cleaned by development in chloroform–methanol (1:1). Samples were applied on plates as 1, 5, and 10 µl aliquots using a 10 µl Drummond digital microdispenser (Broomall, PA). Neutral lipid standard 18-5A (Nu-Check Prep., Elsyian, MN) containing 0.20 µg/µl each of phosphatidylcholine, cholesterol, oleic acid, triolein, and cholesterol oleate was used as a reference for identifying neutral lipid fractions of the samples. The standards were also used in 1, 5, and 10 µl amounts. Plates were developed for 10 cm past the preadsorbent–adsorbent interface in a paper-lined glass chromatography tank containing 100 ml of petroleum ether–diethyl ether–acetic acid (80:20:1). To detect the neutral lipids, plates were sprayed with 5% PMA in ethanol in a fume hood and heated at 110°C in an oven for 10 min.

Results and Discussion. Distinct purple-blue zones were obtained on a yellow background with the PMA reagent. The R_f values of the neutral lipid standards were: cholesterol, 0.22; oleic acid, 0.36; triolein, 0.56; and cholesterol oleate, 0.82. The phosphatidylcholine standard remained at the origin. The major worm neutral lipid fraction had an R_f identical to that of the cholesterol standard and was free sterol. Worm chromatograms also had spots with lesser amounts of free fatty acids, triacylglycerols, and sterol esters, which had R_f values identical to those of the oleic acid, triolein, and cholesterol oleate standards, respectively.

A similar neutral lipid profile has been reported for *E. trivolvis*.[11] A different neutral lipid profile has been reported for *Cotylophoron cotylophorum*, *Gastrothylax crumenifer*, *Gigantocotyle explanatum*, *Echinostoma malayanum*, and *Isoparorchis hypselobagri*, in which triacylglycerol represented the major neutral lipid fraction.[71]

C. PHOSPHOLIPIDS IN *BIOMPHALARIA GLABRATA* SNAILS INFECTED WITH THE INTRAMOLLUSCAN STAGES OF *ECHINOSTOMA CAPRONI*†

This study examined the major phospholipids in the DGG of the snail host (the target organ system for redial development) and determined how the phospholipid composition is altered by larval trematode parasitism.

Sample preparation. *B. glabrata*, 8 to 10 mm in shell diameter, were infected with *E. caproni* miracidia and maintained in aquaria.[22] Phospholipids of infected and uninfected snails were analyzed

---

* Adapted from Horutz and Fried.[23]
† Adapted from Perez et al.[67]

at 6 weeks postinfection. DGG samples (approximately 100 mg wet weight) were extracted in chloroform–methanol (2:1) as described in Section III.C. The extracts were dried under a stream of nitrogen and reconstituted with 500 µl of chloroform–methanol (2:1) prior to chromatography.

Chromatography. Silica gel 20 × 10 cm HPTLC laned plates with preadsorbent application area (Whatman, Clifton, NJ) were used for TLC. Samples and standards were applied to plates with a 10 µl digital microdispenser (Drummond, Broomall, PA). Samples were applied as 2, 4, and 6 µl aliquots. A polar lipid mixture (1 mg/ml) containing equal amounts of cholesterol (CH), phosphatidylethanolamine (PE), phosphatidylcholine (PC), and lysophosphatidylcholine (LPC) (Matreya, Inc., Pleasant Gap, PA) and an additional phosphatidylserine (PS) standard (1 mg/ml) were used. The standards were applied on the plates as 2, 4, 6, 8, and 10 µl aliquots. For optimal separation of the phospholipids, plates were developed four times in a paper-lined tank to a distance of 7 cm from the silica gel–preadsorbent interface, with drying between runs. The mobile phase consisted of chloroform–methanol–isopropanol–0.25% KCl–ethyl acetate (30:9:25:6:18). After the fourth development, the plate was dried with a hair dryer and the phospholipids detected by spraying with a 10% cupric sulfate–8% phosphoric acid solution and subsequent heating in a gravity convection oven at 160°C for 10 min. Specific detection reagents were also used to identify certain classes of phospholipids; for example, plates were sprayed with 0.2% ninhydrin in *n*-butanol saturated with water and heated at 100°C to stain the amino-lipids, PS, and PE.

Results and Discussion. The zones appeared brown on a white background with the cupric sulfate–phosphoric acid detection reagent. Ninhydrin reagent confirmed the presence of PS and PE in the standards and samples. The R_f values of the polar lipid mixture constituents were: LPC, 0.20; PC, 0.36; PE, 0.70; CH, 0.97. The PS standard appeared as three zones with R_f values of 0.38, 0.39, and 0.44. The main phospholipids in the snail DGG were PC, PS, and PE. Densitometric analysis showed that the amount of PS was significantly greater in infected vs. uninfected snails.

D. AMINO ACIDS IN *ECHINOSTOMA TRIVOLVIS* ADULTS*

This study examined the free-pool amino acids in the adult stage of *E. trivolvis*. Amino acid release into a nonnutrient medium was also examined.

Sample Preparation. Worms were obtained from the intestines of experimentally infected hosts (see Section VII.B), rinsed in saline, and pooled to obtain approximately 100 mg of tissue. The free-pool amino acids were extracted in 2 ml of 70% ethanol over 24 h. The ethanol supernatant was removed to a separate tube, and the worm tissue was homogenized in 1 ml of 70% ethanol for 2 min. The homogenate was allowed to stand for 10 min at 4°C before removing the supernatant. The two ethanolic extracts were then pooled, evaporated in air, and reconstituted in 100 µl of 70% ethanol prior to TLC.

Chromatography. Baker-Flex cellulose sheets, 20 × 20 cm (J.T. Baker Chemical Co., Phillipsburg, NJ), were used for TLC as received from the supplier. Disposable micropipets (Drummond Co., Inc., Broomall, PA) of 1, 5, and 10 µl capacities were used to apply standard and sample aliquots to the sheets. Serine, alanine, proline, methionine, and other amino acid standards (Sigma Chemical Co., St Louis, MO) were prepared as 1 µg/µl solutions in deionized water. Preliminary experiments using numerous chromatographic systems showed that these amino acids were the major free-pool ones in *E. trivolvis* adults. The individual amino acids were applied on the sheet as 1 µl aliquots and the samples were applied as 1, 5, or 10 µl aliquots. Sheets were developed for 12 cm past the origin in a vapor-saturated, paper-lined glass chromatographic tank containing 100 ml of *n*-butanol–acetic acid–water (12:3:5) to which 0.5 g of ninhydrin had been added for amino acid detection. After development, the sheets were removed from the tank, air dried, and heated in an oven at 90°C for 5 min.

Results and Discussion. The R_f values of the amino acid standards were: serine, 0.42; alanine, 0.54; proline, 0.62; methionine, 0.75. Worm samples yielded distinct zones with identical mobilities

---

* Adapted from Bailey and Fried.[24]

to the standards. Differences in colors of individual amino acids were noticeable and were compared to the corresponding amino acid standard in addition to identification based on R_f values. Other minor unidentified zones may also be present in experimental samples.

E. SUGARS IN THE HEMOLYMPH OF *BIOMPHALARIA GLABRATA* INFECTED WITH THE INTRAMOLLUSCAN STAGES OF *ECHINOSTOMA CAPRONI**

This study determined the effect of larval *E. caproni* infection on the glucose and trehalose content of the hemolymph of the snail host *B. glabrata*. Glucose and trehalose are the major carbohydrates in the hemolymph of *B. glabrata*.[72]

Sample Preparation. *B. glabrata*, 8 to 10 mm in shell diameter, were infected with *E. caproni* miracidia and maintained in aquaria as described elsewhere[22] (Section VII.C). The hemolymph of infected snails and matched controls was collected at 6 weeks post-infection. Analyses were performed on the hemolymph of pooled samples (five infected snails per pool vs. five uninfected snails per pool). The hemolymph was transferred to a microcentrifuge tube and centrifuged for 2 min at 8000 × g. One hundred µl of plasma (since snail blood has no clotting factor, the term plasma is used over serum) was separated from the amoebocyte–debris pellet and placed in a new centrifuge tube with 500 µl of 70% ethanol. The sample was then centrifuged at 8000 × g for 5 min, and the supernatant transferred to a new vial. This vial was placed in a water bath at 45 to 55°C and the sample evaporated to dryness under air. The residue was reconstituted with 200 µl of 70% ethanol.

Chromatography. TLC was performed on 10 × 10 cm HPTLC silica gel 60 CF_{254} plates with preadsorbent zone and channels (Merck Inc., Rahway, NJ). The plates were pretreated by spraying with $0.1M$ sodium bisulfite solution, drying in air for 10 min, spraying with pH 4.8 citrate buffer (Sigma Chemical Co., St. Louis, MO), and again drying in air for 10 min. Samples and standards were applied with a 10 µl digital microdispenser (Drummond, Broomall, PA). Glucose and trehalose standards (Sigma Chemical Co., St. Louis, MO) were prepared as 1 mg/ml solutions in 70% ethanol. Standards were applied on the plates as 1.2 and 2.4 µl aliquots and samples as 3.5 and 5.0 µl aliquots. For optimal separation of sugars, the plates were developed three times, each time for a distance of 7 cm with acetonitrile–deionized water (85:15) in a paper-lined Camag twin-trough HPTLC chamber that had been preequilibrated with the mobile phase for 10 min. Fresh solvent was used for each run, and between runs, the plate was dried with a hair dryer. Approximately 12 ml of solvent was required for each development, and the average developing time per run was 15 min. After the third development, the plate was dried in an oven at 100°C, and the sugar spots were detected by spraying with an α-naphthol–sulfuric acid reagent (prepared by adding 5 g of α-naphthol to 33 ml of absolute ethanol and then adding 20 ml of sulfuric acid, 127 ml of ethanol, and 13 ml of deionized water). After spraying, the plate was heated for 5 min at 110°C.

Results and Discussion. The R_f values of the standards were 0.19 for trehalose and 0.40 for glucose. The zones appeared as blue-purple spots on a white background. The main sugars in the hemolymph of infected and uninfected snails had mobilities identical to the trehalose and glucose standards. An unknown zone with an R_f of about 0.09 appeared in both samples. Infected snails had significantly reduced levels of glucose and trehalose in the hemolymph as determined by densitometry. Larval schistosomes have been reported to produce a similar effect.

F. CAROTENOID PIGMENTS IN *ECHINOSTOMA TRIVOLVIS* REDIAE†

This study examined the carotenoid pigments in the daughter rediae of *E. trivolvis*. Similar analysis of pigments in rediae or other tissues can be performed, particularly when their presence is known by other means, e.g., gross or microscopic examination.

Sample Preparation. Rediae were dissected from the DGG complex of naturally infected *Helisoma trivolvis* snails. They were rinsed in several changes of saline and pooled to obtain samples

---

* Adapted from Perez et al.[26]
† Adapted from Fried et al.[29]

of 1000 rediae for analysis. The sample was extracted immediately with 3 ml of chloroform–methanol (2:1). The extract was filtered through glass wool, and the supernatant was transferred to a vial and evaporated to dryness under a stream of N_2. The residue was reconstituted in 100 µl of chloroform–methanol (2:1) for chromatography.

Chromatography. Silica gel plates (10 × 20 cm, high performance, LHPKDF) with channels and preadsorbent spotting area were used (Whatman, Clifton, NJ). β-carotene (0.01 mg/ml) and lutein (0.1 mg/ml) standards (Sigma Chemical Co., St. Louis, MO) were prepared in chloroform. Various aliquots of samples and standards (usually 10 or 20 µl) were applied to the plates using a 25 µl digital microdispenser (Drummond, Broomall, PA). The plates were developed in the Mangold solvent system, i.e., petroleum ether–diethyl ether–acetic acid (80:20:1) and air-dried, and the spots were visualized immediately in daylight.

Results and Discussion. The R_f values of the carotenoid standards were 0.17 for lutein and 0.93 for β-carotene. The carotenoid zones were also seen in the rediae sample, a fast-moving one with an R_f identical to the carotene standard (0.93) and a slow-moving one with an R_f identical to the lutein standard (0.17). A second chromatographic system was used to confirm the presence of these pigments in the rediae. This system consisted of a C_{18} reversed phase bonded plate (Whatman, Clifton, NJ) and the mobile phase petroleum ether–acetonitrile–methanol (2:4:4). Rediae extracts were prepared in acetone, and sample aliquots along with standards were applied on the reversed-phase plate. Following development, the two rediae pigments were detected. The R_f for lutein was 0.56 and the R_f for β-carotene was 0.10.

REFERENCES

1. **Bailey, H. H.,** A technique for tentative identification of helminth neutral lipid classes: thin-layer chromatography, in *Experiments and Techniques in Parasitology,* MacInnis, A. J. and Voge, M., Eds., W. H. Freeman, San Francisco, 1970, 192–198.
2. **Fried, B. and Shapiro, I. L.,** Accumulation and excretion of neutral lipids in the metacercaria of *Leucochlordiomorpha constantiae* (Trematoda) maintained *in vitro, J. Parasitol.,* 61, 906, 1975.
3. **Fried, B. and Shapiro, I. L.,** Thin-layer chromatographic analysis of phospholipids in *Echinostoma revolutum* (Trematoda) adults, *J. Parasitol.,* 65, 243, 1979.
4. **Barrett, J., Cain, G. D., and Fairbairn, D.,** Sterols in *Ascaris lumbricoides* (Nematoda), *Macracanthorhynchus hirudinaceus* and *Moniliformis dubius* (Acanthocephala), and *Echinostoma revolutum* (Trematoda), *J. Parasitol.,* 56, 1004, 1970.
5. **Bailey, R. S., Jr. and Fried, B.,** Thin-layer chromatographic analyses of amino acids in *Echinostoma revolutum* (Trematoda) adults, *Int. J. Parasitol.,* 7, 497, 1977.
6. **Fried, B., Beers, K., and Sherma, J.,** Thin-layer chromatographic analysis of beta-carotene and lutein in *Echinostoma trivolvis* (Trematoda) rediae, *J. Parasitol.,* 79, 113, 1993.
7. **Smith, T. M., Brooks, T. J., Jr., and White, H. B., Jr.,** Thin-layer and gas-liquid chromatographic analysis of lipid from cercariae of *Schistosoma mansoni, Am. J. Trop. Med. Hyg.,* 15, 307, 1966.
8. **Fried, B., Lewis, P. D., Jr., and Beers, K.,** Thin-layer chromatographic and histochemical analyses of neutral lipids in the intramolluscan stages of *Leucochloridium variae* (Digenea, Leucochloridiidae) and the snail host, *Succinea ovalis, J. Parasitol.,* 81, 112, 1995.
9. **Fried, B. and Morrone, L. J.,** Histochemical lipid studies on *Echinostoma revolutum, Proc. Helminthol. Soc. Wash.,* 37, 122, 1970.
10. **Fried, B. and Pucci, D. L.,** Histochemical and thin-layer chromatographic analyses of neutral lipids in *Leucochloridiomorpha constantiae* adults, *Int. J. Parasitol.,* 6, 479, 1976.
11. **Fried, B. and Boddorf, J. M.,** Neutral lipids in *Echinostoma revolutum* (Trematoda) adults, *J. Parasitol.,* 64, 174, 1978.
12. **Meyer, F., Meyer, H., and Bueding, E.,** Lipid metabolism in the parasitic and free-living flatworms, *Schistosoma mansoni* and *Dugesia dorotocephala, Biochim. Biophys. Acta,* 210, 257, 1970.
13. **Sharaf, D. M., Clark, S. J., and Downing, D. T.,** Skin surface lipids of the dog, *Lipids,* 12, 786, 1977.
14. **Zibulewsky, J., Fried, B., and Bacha, W. J., Jr.,** Skin surface lipids of the domestic chicken, and neutral lipid standards as stimuli for the penetration of *Austrobilharzia variglandis* cercariae, *J. Parasitol.,* 68, 905, 1982.

15. **Feiler, W. and Haas, W.,** *Trichobilharzia ocellata*: chemical stimuli of duck skin for cercarial attachment, *Parasitology*, 96, 507, 1988.
16. **Keenan, T. W., Huang, C. M., and Zierdt, C. H.,** Comparative analysis of lipid composition in axenic strains of *Blastocystis hominis*, *Comp. Biochem. Physiol.*, 102B, 611, 1992.
17. **Fried, B. and Appel, A. J.,** Excretion of lipids by *Echinostoma revolutum* (Trematoda), *J. Parasitol.*, 63, 447, 1977.
18. **Haseeb, M. A. and Fried, B.,** Chemical communication in helminths, *Adv. Parasitol.*, 27, 169, 1988.
19. **Fried, B.,** Trematoda, in *Methods of Cultivating Parasites in Vitro*, Taylor, A. E. R. and Baker, J. R., Eds., Academic Press, London, 1978, 151-192.
20. **Butler, M. S. and Fried, B.,** Histochemical and thin-layer chromatographic analyses of neutral lipids in *Echinostoma revolutum* metacercariae cultured *in vitro*, *J. Parasitol.*, 63, 1041, 1977.
21. **Fried, B. and Bradford, J. D.,** Histochemical and thin-layer chromatographic analyses of neutral lipids in various host sites infected with *Leucochloridiomorpha constantiae* (Trematoda), *Comp. Biochem. Physiol.*, 78B, 175, 1984.
22. **Fried, B., Schafer, S., and Kim, S.,** Effects of *Echinostoma caproni* infection on the lipid composition of *Biomphalaria glabrata*, *Int. J. Parasitol.*, 19, 353, 1989.
23. **Horutz, K. and Fried, B.,** Effects of *Echinostoma caproni* infection on the neutral lipid content of the intestinal mucosa of ICR mice, *Int. J. Parasitol.*, 25, 653, 1995.
24. **Bailey, R. S., Jr. and Fried, B.,** Thin-layer chromatographic analyses of amino acids in *Echinostoma revolutum* (Trematoda) adults, *Int. J. Parasitol.*, 7, 497, 1977.
25. **Mack, S. R., Samuels, S., and Vanderberg, J. P.,** Hemolymph of *Anopheles stephensi* from non-infected and *Plasmodium berghei*-infected mosquitoes. 2. Free amino acids, *J. Parasitol.*, 65, 130, 1979.
26. **Perez, M. K., Fried, B., and Sherma, J.,** High performance thin-layer chromatographic analysis of sugars in *Biomphalaria glabrata* (Gastropoda) infected with *Echinostoma caproni* (Trematoda), *J. Parasitol.*, 80, 336, 1994.
27. **Hoskin, G. P. and Cheng, T. C.,** Occurrence of carotenoids in *Himasthla quissetensis* rediae and the host, *Nassarius obsoletus*, *J. Parasitol.*, 61, 381, 1975.
28. **Fried, B., Holender, E. S., Shetty, P. H., and Sherma, J.,** Effects of *Echinostoma trivolvis* (Trematoda) infection on neutral lipids, sterols and carotenoids in *Helisoma trivolvis* (Gastropoda), *Comp. Biochem. Physiol.*, 97B, 601, 1990.
29. **Fried, B., Beers, K., and Sherma, J.,** Thin-layer chromatographic analysis of beta-carotene and lutein in *Echinostoma trivolvis* (Trematoda) rediae, *J. Parasitol.*, 79, 113, 1993.
30. **MacInnis, A. J. and Voge, M.,** *Experiments and Techniques in Parasitology*, W. H. Freeman, San Francisco, 1970.
31. **Welsh, J. H., Smith, R. I., and Kammer, A. E.,** *Laboratory Exercises in Invertebrate Physiology*, Burgess, Minneapolis, MN, 1968.
32. **Needham, J. G., Galtsoff, P. S., Lutz, F. E., and Welch, P. S.,** *Culture Methods for Invertebrate Animals*, Dover, New York, 1937.
33. **Loker, E. S. and Hertel, L. A.,** Alterations in *Biomphalaria glabrata* plasma induced by infection with the digenetic trematode *Echinostoma paraensei*, *J. Parasitol.*, 73, 503, 1987.
34. **Fried, B. and Sherma, J.,** Thin-layer chromatography of lipids found in snails, *J. Planar Chromatogr.– Mod. TLC*, 3, 290, 1990.
35. **Mack, S. R. and Vandenberg, J. P.,** Hemolymph of *Anopheles stephensi* from noninfected and *Plasmodium berghei*-infected mosquitoes. 1. Collection procedure and physical characteristics, *J. Parasitol.*, 64, 918, 1978.
36. **Daggett, P. M. and Nerad, T. A.,** Sources of strains for research and teaching, in *Protocols in Protozoology*, Lee, J. L. and Soldo, T., Eds. Allen Press, Lawrence, 1992, E1.1–E1.4.
37. **Taylor, A. E. R. and Baker, J. R.,** *Methods of Cultivating Parasites in vitro*, Academic Press, London, 1978.
38. **Keenan, T. W., Huang, C. M., and Zierdt, C. H.,** Comparative analysis of lipid composition in axenic strains of *Blastocystis hominis*, *Comp. Biochem. Physiol.*, 102B, 611, 1992.
39. **Fried, B.,** Obtaining and handling biological materials and prefractionating extracts for lipid analyses, in *Handbook of Chromatography – Analysis of Lipids*, Mukherjee, K. D. and Weber, N., Eds., CRC Press, Boca Raton, FL, 1993, 1–10.
40. **Fried, B. and Sherma, J.,** *Thin-Layer Chromatography – Techniques and Applications*, 3rd ed., Marcel Dekker, New York, 1994.
41. **Graff, O. J.,** Metabolism of C^{14}-glucose by *Moniliformis dubius* (Acanthocephala), *J. Parasitol.*, 51, 72, 1964.
42. **Sherma, J.,** Sample preparation for quantitative TLC, in *Thin-Layer Chromatography — Quantitative Clinical and Environmental Applications*, Touchstone, J. C. and Rogers, D., Eds., Wiley-Interscience, New York, 1980, 17–35.
43. **Folch, J., Lees, M., and Sloane-Stanley, G. H.,** A simple method for the isolation and purification of total lipids from animal tissue, *J. Biol. Chem.*, 226, 497, 1957.
44. **Beers, K., Fried, B., Fujino, T., and Sherma, J.,** Effects of diet on the lipid composition of the digestive gland-gonad complex of *Biomphalaria glabrata* (Gastropoda) infected with larval *Echinostoma caproni* (Trematoda), *Comp. Biochem. Physiol.*, 110B, 729, 1995.
45. **Maloney, M. D. and Semprevivo, L. H.,** Thin-layer and liquid column chromatographic analyses of the lipids of adult *Onchocerca gibsoni*, *Parasitol. Res.*, 77, 294, 1991.
46. **Mangold, H. K.,** Aliphatic lipids, in *Thin-Layer Chromatography*, 2nd ed., Stahl, E., Ed., Springer-Verlag, New York, 1969, 363–421.

47. **Fried, B. and Appel, A. J.,** Excretion of lipids by *Echinostoma revolutum* (Trematoda) adults, *J. Parasitol.*, 63, 447, 1977.
48. **Shetty, P. H., Fried, B., and Sherma, J.,** Effects of patent *Echinostoma caproni* infection on the sterol composition of the digestive gland-gonad complex of *Biomphalaria glabrata*, as determined by gas-liquid chromatography, *J. Helminthol.*, 66, 68, 1992.
49. **Skipski, V. P., Smolowe, A. F., Sullivan, R. C., and Barclay, M.,** Separation of lipid classes by thin-layer chromatography, *Biochim. Biophys. Acta*, 106, 386, 1965.
50. **Fried, B., Tancer, R. B., and Fleming, S. J.,** *In vitro* pairing of *Echinostoma revolutum* (Trematoda) metacercariae and adults, and characterization of worm products involved in chemoattraction, *J. Parasitol.*, 66, 1014, 1980.
51. **Wagner, H., Horhammer, L., and Wolff, P.,** Thin-layer chromatography of phosphatides and glycolipids, *Biochem. Z.*, 334, 175, 1961.
52. **Rouser, G., Kritchevsky, G., and Yamamota, A.,** Column chromatographic and associated procedures for separation and determination of phosphatides and glycolipids, in *Lipid Chromatographic Analysis*, Vol. 1, Marinetti, G.V., Ed., Edward Arnold, London, 1967, 99–162.
53. **Thompson, S. N., Mejia-Scales, V., and Borchardt, D. B.,** Physiologic studies of snail-schistosome interactions and potential for improvement of *in vitro* culture of schistosomes, *In Vitro Cell. Dev. Biol.*, 27A, 497, 1991.
54. **Maloney, M. D., Semprevivo, L. H., and Coles, G. C.,** A comparison of the glycolipid compositions of cercarial and adult *Schistosoma mansoni* and their associated hosts, *Int. J. Parasitol.*, 20, 1091, 1990.
55. **Giorgio, S., Jasiulionis, M. G., Straus, A. H., Takahashi, H. K., and Barbieri, C. L.,** Inhibition of mouse lymphocyte proliferation response by glycosphingolipids from *Leishmania (L.) amazonensis*, *Exp. Parasitol.*, 75, 119, 1992.
56. **Mack, S. R., Samuels, S., and Vanderberg, J. P.,** Hemolymph of *Anopheles stephensi* from non-infected and *Plasmodium berghei*-infected mosquitoes. 2. Free amino acids, *J. Parasitol.*, 65, 130, 1979.
57. **Mack, S. R., Samuels, S., and Vanderberg, J. P.,** Hemolymph of *Anopheles stephensi* from noninfected and *Plasmodium berghei*-infected mosquitoes. 3. Carbohydrates, *J. Parasitol.*, 65, 217, 1979.
58. **Sherma, J. and Fried, B.,** Preparative thin-layer chromatography, in *Preparative Liquid Chromatography*, Bidlingmeyer, B. A., Ed., Elsevier, Amsterdam, 1987, 105–127.
59. **Weber, N., Vosmann, K., Aitzetmüller, K., Filipponi, C., and Taraschewski, H.,** Sterol and fatty acid composition of neutral lipids of *Paratenuisentis ambiguus* and its host eel, *Lipids*, 29, 421, 1994.
60. **Rouser, G., Siakotos, A. N., and Fleischer, S.,** Quantitative analysis of phospholipids by thin-layer chromatography and phosphorus analysis of spots, *Lipids*, 1, 85, 1966.
61. **McManus, D. P., Marshall, I., and James, B. L.,** Lipids in digestive gland of *Littorina saxatilis rudis* (Maton) and in daughter sporocysts of *Microphallus similis* (Jäg. 1900), *Exp. Parasitol.*, 37, 157, 1975.
62. **Mangold, H. K. and Mukherjee, K. D.,** New methods of quantitation in thin layer chromatography: tubular thin-layer chromatography, *J. Chromatogr. Sci.*, 13, 398, 1975.
63. **Mukherjee, K. K.,** Applications of flame ionization detectors in thin-layer chromatography, in *Handbook of Thin-Layer Chromatography*, Sherma, J. and Fried, B., Eds., Marcel Dekker, New York, 1991, 339–350.
64. **Aisien, S. O., Opute, F. I., Ali, S. N., and Obiamiwe, B. A.,** Lipid composition of adult *Foleyella agamae*, *Int. J. Parasitol.*, 16, 655, 1986.
65. **Sherma, J.,** Planar chromatography, *Anal. Chem.*, 66, 67R, 1994.
66. **Touchstone, J. C.,** Instrumentation for thin-layer chromatography – a review, *J. Chromatogr. Sci.*, 26, 645, 1988.
67. **Perez, M. K., Fried, B., and Sherma, J.,** Comparison of mobile phases and HPTLC qualitative and quantitative analysis, on preadsorbent silica gel plates, of phospholipids in *Biomphalaria glabrata* (Gastropoda) infected with *Echinostoma caproni* (Trematoda), *J. Planar Chromatogr.– Mod. TLC*, 7, 340, 1994.
68. **Lough, A. K., Felinski, L., and Garton, G. A.,** The production of methyl esters of fatty acids as artifacts during the extraction or storage of tissue lipids in the presence of methanol, *J. Lipid Res.*, 3, 478, 1962.
69. **Haseeb, M. A., Fried, B., and Eveland, L. K.,** *Schistosoma haematobium*: neutral lipid composition and release by adults maintained *in vitro*, *Comp. Biochem. Physiol.*, 81B, 43, 1985.
70. **Chitwood, D. J., Lusby, W. R., and Fried, B.,** Sterols of *Echinostoma revolutum* (Trematoda) adults, *J. Parasitol.*, 71, 846, 1985.
71. **Yusufi, A. N. K. and Siddiqi, A. H.,** Lipid composition of *Gastrodiscoides hominis* from pig, *Indian J. Parasitol.*, 1, 59, 1977.
72. **Anderton, C. A., Fried, B., and Sherma, J.,** HPTLC determination of sugars in the hemolymph and digestive gland-gonad complex of *Biomphalaria glabrata* snails, *J. Planar Chromatogr.– Mod. TLC*, 6, 51, 1993.
73. **Fried, B. and Robinson, G. A.,** Pairing and aggregation of *Amblosoma suwaense* (Trematoda: Brachylaimidae) metacercariae *in vitro* and partial characterization of lipids involved in chemoattraction, *Parasitology*, 82, 225, 1981.
74. **Kwong, A. Y. H., Wong, P. C. L., and Ko, R. C.,** Lipids of *Angiostrongylus cantonensis* (Nematoda: Metastrongyloidea): a comparison between young adults and gravid worms, *Comp. Biochem. Physiol.*, 95, 193, 1990.
75. **Ehrlich, I. and Hrzenjak, T.,** Polar lipids in *Ascaris lumbricoides*, *Vetrinarski Archiv*, 45, 129, 1975.
76. **Abbas, M. K., Johnson, W. J. and Cain, G. D.,** Fatty acids of the amoeboid sperm of *Ascaris suum* (Nematoda), *Comp. Biochem. Physiol.*, 80B, 791, 1985.

77. **Beach, D. H., Sherman, I. W., and Holz, G. G., Jr.,** Incorporation of docosahexaenoic fatty acid into the lipids of a cestode of marine elasmobranchs, *J. Parasitol.*, 59, 655, 1973.
78. **Fried, B. and Butler, M. S.,** Histochemical and thin-layer chromatographic analyses of neutral lipids in metacercarial and adult *Cotylurus* sp. (Trematoda: Strigeidae), *J. Parasitol.*, 63, 831, 1977.
79. **Sheriff, O. S., Fakri, E., and Kidwai, S. A.,** Lipids in the hydatid fluid collected from lungs and livers of sheep and man, *J. Helminthol.*, 63, 266, 1989.
80. **Dennis, R. D., Baumeister, S., Irmer, G., Gasser, R. B., and Geyer, E.,** Chromatographic and antigenic properties of *Echinococcus granulosus* hydatid cyst-derived glycolipids, *Parasite Immunol.*, 15, 669, 1993.
81. **Persat, J., Bouhours, J. F., Mojon, M., and Petavy, A. F.,** Analysis of the monohexosglyceramide fraction of *Echinococcus multilocularis* metacestodes, *Mol. Biochem. Parasitol.*, 41, 1, 1990.
82. **Gallo, G. J. and Fried, B.,** Association of particular systems with the release of neutral lipids in *Echinostoma revolutum* (Trematoda) adults, *J. Chem. Ecol.*, 10, 1065, 1984.
83. **Burren, C. H., Ehrlich, I., and Johnson, P.,** Excretion of lipids by the liver fluke (*Fasciola hepatica* L.), *Lipids*, 2, 353, 1967.
84. **Clegg, J. A. and Morgan, J.,** The lipid composition of the lipoprotein membranes on the egg-shell of *Fasciola hepatica*, *Comp. Biochem. Physiol.*, 18, 573, 1966.
85. **Ginger, C. D. and Fairbairn, D.,** Lipid metabolism in helminth parasites. I. The lipids of *Hymenolepis diminuta* (Cestoda), *J. Parasitol.*, 52, 1096, 1966.
86. **Webb, R. A. and Mettrick, D. F.,** The role of glucose in the lipid metabolism of the rat tapeworm *Hymenolepis diminuta*, *Int. J. Parasitol.*, 5, 107, 1975.
87. **Johnson, W. J. and Cain, G. D.,** The selective uptake of cholesterol by the rat tapeworm *Hymenolepis diminuta* (Cestoda), *Comp. Biochem. Physiol.*, 91B, 51, 1988.
88. **Nigam, S. C. and Premvati, G.,** Presence of cholesterol in the neutral lipids of three sheep cestodes, *J. Helminthol.*, 54, 215, 1980.
89. **Siddiqui, J., Siddiqi, A. H., Itoh, T., and Matsumoto, T.,** Characterization of sterols of three digenetic trematodes of buffalo, *Mol. Biochem. Parasitol.*, 15, 143, 1985.
90. **Haseeb, M. A., Fried, B., and Eveland, L. K.,** Histochemical and thin-layer chromatographic analyses of neutral lipids in *Schistosoma japonicum* adults and their worm-free incubates, *Int. J. Parasitol.*, 16, 231-236; 665, 1986.
91. **Smith, T. M. and Brooks, T. J., Jr.,** Lipid fractions in adult *Schistosoma mansoni*, *Parasitology*, 59, 293, 1969.
92. **Fried, B., Imperia, P. S., and Eveland, L. K.,** Neutral lipids in adult male and female *Schistosoma mansoni* and release of neutral lipids by adults maintained *in vitro*, *Comp. Biochem. Physiol.*, 68B, 111, 1980.
93. **Silveira, A. M. S., de Lima Friche, A. A., and Rumjanek, F. D.,** Transfer of [^{14}C] cholesterol and its metabolites between adult male and female worms of *Schistosoma mansoni*, *Comp. Biochem. Physiol.*, 85B, 851, 1986.
94. **Mills, G. L., Coley, S. C., and Williams, J. F.,** Chemical composition of lipid deposits isolated from larvae of *Taenia taeniaeformis*, *J. Parasitol.*, 69, 850, 1983.
95. **Thompson, S. N.,** Effect of *Schistosoma mansoni* on the gross lipid composition of its vector *Biomphalaria glabrata*, *Comp. Biochem. Physiol.*, 87B, 357, 1987.

Chapter 5

THIN-LAYER CHROMATOGRAPHY IN THE STUDY OF ENTOMOLOGY

Richard D. Fell

CONTENTS

I. Introduction 72

II. Carbohydrate Analysis 72
   A. Sugars 72
      1. Analytical Techniques and Approaches 74
      2. Solvent Systems 74
      3. Visualization of Sugars 75
   B. Sugar Alcohols 77
      1. Analytical Techniques for Polyols 78

III. Lipids 80
   A. Analytical Techniques and Approaches 82
   B. Solvent Systems 82
      1. Solvent Systems for Standard TLC 84
      2. Visualization Techniques for Lipids 85

IV. Lipids: Ecdysteroid and Terpenoid Compounds 86
   A. Ecdysones 86
   B. Analytical Techniques and Approaches 87
      1. Solvent Systems 88
      2. Visualization Systems 89
   C. Terpenoids 89
   D. Analytical Techniques and Approaches 90
      1. Solvent Systems 90
      2. Visualization of Terpenoid Compounds 90

V. Biogenic Amines 90
   A. Analytical Techniques and Approaches 91
      1. Solvent Systems 91
      2. Visualization and Quantification of Biogenic Amines 91

VI. Insect Pigments 92
   A. Analysis Techniques and Approaches 93
   B. Solvent Systems 93
      1. Pteridine Pigments 93
      2. Ommochromes 93
      3. Carotenoids 94
      4. Visual Pigment Chromophores 95
   C. Visualization Systems 95

VII. Plant Toxins and Insects 95
    A. Analysis Techniques and Approaches 96
    B. Solvent Systems 96
        1. Terpenoids—Cardenolides 96
        2. Terpenoids—Azadirachtin 97
        3. Terpenoids—Sesquiterpene Lactones 98
        4. Terpenoids—Grayanotoxins 98
        5. Phenolics 98
        6. Pyrrolizdine Alkaloids 99
        7. Azoxyglycosides-Cycasin 99
    C. Visualization Systems 99

References 100

I. INTRODUCTION

The use of thin-layer chromatography for the analysis of insect extracts and tissue samples began in the 1960s. Most of this early work was performed on manually prepared chromatographic plates and involved the separation of materials such as lipids and pigments for qualitative analysis or as a preparatory step for subsequent analysis by other techniques. Frequent examples of such use can be found in the early literature on the isolation and purification of insect sex pheromones.[1-4] TLC was also used in studies of uric acid excretion in cockroaches,[5] to examine excreted tryptophan metabolites,[6] and for the purification of juvenile hormone.[7] The ease of use, speed, and relatively low cost of TLC made it a valuable tool for the qualitative separation of a wide variety of compounds and mixtures extracted from insect tissues.

    Improvements in techniques and equipment, as well as the ready availability of high-quality plates with uniform layers have extended and increased the use of TLC in entomological studies, offering the possibility of quantitative, as well as qualitative analysis. High performance thin-layer chromatography (HPTLC), in particular, has developed to the extent that separation and quantification can provide results that are comparable with other analytical methods such as HPLC.[8] In addition, TLC can offer a number of advantages to the researcher when used for the analysis of insect tissue extracts. Sample preparation, for example, may be simplified in many cases, since contamination of the stationary phase is not generally a problem. Multiple samples can be run simultaneously, a benefit in many studies because of high variability between individual insects. Detection procedures may be simplified as well. The process of detection is separate from that of plate development, such that incompatibility problems between solvent systems and detection methods are not a factor. The process of densitometric scanning can also be manipulated to improve results. Scans can be repeated for all or part of the chromatogram, without the need to rerun samples. This characteristic of TLC permits the use of additional detection techniques between scans that can aid in the identification of unknowns. Finally, one of the most important factors for entomologists is the relative ease with which very small extract solutions obtained from small samples can be handled and quantified.

    In this chapter the use of TLC and HPTLC in the study of insects is reviewed, with a primary emphasis on studies conducted during the last 15 years. The review is divided into sections based on the types of compound involved in the studies.

II. CARBOHYDRATE ANALYSIS

A. SUGARS

The analysis of carbohydrate levels in insect tissues and hemolymph has formed the basis of numerous physiological and metabolic studies. The carbohydrates of primary interest in many of the studies have included glucose and the disaccharide trehalose. Trehalose is the dominant blood sugar in insects, but other saccharides have been identified in the hemolymph. These sugars have included fructose, sucrose, maltose, cellobiose, ribose, and fructomaltose. Some of these sugars are not common and may reflect the consumption of specialized food items in the diet. Other sugars such as galactose are commonly consumed, but apparently are not absorbed through the midgut in an unaltered form. Interest in the measurement of sugars has centered primarily on the saccharide composition of the hemolymph, and a number of chromatographic techniques have been described for the analysis of blood sugars.

The analysis of hemolymph sugar composition has presented problems due to small sample volumes and small amounts of the individual sugars. Various techniques have been used, some of which have provided identification and direct quantification of individual sugars, whereas other methods have provided a measure of total sugars, but only indirect estimates of specific sugars such as trehalose. In addition, in many studies hemolymph samples were combined to increase sample volume and total sugar amounts. Such procedures lead to a loss in individual variability and may not be desirable in some research studies. The development of sensitive techniques utilizing thin-layer chromatography has provided alternate approaches that provide for the rapid qualitative and quantitative analysis of samples from individual insects containing sugars in concentration ranges of 100 ng to several micrograms. Total hemolymph sugar levels in insects generally range from 5 to over 40 µg/µl,[9,10] meaning that sample volumes of 0.5 µl to 1 µl of hemolymph are usually sufficient for analysis.

Some of the earlier studies of hemolymph sugars involved the use of paper chromatography techniques[11,12] or nonquantitative thin-layer techniques.[13–15] Mack et al.[13] examined female *Anopheles* mosquitoes for changes in hemolymph sugars after blood feeding and after infection with the malaria parasite *Plasmodiun berghei*. Glucose and trehalose were the dominant sugars, but maltose, glucuronic acid, and inositol were detected after blood feeding. Magnarelli and Anderson[14] and Magnarelli et al.[15] utilized a thin-layer technique to examine blood sugars and body extracts of tabanid flies (horse and deer flies) for the presence of fructose (and other sugars) in studies of nectar feeding.

Bounias[16,17] analyzed hemolymph carbohydrates in honey bees using a quantitative TLC system and densitometric scanning. The method allows for determinations of common blood sugars (trehalose, glucose, and fructose) in a nanomole range, although the method also works well for a number of other sugars, including melezitose, maltose, ribose, rhamnose, deoxyglucose, and deoxyribose.[16,18] These procedures have been used for studies of hyperglycemia in bees induced by injections of epinephrine, norepinephrine, dopamine, and dopa.[19–22] Fell[23] reported on a technique for the analysis of hemolymph sugars by high performance thin-layer chromatography that also utilized densitometric scanning of developed plates for quantitative determinations of sugars. The technique was used by Stein and Fell[24] in studies of the changes in energy reserves of overwintering bald-faced hornet queens and by Cannon and Fell[25] to determine crop contents of wintering carpenter ant workers and reproductives.

Thin-layer chromatographic techniques have proved effective in studies of the absorption of dietary sugars. Thomas and Nation[26] examined glucose absorption in isolated midgut and hindgut preparations of several cricket species. Aliquots of the incubation medium were anlayzed for radioactively labeled sugars after injection of solutions containing ^{14}C glucose into ligated gut preparations. Several TLC solvent systems were used to show that both the midgut and hindgut absorb glucose *in vitro* and that only about 42% of the absorbed glucose was still in the form of glucose; an almost equal amount of fructose (41%) was found, as well as 10 to 20% of the label in trehalose.

Turunen[27] examined the absorption efficiency of dietary galactose in larvae of the cabbage butterfly *Pieris brassicae*, using an approach somewhat similar to that of Thomas and Nation. Larvae were fed on a diet containing ^{14}C galactose and then hemolymph and gut tissue samples were analyzed for radioactivity. Results from high-pressure liquid chromatography and TLC analyses demonstrated that galactose was rapidly metabolized to other carbohydrates, glycolytic intermediates, and lipids by the midgut cells on absorption. TLC separations of hemolymph carbohydrates, followed by removal of the silica gel from the TLC plates and scintillation counting, indicated that trehalose was the main hemolymph carbohydrate after consumption of galactose. Labeled glucose was present only in small amounts in the hemolymph and no galactose was detected. Turunen suggested that a portion of the galactose is converted to glucose, released into the hemolymph and then converted to trehalose by the fat body. Additional studies[28] using the same techniques were made to better define the metabolic pathways involved in nutrient metabolism in the midgut epithelium.

1. Analytical Techniques and Approaches

Most of the recent TLC methods have relied on silica gel for the stationary phase in the analysis of carbohydrates.[14,16,18,23,27] Bounias[18] used 10 × 15 cm precoated plates from Merck (Ref. 5784), whereas Fell[23] used silica gel 60 HPTLC plates (10 × 10 cm, Merck) and Turunen[27] silica gel G plates. Mack et al.,[13] however, used microcrystalline cellulose sheets.[29]

Precoated plates should be washed before use.[30] Fell[23] used methanol (Baker HPLC grade), and Jork et al.[30] recommended either a mixture of methanol–chloroform (1:1) or use of the mobile phase that is to be employed in the actual separations. When washing plates, the mobile phase should be allowed to run well beyond the point at which the solvent front of the chromatogram will be stopped. Jork et al. also suggested that it is inadvisable to activate plates in a stream of hot or cold air before use, since laboratory air is blown over the plate. Fell,[23] however, reported using a hair dryer to dry plates after washing and before pretreatment and final activation in a drying oven.

The pretreatment of silica gel plates has been used to improve the separation of carbohydrates.[31] Bounias[18] dipped plates in a 0.3 M solution of phosphate buffer (pH 6.5) before use and Thomas and Nation[26] used 0.1 N boric acid impregnated silica gel G plates. Turunen[27] recommended the use of sodium acetate impregnated plates (0.02 M). Fell[23] pretreated plates by spraying with a 0.1 M sodium bisulfite solution, drying, and then spraying with a citrate buffer (1:10 dilution of Sigma citrate buffer–water, pH 4.8), a procedure modified from Ghebregzabher et al.[32] and Pruden et al.[33] After pretreatment, the plates were placed in an oven at 100°C for at least an hour. Following the final heat activation, the plates can be stored in a desiccator until use.

2. Solvent Systems

A number of solvent systems have been used for the separation of sugars and other carbohydrates. Bounias,[16] for example, presented a table of elution solvents for the chromatographic separation of a number of sugars on phosphate buffered plates (see above). However, the best resolution for a mixture of different sugars (containing melibiose, trehalose, sucrose, galactose, glucose, fructose, and ribose) was obtained using multiple developments with a series of three solvent systems. Developments were made by ascending elution in cylindrical jars with half of the jar lined with filter paper saturated with the solvent. The first solvent mixture used in development consisted of n-butanol–n-propanol–acetic acid–water (40:30:15:15, v:v) and was run up the plate 11 to 12 cm. The second solvent mixture — run to a distance of 12 to 13 cm — consisted of the same series of compounds, but in slightly different proportions (45:20:15:20, v:v)). The final elution was run to a height of 13 to 15 cm with a solvent mixture consisting of n-butanol–acetic acid–water (65:15:20). Bounias[16] also presented a two-dimensional system for the separation of complex mixtures of as many as 14 sugars. Silica gel (non-buffered) plates were run in the first direction with pentanol–acetone–water (55:25:20) and propanol–water–pyridine (55:15:30) in the second direction.

Magnarelli and Anderson[14] used a solvent system consisting of butanol–glacial acetic acid–ethyl ether–water (9:6:3:1) for the separation of sugar extracts (glucose, fructose, and sucrose) from tabanid flies on precoated silica gel plates; plates were developed twice in the same direction to improve separations. For the separation of a more complex mixture from aphid honeydew, Magnarelli et al.[15] used an n-butanol–acetone–water mixture (2:14:1) for plate development. Standards used for the identification of unknowns in this system included glucose, fructose, sucrose, maltose, melezitose, raffinose, stachyose, and mannose. Turunen[27] used chloroform–methanol

(6:4, v:v) for the separation of glucose, galactose, galactitol, fructose, trehalose, and sucrose on silica gel G plates, pretreated with sodium acetate (see above).

Fell[23] used an 85:15 mixture of acetonitrile and water[34] for silica gel 60 plates pretreated with sodium bisulfite and citrate. Covered Camag Twin Trough Chambers for 10 × 10 cm plates were used for plate development with 3 to 4 ml of fresh solvent for each run. The plates were run three times in the same direction and to the same point with the solvent allowed to run up 7 cm from the bottom of the plate (6.5 cm from the origin). Multiple runs improved separation and gave the most distinct spots. Plates were dried with a hair dryer for a least 1 min between runs. Incomplete drying caused diffuse spots and a variable solvent front. Separation data for eight sugars using this system are presented in Table 5.1, along with data obtained using a different mobile phase composed of ethyl acetate–acetic acid–methanol–water (60:15:15:10, v:v).[32]

TABLE 5.1
HPTLC of Sugars Separated on Silica Gel 60 Plates Pretreated with 0.1 M NaHSO₃ and Citrate Buffer
(Used with permission from Fell[23])

Sugar	R_f System 1	R_f System 2
Ribose	0.65	0.54
Fructose	0.51	0.43
Glucose	0.46	0.42
Galactose	0.38	0.36
Sucrose	0.30	0.30
Maltose	0.24	0.26
Trehalose	0.20	0.23
Melezitose	0.15	0.18

Note: Solvent Systems: 1. acetonitrile–water (85:15) 2. ethyl acetate–acetic acid–methanol–water (60:15:15:10).

Mack et al.[13] used a two-solvent system for the separation of glucose, maltose, trehalose, glucuronic acid, and inositol on cellulose plates. The first chromatographic run was made with a solvent system consisting of n-propanol–methylethylketone–water–formic acid (10:6:3:1, v:v). Two additional runs were made in the same direction and to the same point using ethyl acetate–pyridine–water (12:5:4, v:v). Plates were allowed to dry completely at room temperature between runs.

3. Visualization of Sugars

A large number of techniques have been developed for the visualization of sugars on thin-layer chromatograms and a number of different reagents and detection methods have been summarized recently by Jork et al.[30] Some techniques are more useful for the identification of different saccharides, as different sugars give different colors on reaction with the reagent. Hansen,[35] for example, described a procedure that gives different colored spots (on a white background) for many common sugars, such as glucose, fructose, and sucrose. This procedure worked well with some sugars, but was not found to give good results with trehalose, the dominant blood sugar in insects. Sugar detection with this method is based on spraying plates (after final development and drying) with a mixture of aniline–diphenylamine–acetone–H_3PO_4 80% (4 ml:4 g:200 ml:30 ml) and heating for 30 min in an oven at 105°C. Colors resulting from the reaction of sugars with the reagent include: glucose, galactose, lactose, maltose — blue; fructose, deoxy ribose — red; sucrose, arabinose, melezitose, ribose, xylose — violet; and fucose, rhamnose — green.

Mack et al.[13] used an aniline–diphenylamine reagent for the detection and identification of sugars on cellulose plates. The reagent consisted of 1% aniline plus 1% diphenylamine in a mixture of 10 parts acetone:1 part 9% phosphoric acid in water.[36] Plates were heated to 90°C for 20 min after spraying. This technique proved useful for the identification of maltose (blue), glucuronic acid (brown), and galacturonic acid (green) in the hemolymph of malaria-infected mosquitoes. Other aniline–diphenylamine reagent mixtures have been described by Gauch et al.[34] and Jork et al.[30]

Turunen[27,28] used orcinol ferric chloride (Bial's reagent, Sigma Chemical Company) for the visualization of sugars. The reagent consists of 0.9% ferric chloride and 0.55% orcinol in acidified ethanol. Plates were sprayed and then heated to 110°C for 10 min.

Bounias[18] described the use of N-(1-naphthyl)ethylenediamine dihydrochloride as a reagent for the visualization and quantification of sugars. The method is well suited for the analysis of insect hemolymph or tissue samples, providing good sensitivity for trehalose as well as a number of other sugars. The reagent is quite specific, reacting with saccharides, amino sugars, cyclitols, and ascorbic acid, but not with amino acids or other organic acids.

The reagent consists of a 6.5 mM solution of N-(1-naphthyl)ethylenediamine dihydrochloride in methanol containing 3% sulfuric acid. Plates were heated at 100°C for 5 min after spraying with the reagent. Spots of variable color, depending on the sugar, appeared on a white background within one to several minutes of heating. Color reactions for different sugars are shown in Table 5.2. Detection limits range as low as 0.02 nanomole. After visualization the plates were scanned on a photodensitometer (Vernon PHI-5 dual beam photometer) using white light, and the resulting peak areas were compared to standard curves constructed from the peak areas of known standards. Peak areas of sugars do not increase linearly with concentration, and Bounias recommended the use of a mathematical calibration process for the accurate quantification of unknowns. The general calibration equation given by Bounias[18] is:

$$(S_i)^x = p\,(C_i)$$

in which S_i and C_i are the peak areas and concentrations, x is the coefficient of linear expansion, and p is the slope of the calibration straight line. The parameters (x) and (p) can be determined from the logarithmic form of the above equation:

$$\ln(C_i) = x \ln(S_i) - \ln p$$

Bounias noted, however, that a small deviation in the calculated value of the intercept $(-\ln p)$ gives a large error in (p); therefore, he suggested that the slope be determined directly after plotting $(S_i)^x$ vs. (C_i), including the zero point (C_o and $S_o = 0$), which gives the final equation given above.

TABLE 5.2
Color Reactions of Sugars on Silica Gel Plates after Visualization with N-(1-naphthyl)ethylenediamine Dihydrochloride (Modified from Bounias[18])

Sugar	Color[1]
Deoxyglucose	Gray-blue
Deoxyribose	Gray-mauve
Fructose	Red-violet
Glucose	Gray-violet
Maltose	Violet
Melezitose	Violet
Ribose	Blue to green-gray
Rhamnose	Red-crimson
Sucrose	Violet
Trehalose	Violet

[1] Color development determined after 4 min

Fell[23] described a different technique for the quantification of hemolymph sugars using ceric sulfate as a visualizing reagent. After final development and drying, the plates were dipped into a solution of ceric sulfate in sulfuric acid and then heated for 15 min at 110°C to char the sugars.[33] The dipping solution was made by diluting 1 part of 0.100 N ceric sulfate in 2 N sulfuric acid (Ricca Chemical Corporation, Arlington, Texas) into 10 parts of 15% sulfuric acid. The sugars (di- and trisaccharides and hexoses) appeared as light to dark brown spots on a white background, unless the plate was overheated. Pentoses appeared as yellow-brown spots. Most amino acids did not react with the reagent, although tryptophan reacted at levels as low as 1 µg/µl. (The R_f of tryptophan was above that of the hexoses and dissaccharides using the procedure described above and does not interfere with most common sugar determinations.) Gentle brushing of the plate with a camel hair brush, after final drying and before dipping, helps to remove lint and dust that char during the heating process. The plates are also easier to handle if excess dipping solution is allowed to run off the plate (bottom of the plate can be touched to a paper towel) and the excess wiped from the back of the plate before heating.

Quantitative measurements can be made by absorbance scanning using a CAMAG TLC Scanner (or similar instrument), interfaced to an integrator. The plates are scanned at a wavelength of 440 nm, using a slit width of 0.2 mm and a slit length of 3 mm. Slit length and width should be selected to reduce signal-to-noise ratios and maximize sensitivity.[37] Integrator parameters are set according to the CAMAG TLC program with a peak width of 1 and a peak threshold of 1000. All scans are made in the direction of plate development, and peak areas are used for quantification. A series of mixed standards for the sugars of interest at 4 to 5 concentrations is run with each plate. Peak area values for the standards are used for the construction of standard curves for each sugar. Curves for common sugars such as glucose, fructose, sucrose, and trehalose are similar and are linear over a range of 125 ng to approximately 2 µg. Figure 5.1 shows scan results for trehalose at different concentrations and the separation of a mixture of trehalose, sucrose, glucose, and fructose.

B. SUGAR ALCOHOLS

Sugar alcohols have been shown to play an important role in insect cold-hardiness by increasing the insect's ability to supercool and avoid the lethal effects of freezing.[38,39] They also play a significant role in cryoprotection in freeze-tolerant insects.[40] A number of different polyols have been found in the hemolymph of overwintering immature and adult insects, as well as in overwintering eggs. The polyols include glycerol, which is the most common sugar alcohol,[38] sorbitol, mannitol, threitol, erythritol,[39-42] and ribitol.[43,44]

Many of the studies, particularly the earlier studies, have relied on paper chromatographic techniques for the detection of glycerol and other polyols. Salt,[38] for example, compared hemolymph glycerol levels in the larvae and pupae of freeze-tolerant and freeze-intolerant insect species. Similarly, glycerol and sorbitol were found in the overwintering eggs of *Bombyx mori*,[45] and Baust and Miller[46] used quantitative paper chromatography to measure variations in glycerol content in an Alaskan carabid beetle. Somme,[47] however, used both paper and thin-layer chromatography to identify and quantify mannitol and glycerol in the overwintering eggs of aphids. Yaginuma and Yamashita[48] used instant thin-layer chromatography to examine the metabolic relationships between glycogen, sorbitol, and glycerol in silkworm eggs after ^{14}C-glucose labeling during oogenesis. Izumiyama et al.[49] used a similar technique to measure changes in glycerol content in cricket (*Gryllus bimaculatus*) eggs during oogenesis and embryonic development. Ring and Tesar[50] used qualitative and quantitative TLC methods to monitor changes in polyols, sugars, and lipids in the larvae and adults of a freeze-tolerant arctic beetle (*Pytho americanus*: Salpingidae) during low temperature acclimation. Glycerol content of both larvae and adults were found to increase over an 18-week period while glycogen, and trehalose to a smaller extent, decreased. Hamilton et al.[44] identified ribitol as an important component of the cryoprotectant system in the wood cockroach, *Cryptocercus puntulatus*, using an HPTLC technique.

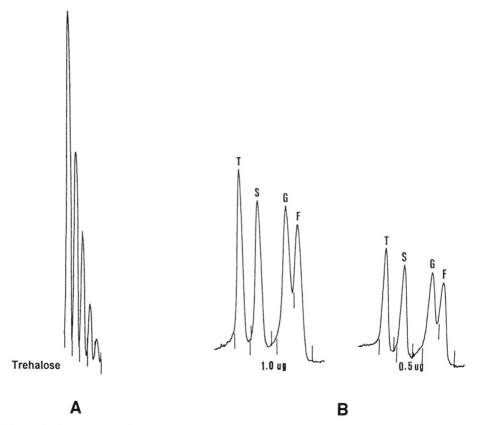

FIGURE 5.1 A, Peaks resulting from densitometric scanning of trehalose at five concentrations: 2.0, 1.0, 0.5, 0.25, 0.125 μg; B, Scanner recordings from the separation of trehalose (T), sucrose (S), glucose (G), and fructose (F) at two concentrations – 1.0 and 0.5 μg. Chromatograms were made on silica gel 60 HPTLC plates, developed in acetonitrile–water (85:15) and visualized by dipping in a ceric sulfate/sulfuric acid solution and heating at 110°C for 15 min.

1. Analytical Techniques for Polyols

Various procedures have been used to extract sugar alcohols from insect tissue or hemolymph. Methods have often involved a modification of van Handel's[51] procedures or the straight homogenation and extraction of tissue with 70 or 80% ethanol. Hemolymph samples are typically small – one to several μl – and can either be combined for a composite sample or analyzed individually. Individual μl samples can be diluted in 5 to 10 μl of 70% ethanol, centrifuged, and the supernatant spotted directly on plates. Larger samples may be centrifuged to remove hemocytes. The plasma fraction can be collected and extracted with ethanol for analysis.

Most TLC methods for the separation of polyols use silica gel plates. Somme[47] used Kieselgel G or Kieselgel G:Kieselgur (2:3) plates for the analysis of mixtures containing glycerol, dulcitol, mannitol, and sorbitol. The Kieselgel plates were impregnated with 0.1 N boric acid and used with one of three solvent systems; (A) ethylmethylketone–acetic acid–methanol (3:1:1, v/v), (B) benzene–acetic acid–methanol (1:1:3, v/v), or (C) n-butanol–acetone–water (4:5:1, v/v). For the mixed Kieselgel:Kieselgur plates, an isopropanol–ethyl acetate–water (83:11:6, v/v) system was used. Most of these systems gave reasonable separations for glycerol, glucose, trehalose, and mannitol. System "B" did not give good separations for glycerol, glucose, and trehalose, and system "A" did not separate mannitol and trehalose. None of the procedures gave acceptable separations of dulcitol and sorbitol.

Ring and Tesar[50] used silica 7G plates for the analysis of sugars and glycerol in tissue extracts. Sample extracts in distilled water were spotted onto plates and developed with a chloroform–methanol–water (64:36:8) solvent system. After drying, the plates were sprayed with a thymol-

H$_2$SO$_4$ reagent[52] and heated at 120°C for 10 min for sugars and 20 min for glycerol. When cool, the spots corresponding to the different sugars or polyols were scraped off the plate. The color was eluted with 4 ml of formic acid, the silica gel was removed by centrifugation and the absorbance of the supernatant read on a spectrophotometer at 350 nm for glycerol and 510 nm for the sugars. R_f values for trehalose, sucrose, glucose, fructose, and glycerol were 0.43, 0.50, 0.56, 0.60, and 0.71, respectively.

Hamilton[43] used silica gel 60 HPTLC plates (Merck) with several different solvent systems for the identification of polyols in hemolymph samples. Standards were mixed at 5 mg/ml in distilled water and spotted in 0.5 µl samples. Sample volumes of 1 µm pooled hemolymph were used after centrifugation to remove hemocytes. Plates were developed in an ascending manner to a point 6 cm from the origin. Separation results for the different solvent systems are shown in Table 5.3. The plates were dried after development and sprayed with one of two visualizing reagents: a nonspecific, but sensitive reagent consisting of 1% potassium permanganate or a more specific bromphenyl blue–boric acid reagent. This reagent consists of 40 mg bromphenyl blue in 100 ml of ethanol containing 100 mg of boric acid and 7.5 ml of 1% aqueous sodium tetraborate and appears to be specific for polyols.[43]

TABLE 5.3
A Comparison of the R_f Values for Polyols Separated on Silica Gel HPTLC Plates Using Different Solvent Systems (Modified from Hamilton[43])

Compound	Solvent System Number			
	1	2	3	4
Glycerol	0.58	.065	0.50	0.69
Threitol	—	0.56	0.35	—
Ribitol	0.35	0.50	0.32	0.57
Arabitol	0.31	0.44	0.30	0.52
Xylitol	0.27	0.41	0.25	0.63
Mannitol	0.21	0.35	0.20	0.41
Sorbitol	0.19	0.28	0.19	0.38
Dulcitol	0.19	0.28	0.19	0.36
Inositol	—	0.06	0.06	—

Solvent system 1: ethyl acetate–methanol–butanol–water (16:4:4:2, v/v)
Solvent system 2: butanol–acetone–water (5:4:1)
Solvent system 3: butanol–water (9:1)
Solvent system 4: acetone–butanol–water (5:3:1)

Cannon and Fell[25] used an HPTLC technique to identify glycerol in tissue extracts of overwintering carpenter ants. Merck silica gel HPTLC plates were used with a solvent system consisting of acetone–butanol–water–acetic acid (40:10:5:1) (K. Tignor, personal communication) to separate a standard series of polyols consisting of glycerol, ribitol, arabitol, sorbitol, mannitol, xylitol, dulcitol, and inositol. Plates were developing to a height of 7 cm in CAMAG twin-trough tanks. After drying, the plates were dipped in 15% sulfuric acid in ethanol and charred at 140°C for 25 min.

Yaginuma and Yamashita[48] described a technique for separation of sugars and sugar alcohols on instant thin-layer plates (chromatomedium, I-TLC SA plates, Gelman Instrument Co., Michigan). Plates were developed six times in the same direction utilizing a solvent system consisting of chloroform–acetic acid–water (50:22:2.5, v/v). Sugars and polyols were visualized by spraying with 1% KMnO$_4$ in 1.0 N NaOH. Good separations were obtained for trehalose, glucose, sorbitol, and glycerol; R_f values were 0.07, 0.30, 0.41, and 0.83, respectively.

An alternate separation and detection method for sugar alcohols was given by Jork et al.[30] Merck HPTLC plates Si 50,000 were prewashed in chloroform–methanol and dried for 30 min at 110°C.

After spotting, plates were developed using a solvent system of 1-propanol and water (18:2) to a height of 7 cm. The visualization process consisted of dipping the dried plates in two solutions and then heating. Solution 1 consisted of 1 ml of a saturated aqueous silver nitrate solution made up to 200 ml with acetone. (Any precipitate was redissolved by adding 5 ml of water and mixing.) Solution 2 was made by dissolving 2 g sodium hydroxide in 2 ml of water with heating, and then made up to a final volume of 100 ml with methanol. The developed plate was dipped in solution 1, dried in a stream of cool air, dipped in solution 2, and then placed in an oven at 100°C for 1 to 2 min. The sugar alcohols appeared as light to dark brown spots on a beige-colored background. (Note: Dark backgrounds can be decolorized by spraying with 5% aqueous ammonia solution and/or with a 5 to 10% sodium thiosulfate in 50% aqueous ethanol.) Quantitative measurements were carried out by reflectance absorbance scanning at 530 nm.

III. LIPIDS

A variety of TLC and HPTLC methods have been used in insect studies, either for the separation and identification of lipids, or for the separation of lipids as a preparative step for subsequent analysis by other techniques such as gas chromatography or GC-MS. Stanley-Samuelson and Dadd,[53] for example, examined changes in arachidonic acid and other tissue fatty acids in the house mosquito, *Culex pipiens*, with changes in the dietary concentration of arachidonic acid. Thin-layer chromatography techniques were used to separate the phospholipid and triacylglycerol fractions from whole body extracts before the analysis of fatty acid composition of each fraction by GC. Subsequent studies were made of the long chain polyunsaturated fatty acids in triacylglycerols and phospholids of lipid extracts of a number of insects, using the same approach.[54,55]

Turunen[56,57] and Turunen and Chippendale[58] examined the role of plant leaf lipids as sources of dietary lipids in several species of leaf-feeding Lepidoptera larvae. Lipids were extracted from leaf tissue, the midgut lumen and midgut tissue, and then fractionated by thin-layer chromatographic techniques. With this approach, Turunen[56] was able to show that a major portion of the lipids in the diet of the cabbage butterfly larva is incorporated into polar lipids (glycolipids) prior to absorption. Turunen and Chippendale[58] were able to show that polar leaf lipids are an important dietary component of phytophagous larvae, with glycoglycerolipids and phospholipids accounting for the majority of the lipid composition. The ensuing study by Turunen[57] showed that these two groups were the major source of essential fatty acids in *Pieris brassicae*, *Inachis io*, and *Anticarsia gemmatilis*. Fischer and Komnick[59] used a similar approach to examine the assimilation of long-chain fatty acids by the midgut in the dragonfly nymph, *Aeshna cyanea*. Nymphs were given oral infusions of lipid emulsions containing the methyl esters of either nervonic acid or erucic acid (or triolein as a control). After the last feeding, hemolymph, midgut, and fat body tissue were isolated, extracted with chloroform–methanol (2:1), and the lipid components separated by TLC. The triacylglycerol and diacylglycerol fractions were isolated from the plates and saponified with methanolic KOH, and the fatty acids were analyzed by HPTLC. The analysis results showed that long-chain fatty acids were shortened on assimilation across the midgut epithelium.

Sickmann et al.[60] looked at the distribution of glycosphingolipids in third instar blowfly larvae (*Calliphora vicina*). These authors used two-dimensional high performance thin-layer chromatography to separate the neutral and acidic glycolipids of different larval organs, including the fatbody, striated muscle, intestinal tract, imaginal discs, and central nervous system. The glycosphingolipids were extracted from the different organs and separated into neutral and acidic fractions by DEAE-Sephadex A-25 column chromatography. The fractions analyzed by HPTLC showed significant differences in the glycolipid components from different organs.

Spates et al.[61] examined the process by which complex blood lipids are metabolized in the midgut of the adult stable fly (*Stomoxys calcitrans*), an obligate blood feeder. TLC analysis of the products of the enzymatic degradation of sphingomyelin and phosphatidylcholine by midgut homogenates indicated that ceramide was one of the major products of sphingomyelin hydrolysis, and that 1,2- and 1,3-diacylglycerols were major products of phosphatidylcholine hydrolysis. Miller et al.[62] examined the effects of dietary ethanol on phospholipase activity in *Drosophila* larvae. *In vitro* and *in vivo* studies were used to examine the hydrolysis of radiolabeled substrates, and TLC was used to fractionate the reaction products. The isolated components were scraped from the TLC plates and analyzed for radioactivity by scintillation counting. Dietary alcohols were found to reduce phosphatidylcholine levels in the insect, partly through an apparent stimulation of a phosphatidylcholine specific phospholipase.

Bounias et al.[63] examined the changes in hemolymph lipids during development of honey bees, separating and quantifying the different lipid fractions by densitometric scanning after separation by TLC and subsequent visualization. Judge[64] and Judge et al.[65] also reported on an HPTLC technique for the analysis of lipids in small hemolymph volumes (0.5 to 10 µl) collected from the moth *Heliothis zea*. Blood samples from individual insects were extracted with chloroform–methanol (2:1) and analyzed with a procedure that provided detection of mono-, di-, and triacylglycerols, cholesterol, cholesterol esters, fatty acids, and alkanes at 50 ng and quantitative measurements at 500 ng. This procedure allowed Judge[64] to monitor diurnal variation in hemolymph 1,2-diacylglycerol concentrations, as well as the effect of flight activity on blood lipids. Stein and Fell[66] used the same procedure to measure changes in the lipid content of eggs of the baldfaced hornet during embryological development.

Several investigators have used thin-layer chromatography techniques in studies of the lipoproteins involved in lipid transport in insect hemolymph. Katagiri[67] used TLC to separate and quantify the phospholipids extracted from a reaction mixture after digestion of locust lipophorin by snake venom phospholipase A_2. Singh and Ryan[68] examined the function of lipid transfer particles in the tobacco hornworm *Manduca sexta*, using TLC to separate the lipids extracted from hemolymph lipoproteins following the injection of radiolabeled ^{14}C-acetate into the moths. Lipid classes separated by the chromatography were subjected to radiochromatogram scanning after visualization.

Petzel et al.[69] examined the fatty acid composition of the Malpighian tubules of female yellow fever mosquitoes and used thin-layer techniques to separate lipid components and analyze the fatty acids in phospholipid and triacylglycerol fractions. Spots corresponding to the different fractions were scraped from the plates into reaction vials and transmethylated to fatty acid methyl esters by refluxing in acidified methanol.[54] The fatty acid methyl esters were then analyzed by gas chromatography and GC-MS.

A number of other researchers have used preparative TLC for the fractionation of lipid extracts from insects. Nelson and Fatland[70,71] used TLC to separate the lipid classes in extracts of internal tissues from pupae of the tobacco hornworm (*Manduca sexta*) in a study of long-chain methyl-branched alcohols and their esters. After initial separation by preparative TLC, the wax ester fraction was rechromatographed, scraped from the plate, and analyzed by GC-MS. Guo et al.[72] conducted a similar study on the long-chain alcohols and their acetate esters in pupae of the southern armyworm, *Spodoptera eridania*.

Juárez et al.[73] and Juárez and Brenner[74] examined the cuticular lipids in kissing bugs, *Triatoma* spp., using TLC to fractionate the lipid extracts of the integument. Quantitative estimates were made of the major lipid classes in the epicuticle and in the entire integument by densitometric scanning of visualized plates. Juárez and Brenner[75] used a similar approach to examine fatty acid biosynthesis in the integument of *T. infestans*. Lipid classes extracted from the integument, after removal of the epicuticular lipids by hexane washes, were separated by TLC. Lipids in the different classes were saponified and the resulting fatty acids esterified and analyzed by GC. More recently, Juárez[76] used thin-layer chromatography to show disruptive effects on the cuticular lipid layer after treating *T. infestans* with sodium trichloroacetate, a compound that inhibited synthesis of the hydrophobic cuticular lipids.

Hamilton et al.[77] examined cuticular extracts of several tick species, including the American dog tick *Dermacentor variabilis*, in a study of the sex pheromone that induces mounting behavior by males. Hexane extracts of female ticks were partitioned into different fractions (acid/base/neutral) and tested for activity. The active neutral lipid fraction was separated further by column chromatography. TLC was used to monitor the elution process and to characterize extracts from different tick species. The thin-layer chromatographic analysis indicated that the main active component was a steryl ester. The identification of cholesteryl oleate as the active pheromone was made by GC-MS. Morse and Meighen[78] also used TLC to examine pheromones and pheromone gland lipids in two budworm species (Tortricidae: *Choristoneura orae*, *C. fumiferana*). Pheromones and other glandular lipids were labeled *in vivo* with ^{14}C acetate, after which the gland components were separated by two-dimensional TLC. The TLC analysis allowed the determination of the blends of the tetradecenyl acetate esters and alcohol, as well as the E–Z ratios of the acetate esters.

Ragab et al.[79] incubated homogenates of reproductive tissues from male and female thysanurans (*Thermobia domestica*) with [U–^{14}C] arachidonic acid to determine if these insects produced prostaglandin-like compounds. The synthesis products were identified by their mobility on thin-layer chromatograms in comparison to prostaglandin, 8- and 5-hydroxyeicosatetranenoic (5-HETE) standards. The incorporation of the radioactive label into metabolites was determined with a Berthold scanner or by autoradiography. The label was found in three main metabolites, 8- and 5-HETE, and a third unidentified metabolite. All of the major metabolites had R_f values different from the prostaglandin standards. In a subsequent study Ragab et al.[80] demonstrated the incorporation of labeled arachidonic acid into phospholipids, particularly phosphatidylcholine and phosphatidylinositol. Thin-layer chromatography procedures were used to separate phospholipids, neutral lipids, and eicosanoids (biologically active oxygenated metabolites of arachidonic acid and certain other C_{20} polyunsaturated fatty acids). Chromatographic spots detected by autoradiography were scraped from the plates and the radioactivity quantified with a scintillation counter (in 10 ml of Picofluor 40 - Packard).

A. ANALYTICAL TECHNIQUES AND APPROACHES

The majority of lipid analysis by thin-layer chromatography has involved the use of silica gel plates. Preparative separations of lipid extracts have generally been made on 20 × 20 cm silica gel G plates (0.250 mm)[69,70,73] or Silica gel 60 plates.[56,79] Stanley-Samuelson and Dadd,[53] however, used silica gel plates with an SiO_2 binder (silica gel H) in their separations of lipid extracts.

The analysis of lipids by high performance thin-layer chromatography has been limited primarily to the use of silica gel 60 HPTLC plates.[59,60,65] Fischer and Komnick[59] analyzed fatty acids by HPTLC and used precoated HPTLC RP–18 plates.

Precoated plates should be washed with chloroform–methanol (2:1)[64] or the mobile phase[77] before use. The solvent should be allowed to run to the top of the plate. After the wash, the plates can be activated by placing them in an oven at 110°C for 1 to 2 hours. Following activation the plates should be allowed to cool in a desiccator before use. Samples can be applied after cooling in volumes of 0.1 μl to 5.0 μl, depending on the type and size of the plate.

B. SOLVENT SYSTEMS

A number of solvent systems have been used for the analysis of lipid mixtures, often with several different solvents used successively to maximize the resolution of different lipid classes. TLC separates lipids according to molecular type with relatively little distinction between chain lengths, and up to 10 lipid classes can be separated on a single plate.[81] Judge,[64] for example, provided a comparison of different solvent systems for their relative effectiveness in separating mixtures of lipids extracted from insect hemolymph. The different systems and their effectiveness in HPTLC separations are shown diagrammatically in Figure 5.2. The mobile phase systems shown in the figure provide a good starting point for the separation of lipids extracted from insect tissue.

Judge[64] found that a two-phase solvent system gave the best results for his analysis of hemolymph lipids in the moth *Heliothis zea*. A solvent system consisting of benzene–diethyl ether–ethanol–(95%)–acetic acid (conc.)) (BEEA – 50:40:2:0.2) was used for the first development and was allowed to migrate 40 mm above the origin. The plate was removed from the chamber and dried under warm, forced air for 5 to 10 min. The second development was made in the same direction and involved the use of a hexane–heptane–diethyl ether–acetic acid solvent (HHEA – 65:18:18:1). The HHEA solvent was allowed to migrate 60 mm from the origin. This system gave good separations

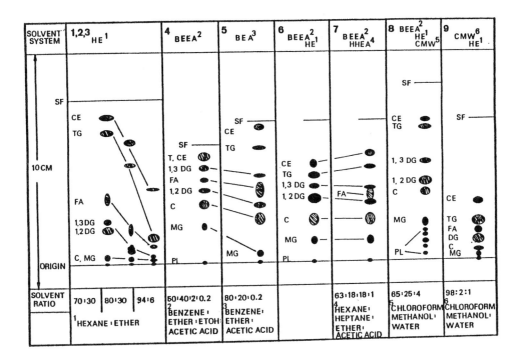

FIGURE 5.2 The separation of lipids by HPTLC using nine different solvent systems. Solvent migration (from the origin) for systems 1, 2, and 3 was 70 mm; 50 mm for system 4, and 60 mm for system 5. For multiphase systems, migration distances were 40 mm, then 60 mm for systems 6 and 7; 55 mm, 75 mm, and 20 mm for system 8; 20 mm and 60 mm for system 9. SF, solvent front; C, cholesterol; CE, cholesterol esters (cholesterol palmitate); FA, fatty acids (palmitic acid); PL, phospholipids (phosphatidylcholine, phosphatidylinositol, phosphtidylethanolamine, and phosphatidylserine); MG, DG, TG, mono-, di-, and triacylglycerols. (From Judge, D. N., M.S. Thesis, 1988. With permission.)

of cholesterol and cholesterol esters, mono-, di-, and triacylglycerols in a range of 0.5 µg to 10 µg. Free fatty acids separate between the two isomeric forms of diacylglycerols, but were not present in detectable amounts in the hemolymph of this insect, and thus were not of concern. One of the other solvent systems (No. 5) can be used if free fatty acids are present in the extracts to be analyzed.

Judge[64] also described a three-phase system (No. 8 in Figure 5.2) to separate phospholipids from neutral lipids on the same plate, if required. BEEA is used as the first solvent and is allowed to migrate a distance of 55 mm from the origin. The second development is made with hexane–ether–acetic acid (HE3 – 96:4:1) to a distance of 75 mm. The third solvent, chloroform–methanol–water (CMW – 65:25:4; modified from Pernes et al.[82]) is run 20 mm from the origin and separates phospholipid groups between the origin and the monoacylglycerols. This system can be used for the separation of phosphatidylcholine, phosphatidylinositol, phosphatidylethanolamine, and phosphatidylserine.

The separations obtained by Judge[64,65] are partially dependent on the type of developing tank used for the analysis. The data presented in Figure 5.2 are based on the use of a horizontal sandwich developing chamber (CAMAG) or unsaturated type S-chamber,[65] and not a vertical twin-trough-type tank. Differences in chamber saturation will affect R_f values, and good lipid separations obtained in sandwich type developing chambers may not be reproducible in N-type chambers.

Hamilton et al.[77] separated extracts of ticks (*Dermacentor variabilis*) on HPTLC using a two-stage development; hexane followed by hexane–diethyl ether (80:20). R_f values for cholesterol, free fatty acids, steryl esters, and hydrocarbons were 0.37, 0.65, 0.84, and 0.99, respectively.

Sickmann et al.[60] used two-dimensional HPTLC systems for the separation of neutral and acidic glycosphingolipids in vertical tanks with a saturated atmosphere. Plates were pre-run in chloroform–acetone (1:1) before use. Neutral glycosphingolipids were separated by running plates the first direction in chloroform–methanol–0.2% aqueous $CaCl_2$ (50:40:10) and in chloroform–methanol–0.2% NH_3 (50:40:10) the second direction. Acidic glycosphingolipids were separated with 1-propanol–0.2% aqueous $CaCl_2$ (7:3) in the first dimension and n-butanol–acetic acid–water (2:1:1) in the second dimension.

Fischer and Komnick[59] separated fatty acids on reversed phase C_{18} HPTLC plates. The mobile phases used for the separation consisted of two solvent systems and three developments in the same direction: (1) acetonitrile–tetrahydrofuran (85:15) to 25 mm; (2) same solvent system to 40 mm; (3) methanol + 0.25% water to 70 mm. Good separations were obtained for $C_{14:1}$ to $C_{24:1}$ fatty acids.

1. Solvent Systems for Standard TLC

Numerous solvent systems have been described for the separation of lipids on standard TLC plates, many of which have been used as a preparative step for subsequent analysis by GC or GC-MS. The following mobile systems have been used for the separation of chloroform–methanol (2:1) or hexane extracts of insects or insect tissue containing neutral and /or polar lipids.

a. Neutral and General Lipid Separations
1. Juarez et al.[73] — purpose: separate lipid classes
    Hexane–ethyl ether–acetic acid (80:20:2) — up 1/2 of plate
    Hexane–ethyl ether (95:5) — to the top of the plate
    Hexane — run to the top of the plate
    Alternate procedure:
    Benzene–ethyl ether–ethanol–acetic acid (50:40:2:0.2) — up 2/3 of plate
    Hexane–ethyl ether (94:6) — to top of plate
2. Singh and Ryan[68] — purpose: separate lipid classes
    Hexane–ethyl ether–acetic acid (70:30:1)
3. Stanley-Samuelson and Dadd[53] — purpose: separate lipid classes
    Petroleum ether–ethyl ether–glacial acetic acid (90:10:1)
4. Petzel et al.[69] — purpose: separate lipid classes for analysis of triacylglycerols and phospholipids
    Petroleum ether–ethyl ether–acetic acid (80:20:1)
5. Nelson and Fatland[70] — purpose: separate wax esters and hydrocarbons
    Hexane–ethyl ether (95:5), developed two times in the same direction, wax ester fraction eluted with chloroform and separated by benzene–hexane (1:1)
6. Guo et al.[72] — purpose: separate wax ester fraction from other lipid classes
    Hexane–ethyl ether–formic acid (80:20:2)
7. Spates et al.[61] — purpose: separate polar and neutral lipid hydrolysis products of sphingomyelin and phosphatidylcholine
    I. Hexane–ethyl ether–acetic acid (30:20:1)
    or
    II. Benzene–ethyl ether–ethanol–acetic acid (50:40:2:0.2)
    or
    III. Chloroform–methanol–acetic acid–water (60:50:5:2)

b. Separation of Polar Lipids
8. Turunen[56] — purpose: separate polar lipids and neutral lipids in tissue extracts
    Chloroform–methanol–acetic acid–water (65:25:8:4 or 25:15:4:2), develop to 12 cm to separate polar lipids, air dry and redevelop with
    Hexane–ethyl ether (4:1), to 18 cm to separate neutral lipids
9. Petzel et al.[69] — purpose: separate phospholipids
    Chloroform–methanol–water (65:25:4)
10. Katagiri[67] — purpose: phospholipid separations
    Chloroform–methanol–acetic acid–water (25:15:4:2)

The solvent systems listed above generally give good separations for lipid classes and can be used for quantitative estimates of lipids after visualization, as well as for qualitative or preparative separations. Some care must be used in interpreting the results of separations to be sure that some fractions do not migrate to the same point. Petzel et al.[69] for example, noted that phospholipids do not separate into pure fractions on one-dimensional TLC. Phosphatidylinositol and phosphatidylserine tend to migrate together, and the phosphatidylcholine fraction may contain phosphatidyl glycerol. Spates et al.[61] found that 1,2 diacylglycerols and cholesterol had the same R_f (0.21) value with system I (see 7 above). System II gave better separations for some of the lipid classes (R_f values for ceramide = 0.2, 1,2 DG = 0.57, 1,3 DG = 0.61, monoacylglycerol = 0.16, cholesterol = 0.28). System III was used to separate a group of polar compounds that remained at the origin and did not separate in either of the other systems, including sphingomyelin (R_f = 0.06), sphingosine (R_f = 0.4) and ceramide (R_f = 0.72). Stanley-Samuelson and Dadd[53] also noted that phospholipids remain at the origin with their solvent system (see 3 above), and that the spot may contain some monoacylglycerols, as well.

2. Visualization Techniques for Lipids

A variety of methods can be used for the detection of lipids. Some methods are useful in the qualitative evaluation of the lipid content of an extract, whereas others can be used for the accurate quantification of lipids with densitometric or fluorescent scanning. A number of detection methods for lipids have been summarized by Judge et al.,[65] along with the instrument parameters necessary for quantification. Additional reagents that can be used for lipid visualization have been presented by Jork et al.[30]

One of the most common and universal detection reagents for lipids is iodine.[79,83,84] A simple vapor chamber can be made by placing a few crystals of iodine in the bottom of a covered twin-trough chamber and allowing the chamber to become saturated with iodine vapor. Solvent-free plates can be placed in the chamber for a short period (30 sec to several min) after final development. The chromatogram can be observed and marked after the spots become visible. The reaction is reversible so that additional analyses can be performed on the lipids after the iodine evaporates.[30]

Several charring systems have been used for the detection and quantification of lipids.[65] Developed plates can be sprayed with sulfuric acid (50%) (or dipped in a sulfuric acid solution) and heated at 100 to 110°C for 15 min or until spots become visible. Lipids can be quantified by absorbance scanning at 350 nm. Detection limits are 1 to 2 µg per lipid. An alternate technique is to dip developed plates into a solution of copper sulfate and phosphoric acid and then heat the plates at 180°C for 20 min. Judge[64] found that 10% copper sulfate in an 8% phosphoric acid solution provided the best results. Detection limits were approximately 0.5 µg with absorbance scanning at 350 nm. Singh and Ryan[68] used a slightly different dipping reagent, immersing plates in a 3% cupric acetate–8% phosphoric acid solution, and heating at 180°C for 15 min.

Fluorogenic reagents provide an alternative to charring methods and can increase both detection sensitivity and quantification accuracy. One of the more sensitive techniques involves the use of 6-p-toluidino-2-napthalenesulfonic acid (TNS) with HPTLC.[65] The TNS reagent is made up as a 1% solution in methanol (or ethanol). Plates are dipped into the solution and then rapidly dried with forced air (<35°C). Lipids can be quantified with fluorescence scanning at 366 nm. Detection limits are approximately 0.25 µg, but sensitivity is partially dependent on plate dryness and can be improved by allowing plates to dry in the dark for several hours. TNS dipped plates are UV light sensitive and exposure to natural light rapidly reduces fluorescence. Light exposure during handling should be minimized, but plates can be stored in aluminum foil in the freezer for several months with little loss of fluorescence. The method provides a linear response with lipid concentration over a range of 1 to 10 µg for mono-, di-, and triacylglycerols. The fluorescent response of cholesterol and cholesterol esters is slightly curvilinear over the same range.[65] Fluorescent intensities for the different lipid classes are variable, such that a series of mixed standards at different concentrations should be run on each plate for quantitative work.

Free fatty acids can be visualized by a prechromatographic dansylation process.[59,85] The fatty acids are converted to fluorescent dansyl derivatives and appear as blue-green spots under UV light (365 nm).

Other fluorogenic reagents used for the detection of lipids include rhodamine B and 6G[30] and 2,7-dichlorofluorscein (DCF).[53,64,83] The rhodamine 6G reagent can be made as either a dipping or a spray solution by dissolving 50 mg (100 mg for spray) rhodamine 6G in 100 ml ethanol (95%). A rhodamine 6G spray in methylene chloride is also available commercially. Lipids on treated plates appear as pink-colored spots and fluoresce a deep yellow orange. Fluorescent scanning at 365 nm can be used for quantitative measurements. Detection limits for free fatty acids are approximately 100 ng.[30] The DCF reagent can be made as a 0.2% solution in 95% ethanol. Detection limits for the DCF reagent are approximately 1 µg with fluorescent scanning at 300 nm.[65] Background fluorescence is often a problem with this reagent, especially if solvent residues remain on the plate.

Several chromogenic reagents can also be used for the visualization of lipids. Miller et al.[62] sprayed plates with Dragendorff's reagent (0.11 M potassium iodide and 0.6 mM bismuth subnitrate in 3.5 M acetic acid) and molybdenum blue reagent (1.3% molybdenum oxide in 4.2 M sulfuric acid) for the detection of phospholipids and phospholipid (phosphatidylcholine) hydrolysis products. Dragendorff's reagent is used for the detection of alkaloids and quaternary nitrogen compounds and the molybdenum reagent for phospholipid detection. An orcinol–sulfuric acid reagent may be used for the detection and quantification of glycolipids.[56,60,86] The reagent can be made by dissolving 200 mg of orcinol in 100 ml of 75% sulfuric acid; plates are sprayed with the reagent and then placed in an oven at 100°C for 10 to 15 min. All of these reagents are available commercially.

IV. LIPIDS: ECDYSTEROID AND TERPENOID COMPOUNDS

A. ECDYSONES

Since the chemical identification of ecdysone in the 1960s, numerous studies on the ecdysteroids of insects have been made. Inasmuch as ecdysones are found in small quantities in insects and other arthropods (on the order of 10^{-5} to 10^{-7}g per gram of tissue[87]), methods for the microdetermination of these hormones have played an important role. Several methods can be used for the qualitative and quantitative analysis of ecdysteroids, and much of the work has relied on the use of radioimmunoassay techniques and high performance liquid chromatography. Thin-layer chromatography has been used for both preparative separations of insect ecdysones and for quantitative determinations, but many of the earlier methods did not give good resolution or sensitivity.[87] The development of reversed-phase, thin-layer chromatography has improved the usefulness of the technique for the analysis of insect ecdysones.

Much of the early work on the use of TLC in insect ecdysteroid studies has been summarized by Morgan and Poole,[87] including several schemes for the extraction of ecdysones from insect tissue, as well as a number of solvent systems. TLC (on silica gel) was used as a purification step in a number of isolation procedures, but often presented problems with recovery because of irreversible adsorption on the silica. Wilson et al.[88] found that reversed-phase TLC plates (C_2, C_8, and C_{18}) not only provided good separations of ecdysteroids with different solvent systems, but that the recovery of ecdysones could be improved. Subsequent studies showed that C_{12} bonded and paraffin-coated silica gel plates also provided good resolution and recovery of ecdysteroids.[89] Later studies with C_{12} and C_{18} nonhydrophobic TLC plates indicated that these plates can give better resolution of ecdysone and 20-hydroxyecdysone than some more hydrophobic C_{18} RP-HPTLC plates.[90]

TLC has been used in a number of studies for the isolation of ecdysteroids from hemolymph and tissue extracts. Wilps and Zöller,[91] for example, used thin-layer chromatography for the separation of ecdysteroids from hemolymph and ovary extracts of the blowfly *Phormia terraenovae* as part of a study on ecdysteroid production during oogenesis. Bands corresponding to ecdysteroids were scraped from developed plates, extracted, and then further analyzed by RIA and HPLC to show changes in ecdysone titers. O'Reilly et al.[92] used TLC in their study of the effects on ecdysteroid metabolism of infection of the fall armyworm by a nuclear polyhedrosis baculovirus (*Autographa californica*). TLC of the products of *in vitro* reactions of hemolymph from infected larvae and radiolabeled ecdysone was used to show that the ecdysteroid UDP–glucosyltransferase released into the hemolymph by the virus catalyzes the conjugation of ecdysteroids with galactose and keeps ecdysteroid levels low.

TLC has also provided a mechanism to examine the products of ecdysone metabolism. Wilson and Lafont[93] examined the metabolites of [^{3}H]20-hydroxyecdysone with TLC in the feces of the locust *Locusta migratoria* and found that most of the metabolites could be resolved by the use of two solvent systems and two-dimensional plate development. One of the problems associated with studies of ecdysteroid metabolism has been the range of polarities exhibited by the products. This problem was at least partially solved with the use of multiple development and solvent systems of different properties.[93] The results from TLC and HPTLC were similar to those obtained with HPLC, although a better resolution of the 2- and 3-acetates of 20-hydroxyecdysone was obtained with HPLC. The use of radio-thin-layer chromatogaphy can provide advantages in the study of ecdysteroid metabolism, however. Wilson et al.[94] found that the presence of labeled nonpolar fractions may be more apparent after radioscanning of developed plates, than if samples are simply run on HPLC columns optimized for the separation of the polar metabolites.

The use of TLC has proved beneficial in the analysis of ecdysteroid conjugates, such as the ecdysteroid acetates and long-chain fatty acyl esters. Dinan[95] has reported on a series of methods that can be used for the initial analysis of ecdysteroid apolar esters with HPTLC. A combination of normal and reversed-phase plates can be used to determine if ecdysone metabolites are esters of long-chain fatty acids, if ecdysone acetates are present, and the possible position(s) of conjugation. Whiting and Dinan[96] used TLC to separate ecdysteroid conjugates from extracts of eggs of the house cricket, *Acheta domesticus*. They also used this approach for the initial identification of the ecdysteroid conjugates that were then purified on HPLC and positively identified by mass-spectrometry.[97]

B. ANALYTICAL TECHNIQUES AND APPROACHES

The separation of ecdysteroids by thin-layer chromatography can be accomplished on normal silica gel TLC plates, HPTLC plates, reversed phase TLC and HPTLC plates, and on paraffin-coated silica gel plates. Plates with or without a fluorescent indicator may be used, although plates with an indicator (F_{254}) provide the option of ecdysteroid detection by visual observation under UV light (fluorescence quenching). Reasonable separations can be obtained on standard TLC plates, but sensitivity tends to be poor. The addition of a paraffin coating to plates can improve both the separation and the recovery of ecdysteroids. The paraffin coating can be added by developing plates in a solvent system of dichloromethane containing 5 to 7.5% refined paraffin oil (Nujol). The solvent front is allowed to run to the top of the plate; the plates are then allowed to dry in a fume hood. Preliminary washing of plates before paraffin coating is advisable.

Reversed-phase TLC plates have proven useful for the separation of free ecdysteroids as well as ecdysteroid conjugates. Ecdysone and 20-hydroxyecdysone can be separated adequately on C_2, C_8, and C_{18} RP-plates with the appropriate solvents.[88] Excellent separations between these two materials can be obtained with C_2 bonded plates and an isopropanol–water solvent system (20:80; see below). C_{18}-bonded silica gel HPTLC with a methanol–water solvent system provides a good general system for the separation of ecdysteroids. RP-8 HPTLC plates have proven useful for the separation of ecdysone long-chain fatty acyl esters.[95] RP plates are also well suited for the recovery of separated ecdysteroids from the silica.

Samples and standards can be applied in methanol using volumes of approximately 1 µl. Ecdysteroid concentrations should normally be in the range of 100 ng to several µg, depending on the detection system.

1. Solvent Systems

A large number of solvent systems have been used for the separation of ecdysteroids by thin-layer chromatography. The varying polarities of ecdysone, ecdysone metabolites, and ecdysteroid conjugates mean that no solvent system is capable of resolving all of these components in a single run. Solvent systems must select for the particular class of ecdysteroids to be separated and the type of plate to be used, if optimal resolution is to be obtained. Some of the common solvent systems for ecdysteroid separation are given below.

Normal phase TLC and HPTLC with silica gel plates

1. Chloroform–96% ethanol (4:1)
    Most commonly used solvent for ecdysteroid separation[87] with numerous variations. R_f values for ecdysone and 20-hydroxyecdysone vary between approximately 0.21 (ecdysone) and 0.15 (20-hydroxyecdysone) for 20% EtOH, to 0.63 and 0.56 for 40% EtOH, respectively.[90] R_f values for a number of ecdysteroids run with this solvent system are also given by Horn and Bergamasco[98] in table format. For normal phase separations, tanks should be presaturated.
2. n-butanol–acetone–glacial acetic acid–ammonia (30%)–water (70:50:18:1.5:60)[92,99]
    Separation of ecdysone and ecdysone glucosides.
3. Solvent system I: Ethanol–ethyl acetate–water (80:20:5)
    Solvent system II: Chloroform–ethanol (4:1)
    Develop plates 1/3 to 1/2 with solvent system I, dry and develop whole plate with solvent system II. If HPTLC plates are used, double development with each solvent can improve resolution. Separation of 20-hydroxyecdysone and metabolites (e.g., 20-hydroxyecdysone acetates, 20-hydroxyecdysonoic acid).[93]
4. Chloroform–ethanol (9:1)
    Separation of ecdysone, ecdysteroid acyl esters, ecdysteroid acetates
    R_f values for ecdysone – 0.07, ecdysone long chain fatty acyl esters – 0.13–0.15, ecdysone 22-acetate – 0.09, ecdysone 2(3)-acetate – 0.25; additional R_f values for other ecdysteroids are given by Dinan.[95]

Paraffin (7.5%) coated silica gel HPTLC plates

1. Methanol–water (50:50)
    Separation of ecdysteroids, R_f values for ecdysone and 20-hydroxyecdysone = 0.35 and 0.52, respectively.[89]

Reversed phase C_2, C_8, and C_{18} silica gel plates (see Wilson et al.[88] for additional information on solvent systems and ecdysteroid separations on RP plates)

1. Methanol–water (50:50) for C_2 and C_8 RP-plates
    Separation of ecdysteroids, R_f values for ecdysone and 20-hydroxyecdysone = 0.38 and 0.53 respectively on RP-2 and 0.28 and 0.44 on RP-8. Methanol can be increased to 60% with an increase in R_f of approximately 0.2 for both ecdysones. MeOH–water (60:40) with RP-18 plates gives R_f values of 0.39 and 0.54.[88]
2. Isopropanol–water (20:80) for C_2, C_8, C_{18} plates
    Separation of ecdysteroids, R_f values for ecdysone and 20-hydroxyecdysone = 0.18 and 0.46 respectively on RP-2; 0.19 and 0.46 on RP-8; and 0.24 and 0.46 on RP-18.
3. Methanol–water (19:1) for C_8, C_{18} plates
    Separation of ecdysone, ecdysone acetate, and ecdysone long chain fatty acyl esters. R_f values for ecdysone ≈ 0.93, for C_{22} long chain fatty acyl esters varied from ≈ 0.19 for

C_{20} fatty acids to 0.46 for C_{12} fatty acids. Additional information on separation can be found in Dinan.[95]

2. Visualization Systems

Ecdysteroids can be detected on TLC plates by several means. One of the simplest and most commonly used is by fluorescent quenching. Ecdysteroids stongly absorb UV light in the range of 240 nm and will absorb the emissions from a fluorescent indicator incorporated into the chromatographic layer. When developed plates with an indicator (F_{254}) are viewed under UV light (at 254 nm), ecdysteroids appear as dark spots on a fluorescent background. Quantitative analyses can be made by scanning plates with a TLC scanner and comparing peak height / areas to standards. Detection limits for ecdysteroids with fluorescent quenching are approximately 100 ng.[88]

Ecdysteroids can also be detected with the use of a vanillin–sulfuric acid reagent. The reagent can be made by adding 2 ml of concentrated sulfuric acid to 3 g of vanillin dissolved in 100 ml of absolute ethanol.[95] (Several formulas are available for this reagent; Wilson et al.[88] dissolved vanillin in concentrated sulfuric acid and 95% ethanol [5:70:20, w/v/v].) The reagent is sprayed onto developed chromatograms, and the plates are then heated at 110°C for 5 to 10 min. Ecdysteroids appear as brown or bluish spots,[95] although the color can vary with ecdysteroid structure. The reagent should be stored in the dark and at a low temperature to prevent deterioration. The vanillin–sulfuric reagent can react with other compounds and thus may not be suitable for the analysis of relatively crude extracts. Detection limits for the vanillin sulfuric reagent are on the order of 100 ng.[88]

Fluorescent derivatives of ecdysteroids can also be formed and used to increase detection limits to about 50 ng. Derivatization can be accomplished by heating developed chromatograms in an atmosphere of ammonium carbonate at 100°C for 1 h. Plates are viewed under UV light at 366 nm; ecdysteroids appear as blue-white fluorescent spots. The reaction is relatively nonspecific and is not recommended for partially purified extracts.[88]

C. TERPENOIDS

The terpenoids are a large group of structurally diverse lipid-like molecules, formed from a variable number of isoprene units.[100] Entomological interest in the terpenoids is primarily in the juvenile hormones, which play a key role in insect development in both the immature and adult stages. Juvenile hormones (JH) are sesquiterpenoids with an epoxide group near one end and a methyl ester on the other end. At least five structurally related JH forms occur, with a bisepoxide form of JHIII also identified.[101,102] Monoterpenes have been identified as pheromone components[103,104] and as components of defensive secretions.[105] The terpenoids also form an important group of plant toxins, a number of which insects may sequester for defensive purposes (see Section VII).

The primary use of thin-layer chromatography for terpenoid compounds has been in studies of juvenile hormone biosynthesis and metabolism. Lefevere et al.[106] examined the biosynthesis of the juvenile hormone III bisepoxide with *in vitro* assays of the corpus allatum–corpus cardiacum complex (CA-CC) from blowfly larvae (*Lucilia cuprina*) and radiolabeled precursors. The major synthesis product found by TLC comigrated with the JH bisepoxide and accounted for 85% of the total radioactivity recovered by scintillation counting. Casas et al.[107] examined the *in vitro* metabolism of JH III and the JH III bisepoxide with cell fractions of *Drosophila melanogaster*. TLC plates were scanned for radioactivity after development to detect labeled metabolites. Similar approaches have been used to characterize JH esterase activity in *D. melanogaster*[108] and epoxide hydrolase activity in the hornworm *Manduca sexta*.[109] Ottea et al.[110] developed a TLC assay for separating the major metabolites of JH hydrolysis, and Meyer and Lanzrein[111,112] looked at JH degradation and the methylation of JH acid by the CA-CC complex of the cockroach *Nauphoeta cinerea*. TLC was used to separate JH metabolites before scintillation counting of plate extracts. Other TLC-based studies include those of Hammock et al.[113] on the influence of hemolymph binding proteins on JH stability and the use of a TLC assay developed by Hammock and Roe[114] for monitoring JH esterase in *M. sexta* and *Trichoplusia ni*.

D. ANALYTICAL TECHNIQUES AND APPROACHES

Reviews by Croteau and Ronald,[115] Coscia,[100] and Fried and Sherma[83] discuss thin-layer chromatography of terpenoid compounds and provide approaches for the analysis of compounds in different terpenoid classes. The review presented here will be limited primarily to the TLC analysis of sesquiterpenoid compounds.

The separation of sesquiterpenoids by TLC has been based on the use of standard silica gel plates (20 × 20) or silica gel 60 plates with fluorescent indicator (F_{254}). Pretreatment of plates generally is not required, although Croteau and Ronald[115] indicated that separation of sesquiterpenoid hydrocarbons can be improved by pretreating plates with 6 to 8% $AgNO_3$.

1. Solvent Systems

Several mobile phases have been used for the separation of the juvenile hormones and JH metabolites. Most solvent systems do not produce clearly defined spots, but adequate separation for the identification of JH, JH acid, JH diol, and JH diol acid, and quantification, if radiolabeled materials are used.[113,114] The most common systems use hexane and ethyl acetate in varying ratios. Standard solvent systems include the following.

1. Hexane–ethyl acetate (2:1) or toluene–propanol (20:1)
   (Hammock and Roe[114])
2. Petroleum ether–ethyl ether (10:1) or benzene–*n*-propanol (4:1 or 10:1)
   A two-dimensional analysis can be used to separate metabolites, using hexane–ethyl acetate in the first direction, followed by benzene–propanol (4:1) 1/3 of the plate in direction two, then benzene–propanol (10:1) to the top of the plate in the same direction (Hammock et al.[113]).
3. Hexane–ethyl acetate (7:4 or 1:1), tanks allowed to equilibrate 60 min before development (Casas et al.[107]).
4. Solvent system 1 — hexane–ethyl acetate (7:4)
   Solvent system 2 — hexane–ethyl acetate–acetic acid (7:4:1)
   Plates were developed with solvent 1, dried and developed in solvent 2. The addition of acetic acid to the solvent reduces tailing with the JH acid and helps resolve the metabolites. Interference with the acid-diol can be eliminated by using plastic backed plates and cutting off the origin where the JH acid-diol remains after development in solvent system 1. Ottea et al.[110] also provided a procedure for derivatizing the metabolites with *n*-butaneboronic acid to increase mobility and aid in identification. R_f values for the JH diol, JH acid, JH III, and JH diol + BBA on plates developed in solvent I were 0.14, 0.29, 0.64, and 0.71, respectively.

2. Visualization of Terpenoid Compounds

The primary method for the detection and quantification of sesquiterpenoids has been by fluorescent quenching and either radiometric scanning of plates or spot removal and scintillation counting. Known standards should be run on plates for the identification of metabolites and JH. Additional procedures for the detection of terpenoid compounds can be found in Coscia.[100]

V. BIOGENIC AMINES

Several TLC techniques have been developed for the analysis of biogenic amines in animal tissue extracts, with at least one method designed specifically for use with insect tissue. Eaton and Mullins[116] developed a method for the separation, detection, and quantification of dansyl derivatives of biogenic amines in insect brain and fat body extracts. One- and two-dimensional chromatographic separations by HPTLC allow for the identification and quantification of dopamine, norepinephrine, octopamine, and serotonin on a single plate.

A. ANALYTICAL TECHNIQUES AND APPROACHES

The analysis of tissue extracts for biogenic amines is made on silica gel 60, 10 × 10 cm HPTLC plates. Plates are prewashed in methanol and activated in an oven (110°C) for at least 1 h before use. Brains and tissue extracts were homogenized in 100 μl of 0.1 n-perchloric acid, followed by several cycles of freezing and rehomogenizing. Sample extracts were centrifuged to remove particulate matter and the supernatant was used for analysis. Derivatizations were made on 100 μl samples (in 0.1 n-perchloric acid) of standard biogenic amine mixtures, tissue extracts, and on standards plus extracts. A 250 μl portion of 1.2 M bicarbonate buffer was added to each sample to adjust the pH to 9.0. One μmol of dansyl chloride in 700 μl of acetone (0.53 mg/ml) was added to the solutions, mixed thoroughly, and incubated for 20 min at 40°C. (The dansyl chloride concentration must be tenfold greater than the amines for complete derivatization and at least 5 mM.) After incubation, 100 μl of a proline solution (10 μl per 100 μl water) was added to remove excess dansyl chloride. The resulting mixtures were incubated a second time (20 min, 40°C), and then the excess acetone was removed by directing a stream of nitrogen over the reaction vials for 4 min. The derivatized amines were extracted from the aqueous mixtures with three 250-μl volumes of toluene. The toluene extracts were combined and evaporated to dryness in a warm water bath under a stream of dry nitrogen. The derivative residues were then taken up in 100 μl of toluene, although volumes may need to be adjusted. The optimal concentration range for derivatized amines was 1.0 to 50 pmol. Sample mixtures and standards were spotted on plates in 1 μl volumes. Reagent blanks should also be spotted on plates since DANS-dimethylamide and DANS-NH_2 are formed in side reactions.

1. Solvent Systems

Chromatograms were developed in twin-trough tanks using one of four solvent systems: A, butyl acetate–ethyl acetate (5:1); B, chloroform–butyl acetate–ethyl acetate (3:3:1); C, chloroform–butyl acetate–ethyl acetate–triethylamine (6:1:2:0.5); D, butyl acetate–chloroform (1:1). Two-dimensional separations were made using solvent B for the first development and then solvent C in the second direction. Developed plates were air-dried in a fume hood for at least 4 min between runs or before preservation. Plates were preserved by dipping in a solution of 10% paraffin oil in cyclohexane and drying for 30 sec at 100°C. Separations obtained by this method are shown in Table 5.4 and Figure 5.3.

TABLE 5.4
A Comparison of the Separations of Derivatized Amines by Three Solvent Systems on Silica Gel 60 HPTLC Plates (Modified from Eaton and Mullins[116])

	R_f values for solvent systems[1]		
Amine derivative	A	B	C
Dopamine	0.70	0.69	0.57
Norepinephrine	0.55	0.40	0.20
Octopamine	0.58	0.45	0.25
Serotonin	0.68	0.54	0.24

[1] Solvent systems: A, butyl acetate–ethyl acetate (5:1); B, chloroform–butyl acetate–ethyl acetate (3:3:1); C, chloroform–butyl acetate–ethyl acetate–triethylamine (6:1:2:0.5).

2. Visualization and Quantification of Biogenic Amines

The dansyl derivatives of biogenic amines fluoresce under UV light at 366 nm. Plates can be scanned in TLC scanner using a fluorescence mode. An excitation wavelength of 366 nm is used with a cut-off filter of 450 nm to prevent light of lower wavelengths from reaching the photocell detector. Peak areas for standards can be used to construct standard curves for the quantification of biogenic amines in sample extracts. Standard curve data are linear in the range of 1.0 to 50 pmol with r^2 values of 0.98 or better.[116]

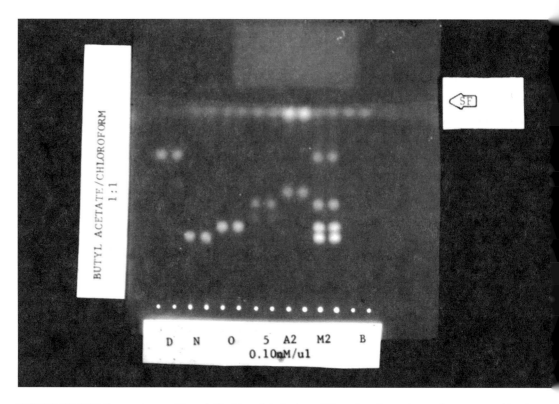

FIGURE 5.3 HPTLC separation on silica gel 60 of dansyl derivatives of biogenic amines using a solvent system of butyl acetate and chloroform (1:1). D, dopamine; N, norepinephrine; O, octopamine; S, serotonin; M, mixed standards; B, reagent blank. (Used with permission of J. L. Eaton.)

VI. INSECT PIGMENTS

Insects display a tremendous diversity of colors. Some insect colors are due to the interaction of incident white light with structural features of the insect, with different colors resulting from light scattering, interference, or diffraction phenomena. Other colors are due to the presence of pigments. Many types of natural pigments have been identified in insects; some, such as the ommochromes, melanins, porphyrins, papiliochromes, and pteridines, are produced metabolically. Other pigments cannot be synthesized and are sequestered from plants, often as precursors that are subject to later modification by the insect. Common pigments of plant origin include the carotenoids and the flavonoids. Numerous studies have been made on the pigments found in insects and the subject has been reviewed by Kayser.[117]

Eye color pigments have attracted the attention of a number of researchers, and thin-layer chromatography has been widely used to separate eye pigments and their precursors.[118,119] Ferré[119] and Ferré et al.[120] examined extracts of eye color mutants in *Drosophila melanogaster* by two-dimensional TLC in studies of the biosynthetic pathways of xanthommatin and pteridines. Ferré[119] showed that kynurenic acid accumulated in the eyes of cinnabar mutants and Ferré et al.[120] provided extensive data on the chromatographic separation of pteridines and their fluorescent intermediate metabolites in both wild type and mutant flies. Ferré et al.[120] also provided quantitative data on the xanthommatins and pteridines in eye color mutants. Reaume et al.[121] examined the effects of a mutation at the rosy gene (which codes for the enzyme xanthine dehydrogenase) in *D. melanogaster* with respect to eye color pigments. Separation and identification by TLC of the eye pigments in extracts indicated that the brown eye color of mutant adults was due to a deficiency of red pteridine pigments. Eye pigmentation in insects has been reviewed by Summers et al.[122]

Similar approaches have been used in the study of photopigments of the insect eye. Pepe and Cugnoli,[123] for example, used TLC in their study of a water-soluble photopigment from the compound eye of the honey bee. Results of a TLC separation of extracts of the eye pigment protein complex indicated that all-*trans*-retinal is bound to the protein and is isomerized to 11-*cis*-retinal in the light. Shukolyukow and Denisova[124] examined the effects of carotenoid levels on the synthesis of visual pigments in the eye of the blowfly, *Calliphora erythrocephala*. TLC was used for the separation of retinals extracted from the visual pigments and for the isolation of carotenoids from eye extracts. A recent review of insects visual pigments has been presented by White.[125]

A number of studies have also been made on pigments in other parts of the insect body. Hori and Riddiford[126] used TLC to isolate and identify ommochromes from the epidermis of the tobacco hornworm. Koch[127] examined the incorporation of ^{14}C-labeled pigment precursors (such as tryptophan and 3-hydroxy-kynurenine) into ommatins in the red wing scales of the butterfly *Araschinia levana*. TLC was used for the separation of the ommatin pigments extracted from the wings and for the determination of which pigments were labeled by scanning plates for radioactivity. Melber and Schmidt[128] used TLC for the isolation and partial identification of pterin pigments and uric acid from several species of pyrrhocorid bugs (*Dysdercus* spp.–cotton stainers).

A. ANALYSIS TECHNIQUES AND APPROACHES

The separation of insect pigments is most commonly performed on microcrystalline cellulose plates, 20 × 20 cm in size, 0.25 mm in thickness. Silica gel plates have been used for some separations involving the chromophores of visual pigments. Pretreatment of plates is not required, although a prewash with the developing solvent may be recommended. Silica gel plates should be activated before use by heating to 110°C for at least 1 h. Plate activation, however, is not recommended for cellulose plates.

B. SOLVENT SYSTEMS

Different solvent systems are required for the separation of different pigment classes. The following solvent systems are listed with respect to pigment classes and plate type.

1. Pteridine Pigments

Solvent system I: Isopropanol–2% aqueous ammonium acetate (1:1)

Solvent system II: 3% (w/v) aqueous ammonium chloride

This system was designed for a two-dimensional separation on cellulose with plates run in direction one for 14 cm with solvent I, then for the same distance in direction two with solvent system II.[119,120] Plates are allowed to dry between runs at room temperature for several hours. The chromatography procedures should be performed under low red light or in the dark to avoid photodegradation of the pigments. Separations of pterin pigments from the eyes of *Drosophila* are shown in Figure 5.4 and Table 5.5. Melber and Schmidt[128] presented several other solvent systems for pteridine pigments.

2. Ommochromes

(a) 80% formic acid–methanol–HCl (80: 15: 0.5)
(b) butanol–acetic acid–water (4: 1: 2)
(c) propanol–88% formic acid–water (1: 1: 4)
(d) collidine–water (3: 1)
(e) propanol–water (8: 3)
(f) propanol–NH_4OH–water (55: 10: 35)
(g) pyridine–butanol–water (1: 1: 1)

Solvent systems for ascending chromatography on cellulose plates of ommochrome pigments. The addition of 2-mercaptoethanol (at 0.1% v/v) is recommended for the chromatography of reduced state ommochromes.[126] R_f values for some pigments in the different mobile phases are shown in Table 5.6.

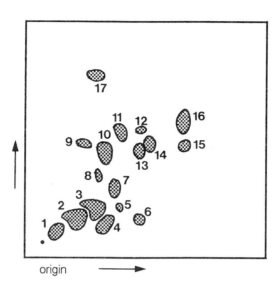

FIGURE 5.4 Two-dimensional thin-layer chromatogram of an extract of the eye pigments of adult wild-type *Drosophila* flies. The plate was developed using the two solvent systems described in the text. The spots are identified in Table 5.5; arrows show direction of chromatographic development. (From Ferré, J., et al., *Biochem. Genetics*, 24, 569, 1986. With permission.)

TABLE 5.5
Identification of the Chromatographic Spots Shown in Figure 5.4
(Modified from Ferré et al.[120])

Spot No.	Compound	Color
1	Neodrosopterin	Red
2	Drosopterin	Orange-red
3	Isodrosopterin	Orange-red
4	Aurodrosopterins	Orange-red
5	Probably oxidized drosopterins	Red
6	"	Red
7	Isoxanthopterin	Colorless (purple under UV)[1]
8	7,8-dihydro-6-formylpterin ?	Yellow (greenish yellow under UV)
9	Deoxysepiaterin	Yellow
10	Sepiapterin	Yellow
11	Acetyldihydrohomopterin	Brown
12	Probably oxidized for of 11	Brown
13	7,8-Dihydrobiopterin	Blue fluorescence under UV
14	Pterin	Blue fluorescence under UV
15	Mix of Neopterin and Biopterin	Blue fluorescence under UV
16	Biopterin	Blue fluorescence under UV
17	Xanthurenic acid	Light blue fluorescence under UV

[1] Visualization under UV light at 365 nm

3. Carotenoids

Hexane–2-propanol (10: 0.1) — 1 time

Dichloromethane–butyl acetate (10: 0.2) — 2 times

Separations were performed on silica gel plates. After development the plates were viewed under dim blue light, the pigments were marked and then scraped off the plates for spectroscopic measurements.[124] Carotenoids are sensitive to oxygen, light, and strong acids; therefore, care must be taken in extraction and handling (see Kayser[117]).

TABLE 5.6
Separation of Ommochrome Pigments and a Precursor Compound on Cellulose
Thin-Layer Chromatography Plates (Modified from Hori and Riddiford[126])

Compound	R_f values				
	Solvent 1	Solvent 2	Solvent 3	Solvent 4	Solvent 5
D-dihydroxanthommatin	0.31	0.06	0.07	0.30	—
L-dihydroxanthommatin	0.43	0.12	0.15	0.36	—
L-xanthommatin	0.88	0.11	—	—	—
3-hydroxy-L-kynurenine	—	0.49	—	0.43	0.40
Xanthurenic acid	—	0.65	—	0.74	0.59

Solvent systems: 1, 80% formic acid–methanol–HCl (80:15:0.5); 2, butanol–acetic acid–water (4:1:2); 3, propanol–88% formic acid–water (1:1:4); 4, collidine–water (3:1); 5, propanol–water (8:3).

4. **Visual Pigment Chromophores**
Hexane–ether (85: 15)
[Hexane–acetone (100: 3, with plate developed 3 to 4 times)]
   Separation of all-*trans*-retinal and *cis*-retinal on silica gel plates (11-*cis*-retinal and 13-*cis*-retinal cannot be completely resolved by this method).

C. VISUALIZATION SYSTEMS

The detection of pteridine pigments is made by UV light at 366 nm. Ommochromes can be detected by their natural color or by UV absorption at 254 or 375 nm.[126] Carotenoids can also be detected by their natural color (typically yellow to red). The spots of the retinal chromophores can be detected by exposure of plates to iodine vapors or by spraying the plates with $SbCl_3$.[124]

VII. PLANT TOXINS AND INSECTS

Plants produce a tremendous diversity of secondary compounds that serve for chemical defense against herbivores. Such defensive chemicals are often placed into major classes that include terpenoids, phenolics, and nitrogen-containing compounds such as alkaloids, amines, and nonprotein amino acids (see Harborne[129]). Phytophagous insects must adapt to the presence of these allelochemics, and many insects have not only developed the ability to detoxify toxic compounds, but also utilize them as attractants or feeding stimulants. Some insects may also sequester toxins within their bodies as part of their own chemical defense system against predators. Numerous studies have been made on the sequestration of plant toxins by insects (see Duffey[130] for a review of this topic), and many of these studies have used thin-layer chromatographic techniques for the analysis of sequestered plant compounds in insect tissues.

The best-known examples of insects sequestering and utilizing plant toxins for their own defense come from studies on the Monarch butterfly (*Danaus plexippus*) and other insects feeding on milkweeds. Monarch butterflies sequester cardiac glycosides, and the variations in cardenolide content found in these insects reflect the plant sources. Roeske et al.,[131] for example, showed that different butterfly populations may show different TLC cardenolide profiles, and Brower et al.[132–135] showed that the cardenolide TLC fingerprints of wild butterflies derive from their specific *Asclepias* food plants. Lynch and Martin[136] found similiar correlations in the cardenolid TLC profiles for Monarch butterflies and their larval host plants in Louisiana, and Cohen and Brower[137] found that the cardenolides sequestered by the Dogbane Tiger Moth (*Cycnia tenera*; Arctiidae) were directly related to those of their larval host. In a similar study, Moore and Scudder[138] used TLC to examine the distribution of sequestered cardenolides in the tissues of the milkweed bug, *Oncopeltus fasciatus*.

The uptake and metabolism of another toxin in the terpenoid group, azadirachtin (tetranortriterpenoid isolated from the neem tree, *Azadirachta indica* A. Juss), has been examined by TLC. Barnby et al.[139] injected a tritiated derivative (22; 23-dihydroazadirachtin) into larvae of the tobacco budworm (*Heliothis virescens*), extracted feces and hemolymph from the larvae, and then analyzed the extract by TLC for the presence of radiolabeled metabolites and 22,23-dihydroazadirachtin. A number of other plant terpenoid compounds, such as the terpenes of Douglas fir, have also been identified as important antifeedants for insects.[140] Chou and Mullin[141] used preparative TLC for the purification of antifeedant sesquiterpene lactones from sunflower and then quantified them by HPTLC.

The effects of a number of plant phenolic compounds have been studied with the aid of TLC. Center and Wright[142] used TLC and HPLC to characterize the phenolic compounds in leaves of the water hyacinth and related these levels to feeding preferences of the water hyacinth weevil (*Neochetina eichhornia*). Shapiro[143] used a fluorescent analog of coumarin (7-amino-3-phenyl coumarin) to study the uptake of ingested xenobiotics into the hemolymph of larvae of the weevil *Diaprepes abbreviatus*. TLC was used to identify and quantify (by scanning fluorescence densitometry) the fluroescent products bound to hemolymph proteins. Other studies utilizing TLC have been made on the pyrrolizidine alkaloids of plants and their attractiveness to Arctiidae moths,[144] and the acquisition of cycasin, an azoxyglycoside, by larvae of the lycaenid butterfly, *Eumaeus atala*, feeding on cycads.

A somewhat different consequence of plant toxins can occur from their incorporation into insect products such as honey. Flowers of a number of plant species may contain toxic compounds that lead to the poisoning of honey bees.[145] In some cases a toxic honey can be produced that causes problems to humans after consumption, but does not seem to have significant effects on the bees. Scott et al.[146] analyzed a toxic honey sample from British Columbia, Canada by TLC and found that it contained grayanotoxins II and III and desacetyl pieristoxin B. Grayanotoxins are diterpenoid constituents of plants in the family Ericaceae. More recently Fell[147] analyzed a toxic honey from southwest Virginia using similar techniques and found significant levels of grayanotoxins I and III. The presumed source of this latter honey was mountain laurel, *Kalmia latifolia*. Sütlüpinar et al.[148] have also reported on toxic honey in Turkey produced by honey bees from *Rhododendron* (*R. ponticum* and *R. luteum*).

A. ANALYSIS TECHNIQUES AND APPROACHES

The analysis of secondary plant compounds is performed on silica gel, using either silica gel 60, 20 × 20 cm plates or silica gel 60 HPTLC plates. The larger plates are preferred for the separation of complex mixtures of cardenolides or phenolics when constructing plant or insect profiles. Chou and Mullin[141] used HPTLC silica gel 60 F_{254} plates for the separation and quantification of sesquiterpenoid lactones. HPTLC plates also give good resolution of grayanotoxins and improved results with densitometric scanning. Plates should be prewashed and activated as described previously.

B. SOLVENT SYSTEMS

Solvent systems for the assay of different plant compounds are described according to their major groups.

1. Terpenoids — Cardenolides

Chloroform–methanol–formamide (90:60:1, v/v/v); developed 4 times

or

Ethyl acetate–methanol (97:3), developed 2 times

Extracts containing 25 to 100 μg of cardenolides were spotted in 20 μl of chloroform on silica gel 60 F_{254} plates. Digitoxin and digitoxigenin (25 μg) were spotted as reference standards. Plates were developed in saturated, filter-paper lined chambers and air-dried after each run. The chloroform-based solvent generally gives better separations and cardenolide fingerprints (Figure 5.5). Cardenolides are extracted from plant and insect tissue with 95% ethanol, but must be cleaned of pigments before TLC. The basic procedure described by Brower et al.[132] is as follows: evaporate 3 ml of extract to dryness, then redissolve in 1 ml 95% ethanol and add 2 ml of 5% aqueous lead acetate solution to precipitate pigments. Mix, chill in an ice bath for 20 min, and centrifuge at moderate

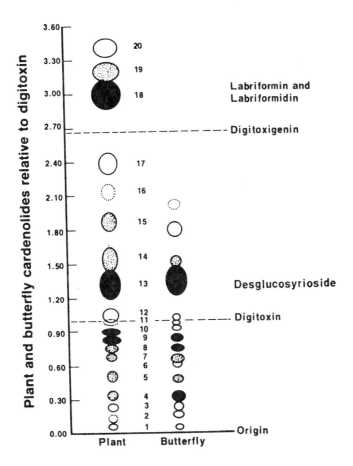

FIGURE 5.5 Thin-layer chromatogram showing a cardenolide fingerprint profile for monarch butterflies reared on milkweed (*Asclepias eriocarpa*). The chromatogram was developed in chloroform–methanol–formamide (90:60:1). Spot intensity (darkness) indicates the relative concentrations of the compounds. Mean migration distance digitoxin was 31.2 mm. (From Brower, L. P., et al., *J. Chem. Ecol.*, 8, 570, 1982. With permission.)

speed for 10 min. Decant and save the supernatant to a vial with 0.5 ml of granular ammonium sulfate. Reextract the lead precipitate with 2 ml of 1:2 95% ethanol and water, mix, centrifuge, and combine the supernatant with the first. Mix the combined extract and centrifuge (to remove the ammonium sulfate), save as supernatant-2, and rewash the residue with an additional 2 ml of ethanol and water. Combine the rinse after centrifuging with supernatant-2 and extract with two separate 2-ml portions of chloroform using a Pasteur pipet. The chloroform extracts can be dried by passing through anhydrous sodium sulfate on top of glass wool in a small funnel. The filter should be rinsed with 2 ml of chloroform after the second extract and evaporated to dryness with N_2. Resuspend in chloroform for TLC analysis.

Methylene chloride–methanol–formamide (105:15:1)

Silica gel G plates were activated after spotting over concentrated H_2SO_4 for 24 h prior to and 12 h after spotting. Plates were developed in saturated chambers (filter paper lined) to a height of 15 to 16 cm.[138]

2. Terpenoids — Azadirachtin

Isopropanol–n-hexane

Separations were made on Polygram, flexible polyester-backed, silica gel G plate, 4 × 8 cm (Macherey-Nagel) that could be cut into sections for scintillation counting.

3. **Terpenoids — Sesquiterpene Lactones**
   Chloroform–ethyl acetate–isopropanol (45:45:10)
   Separations were made on 10 × 10 cm HPTLC plates for the quantification of sesquiterpene lactones; the development of plates was carried out in closed tanks.[141] Sesquiterpene lactones can also be separated in a solvent system of chloroform–acetone (6:1).[149]

4. **Terpenoids — Grayanotoxins**
   Benzene–methanol–acetic acid (18:6:1)
   or
   Toluene–ethyl acetate–90% formic acid (5:4:1)
   or
   Acetone–chloroform (1:1)
   Separations can be run on either standard 20 × 20 cm plates or HPTLC plates. Extracts and standards were applied in chloroform or chloroform–acetone (85:15). The honey was extracted by first mixing 50 gm of honey with 100 ml of methanol–water (1:3) and then adjusting the pH to 6.5 with a dilute NaOH solution. The sample was filtered through Whatman No. 40 filter paper and then extracted three to five times with 150 ml of chloroform. The chloroform extracts were combined and evaporated to dryness. The residue was redissolved in 0.5 ml chloroform and analyzed. The HPTLC using the first solvent system procedure gives good results in the range of approximately 0.25 μg to 5 μg.

5. **Phenolics**
   Benzene–acetic acid–water (10:7:3)
   Separations were made on silica gel 60 F_{254} plates. Standards for identification included hydrobenzoic, coumaric, vanillic, chlorogenic, ferulic, o-coumaric, caffeic, shikimic, and tannic acids, plus pyrogallol and hydroquinone (10 μg each). Work was performed in a nitrogen-filled glove box to prevent oxidation of extracts. R_f values for the different compounds are shown in Table 5.7.[142]

TABLE 5.7
Separation of Phenolic Compounds by Thin-Layer Chromatography; R_f Values Resulting from a Solvent System of Benzene–Acetic Acid–Water (10:7:3) (Modified from Center and Wright[142])

Compound	R_f value
Gallic acid, chlorogenic acid	0.0 - (origin)
Catechol	0.02
Protocatechuic acid	0.07
Pyrogallol	0.07
p-Coumaric acid	0.13
Caffeic acid	0.14
p-Hydroxybenzoic acid	0.39
Ferulic acid	0.55
Vanillic acid	0.56
Shikimic acid	0.62
Coumarin	0.68
Vanillin	0.76

Chloroform–ethyl acetate (4:1)
   Extracts with coumarin-10 (7-amino-3-phenyl coumarin) were run on silica gel 20 × 20 cm plates.[143] Coumarin-10 is a fluorescent analog of natural coumarin and can be detected and quantified by fluorescence scanning.

6. Pyrrolizidine Alkaloids
Methanol–chloroform–ammonia (14.5:85:0.5)
Samples were run on silica gel G plates.[144]

7. Azoxyglycosides — Cycasin
n-butanol–acetic acid–water (4:1:1)
Extracts were run on silica gel plates with standards of cycasin ($R_f = 0.32$) and macrozamin ($R_f = 0.19$).[150]

C. VISUALIZATION SYSTEMS

Cardenolides — The standard detection procedure used by Brower et al.[132] for cardenolide compounds uses two spray solutions applied after final development. Solution 1 consists of a saturated solution of 2.2',4,4'–tetranitrodiphenyl (TNDP) in benzene. Spray solution 2 consists of 10% KOH in 50% aqueous methanol and is applied after the plate has dried from spraying with solution 1. The blue color formed by the TNDP–cardenolide complex fades rapidly, so plates are photographed to provide a record for comparison.

Sesquiterpene lactones — Quantitative determinations of separated compounds were made by UV-reflectance scanning at 220 nm with comparisons to standard curves of authentic compounds.[141] Sesquiterpene compounds can be identified by the color formed after spraying with a solution of 0.5 g p-dimethyl-aminobenzaldehyde, 0.5 ml concentrated H_2SO_4, three drops of glacial acetic acid, and 9.0 ml of 95% ethanol.[149] Plates are slowly heated at 70°C after spraying. Colors are somewhat unstable and may change on long standing. Sensitivity for sesquiterpene lactones is approximately 0.05 to 0.1 μg. Picman et al.[149] also presented a table of colors for over 70 compounds.

Grayanotoxins — Several detection reagents have been used for visualizing these diterpene toxins. Godin's reagent was originally used for polyols, but colors grayanotoxins a gray-blue. The reagent is prepared by mixing one volume of a 1% vanillin solution in ethanol with one volume of 3% perchloric acid in water just before spraying onto developed plates. The plates are then heated at 85°C for 3 to 5 min.[151] A second spray reagent can be made from a mixture of anisaldehyde (0.5ml), acetic acid (10 ml), concentrated sulfuric acid (5 ml), and methanol (85 ml). Plates are heated after spraying at 110°C.[146]

Phenolics — Many phenolic compounds can be detected by spraying developed plates with the Folin and Ciocalteu reagent or by quenching under UV light at 254 nm (with F_{254} plates).[142]

Pyrrolizidine alkaloids — Bogner and Broppre[144] recommend a procedure adapted from Mattocks.[152] Developed plates are placed in a saturated iodine atmosphere for 15 min and then sprayed with a 10% solution of 4-dimethylamino-benzaldehyde in concentrated hydrochloric acid that is diluted 1:4 with acetone just before use. For a selective visualization of dihydropyrrolizines, the plates are sprayed with the reagent without the initial iodine treatment.

Azoxyglycoside — Cycasin and macrozamin can be visualized by spraying plates with a reagent consisting of 1 g of 4,5-dihydroxy-2,7-naphthalene disulfonic acid disodium salt dissolved in 10 ml of water and diluted to 500 ml with 12.5 M sulfuric acid. Plates are heated to 110°C for 10 to 15 min after spraying.[150]

ACKNOWLEDGMENTS

I would like to thank Andrea Dvorak for her assistance with the library research and Dr. D. E. Mullins for his review of parts of the manuscript.

REFERENCES

1. **Berger, R. S., McGough, J. M., Martin, D. F., and Ball, L. R.,** Some properties and the field evaluation of the pink bollworm sex attractant, *Ann. Entomol. Soc. Am.*, 57, 606, 1964.
2. **Devakul, V. and Maarse, H.,** A second compound in the odorous gland liquid of the giant water bug *Lethocerus indicus* (Lep. and Srv.), *Anal. Biochem.*, 7, 269, 1964.
3. **Tashiro, H. and Chambers, D. L.,** Reproduction in the Califormia red scale, *Aonidiella aurantii* (Homoptera: Diaspididae). I. Discovery and extraction of a female sex pheromone, *Ann. Entomol. Soc. Am.*, 60, 1166, 1967.
4. **Jacobson, M.,** *Insect Sex Pheromones*, Academic Press, New York, 1972.
5. **Mullins, D. E. and Cochran, D. G.,** Nitrogen excretion in cockroaches: uric acid is not a major product, *Science*, 177, 699, 1972.
6. **Mullins, D. E. and Cochran, D. G.,** Tryptophan metabolite excretion by the American cockroach, *Comp. Biochem. Physiol.*, 44B, 549, 1973.
7. **Johnson, R. A. and Hill, L.,** The activity of the corpora allata in the fourth and fifth larval instars of the migratory locust, *J. Insect Physiol.*, 19, 1921, 1973.
8. **Trieber, L. R.,** Evaluation of TLC as an analytical tool; comparison of TLC with other methods, *J. Chromatogr. Sci.*, 24, 220, 1986.
9. **Florkin, M. and Jeuniaux, C.,** Hemolymph composition, in *The Physiology of Insects*, Vol. 5, Rockstein, M., Ed., New York, Elsevier, 1974, 255–307.
10. **Mullins, D. E.,** Chemistry and physiology of the hemolymph, in *Comprehensive Insect Physiology, Biochemistry and Pharmacology*, Vol. 3, Kerkut, G. and Gilbert, L., Eds., Pergamon Press, Oxford, 1985, 355–400.
11. **Maurizio, A.,** Untersuchungen uber das Zuckerbild Hamolymphe der Honigbiene (*Apis mellifica* L.) I. Das Zuckerbild des Blutes Erwachsener Bienne, *J. Insect Physiol.*, 11, 745, 1965.
12. **Caldes, G. and Prescott, B.,** A simple method for the detection and determination of trehalose by spot elution paper chromatography, *J. Chromatogr.*, 111, 466, 1975.
13. **Mack, S. R., Samuels, S., and Vanderberg, J. P.,** Hemolymph of *Anopheles stephensi* from noninfected and *Plasmodium berghei*-infected mosquitoes. 3. Carbohydrates, *J. Parasitol.*, 65, 217, 1979.
14. **Magnarelli, L. A. and Anderson, J. F.,** Follicular development in salt marsh Tabanidae (Diptera) and incidence of nectar feeding with relation to gonotrophic activity, *Ann. Entomol. Soc. Am.*, 70, 529, 1977.
15. **Magnarelli, L. A., Anderson, J. F., and Thorne, J. H.,** Diurnal nectar-feeding of salt marsh Tabanidae (Diptera), *Environ. Entomol.*, 8, 544, 1979.
16. **Bounias, M.,** Micro-analyse quantitative de quelques métabolites de l'hémolymph d'insectes — I. Les glucides libres, *Analysis*, 4, 87, 1976.
17. **Bounias, M.,** A comparison of hemolymph levels and inter-relations of trehalose, glucose and fructose in worker bees from different races and hybrids, *Comp. Biochem. Physiol.*, 69B, 471, 1981.
18. **Bounias, M.,** N-(1-Naphthyl)ethylenediamine dihydrochloride as a new reagent for nanomole quantification of sugars on thin-layer plates by a mathematical calibration process, *Anal. Biochem.*, 106, 291, 1980.
19. **Bounias, M.,** Cinétique d'action de la noradrénaline, de la dopa et de la dopamine sur la glycémie de l'abeille *in vivo*, *Arch. Int. Physiol. Biochem.*, 88, 93, 1980.
20. **Bounias, M. and Pacheco, H.,** Effets hyperglycémiants de l'adrenaline injectée chez l'Abeille *in vivo*, *C.R. Acad. Sci., Paris*, 289D, 33, 1979.
21. **Bounias, M. and Pacheco, H.,** Sensibilité de la glycémie de l'Abeille à l'action de d'insuline et du glucagon injectés *in vivo*, *C.R. Acad. Sci., Paris*, 289D, 201, 1979.
22. **Bounias, M. and Kruk, I.,** Adrenochrome as competitive inhibitor of epinephrine effects: A comparative study on trehalose and glucose levels in the honey bee hemolymph, *Comp. Biochem. Physiol.*, 74C, 143, 1983.
23. **Fell, R. D.,** The qualitative and quantitative analysis of insect hemolymph sugars by high performance thin-layer chromatography, *Comp. Biochem. Physiol.*, 95A, 539, 1990.
24. **Stein, K. J. and Fell, R. D.,** Seasonal comparison of weight, energy reserve, and nitrogen changes in queens of the baldfaced hornet (Hymenoptera: Vespidae), *Environ. Entomol.*, 21, 148, 1992.
25. **Cannon, C. A. and Fell, R. D.,** Cold hardiness of the overwintering black carpenter ant, *Physiol. Entomol.*, 17, 121, 1992.
26. **Thomas, K. K. and Nation, J. L.,** Absorption of glucose, glycine and palmitic acid by isolated midgut and hindgut from crickets, *Comp. Biochem. Physiol.*, 79A, 289, 1984.
27. **Turunen, S.,** Efficient use of dietary galactose in *Pieris brassicae*, *J. Insect Physiol.*, 38, 503, 1992.
28. **Turunen, S.,** Metabolic pathways in the midgut epithelium of *Pieris brassicae* during carbohydrate and lipid assimilation, *Insect Biochem. Molec. Biol.*, 23, 681, 1993.
29. **Sammuels, S. and Fisher, C.,** Direct quantification of micro-thin-layer chromatograms, *J. Chromatogr.*, 71, 297, 1972.
30. **Jork, H., Funk, W., Fischer, W., and Wimmer, H.,** Thin-layer chromatography: reagents and detection methods, Volume 1a – *Physical and Chemical Detection Methods: Fundamentals, Reagents* I, VCH Publishers, New York, 1990.
31. **Gocan, S.,** Stationary phases in thin-layer chromatography, in *Modern Thin-Layer Chromatography*, Grinberg, N., Ed., Marcel Dekker, New York, 1990, 5–137.

32. **Ghebregzabher, M., Ruffini, S., Monaldi, B., and Lato, M.,** Thin layer chromatography of carbohydrates, *J. Chromatogr.*, 174, 195, 1976.
33. **Pruden, B. B., Pineault, G., and Loutfi, H.,** A thin-layer chromatographic method for the quantitative determinaton of D-mannose, D-glucose, and D-galactose in aqueous solution, *J. Chromatogr.*, 115, 477, 1975.
34. **Gauch, R., Leuenberger, U., and Baumgartner, E.,** Quantitative determination of mono-, di-, and trisaccharides by thin-layer chromatography, *J. Chromatogr.*, 174, 195, 1979.
35. **Hansen, S. A.,** Thin-layer chromatographic method for the identification of mono-, di- and trisaccharides, *J. Chromatogr.*, 107, 224, 1975.
36. **Bassal, T. T. M.,** Detection on silicic acid chromatograms of trehalose and other sugars in arthropod fluids, *J. Med. Entomol.*, 10, 228, 1973.
37. **Poole, C. F., Coddens, M. E., Butler, H. T., Schuette, S. A., Ho, S. S. J., Khatib, S., Piet, L., and Brown, K. K.,** Some quantitative aspects of scanning densitometry in high-performance thin layer chromatography, *J. Liq. Chromatogr.*, 8, 2875, 1985.
38. **Salt, R. W.,** Natural occurrence of glycerol in insects and its relation to their ability to survive freezing, *Can. Entomol.*, 84, 491, 1957.
39. **Story, K. B. and Story, J. M.,** Biochemistry of freeze tolerance in terrestrial insects, *Trends Biochem. Sci.*, 8, 242, 1983.
40. **Danks, H. V.,** Modes of seasonal adaptation in the insects, *Can. Entomol.*, 110, 1167, 1978.
41. **Baust, J. G. and Lee, R. E.,** Population differences in antifreeze/cryoprotectant accumulation patterns in an antarctic insect, *Oikos*, 40, 120, 1983.
42. **Miller, L. K. and Smith, J. S.,** Production of threitol and sorbitol by an adult insect: association with freezing tolerance, *Nature*, 258, 519, 1975.
43. **Hamilton, R. L.,** Cold-acclimation of the wood cockroach *Cryptocercus punctulatus* (Scudder) (Dictyoptera: Cryptocercidae), M.S. Thesis, Virginia Polytechnic Institute and State University, Blacksburg, VA, 1985.
44. **Hamilton, R. L., Mullins, D. E., and Orcutt, D. M.,** Freezing-tolerance in the woodroach *Cryptcercus punctulatus* (Scudder), *Experientia*, 41, 1535, 1985.
45. **Chino, H.,** Conversion of glycogen to sorbitol and glycerol in the diapause egg of the Bombyx silkworm, *Nature*, 180, 606, 1957.
46. **Baust, J. G. and Miller, L. K.,** Variations in glycerol content and its influence on cold hardiness I the Alaskan carabid beetle, *Pterostichus brevicornis*, *J. Insect Physiol.*, 16, 979, 1970.
47. **Somme, L.,** Mannitol and glycerol in overwintering aphid eggs, *Norsk Entomol. Tidssk.*, 16, 107, 1969.
48. **Yaginuma, T. and Yamashita, O.,** Polyol metabolism related to diapause in *Bombyx* eggs: different behavior of sorbitol from glycerol during diapause and post-diapause, *J. Insect Physiol.*, 24, 347, 1978.
49. **Izumiyama, S., Suzuki, K., and Miya, K.,** Glycerol in the eggs of the two-spotted cricket, *Gryllus bimaculatus* De Geer, *Appl. Ent. Zool.*, 18, 295, 1983.
50. **Ring, R. A. and Tesar, D.,** Cold-hardiness of the arctic beetle, *Pytho americanus* Kirby Coleoptera, Pythidae (Salpingidae), *J. Insect Physiol.*, 26, 763, 1980.
51. **Van Handel, E.,** Microseparation of glycogen, sugars, and lipids, *Anal. Biochem.*, 11, 266, 1965.
52. **Kartnig, T. and Wegschaider, O.,** Eine Möglichkeit zur Identifizierung von Zuckern auskleinsten Meugen von Glycosiden oder aus Zucher gemischen, *J. Chromatogr.*, 61, 375, 1971.
53. **Stanley-Samuelson, D. W. and Dadd, R. H.,** Arachidonic and other tissue fatty acids of *Culex pipiens* reared with various concentrations of dietary arachidonic acid, *J. Insect Physiol.*, 27, 571, 1981.
54. **Stanley-Samuelson, D. W. and Dadd, R. H.,** Long-chain polyunsaturated fatty acids: patterns of occurrence in insects, *Insect Biochem.*, 13, 549, 1983.
55. **Stanley-Samuelson, D. W. and Loher, W.,** Arachidonic and other long-chain polyunsaturated fatty acids in spermatophores and spermathecae of *Teleogryllus commodus*: significance in prostaglandin-mediated reproductive behavior, *J. Insect Physiol.*, 29, 41, 1983.
56. **Turunen, S.,** Uptake of dietary lipids: a novel pathway in *Pieris brassicae*, *Insect Biochem.*, 18, 499, 1988.
57. **Turunen, S.,** Plant leaf lipids as fatty acid sources in two species of Lepidoptera, *J. Insect Physiol.*, 36, 665, 1990.
58. **Turunen, S. and Chippendale, G. M.,** Relationship between dietary lipids, midgut lipids, and lipid absorption in eight species of Lepidoptera reared on artificial and natural diets, *J. Insect Physiol.*, 35, 627, 1989.
59. **Fischer, R. and Komnick, H.,** Peroxisomal acyl-CoA oxidase and chain-shortening of dietary fatty acids in the midgut of dragonfly larvae, *Aeshna cyanea*, *Insect Biochem. Molec. Biol.*, 22, 793, 1992.
60. **Sickmann, T., Weske, B., Dennis, R. D., Mohr, C., and Wiegandt, H.,** Chemical distribution of glycosphigolipids in third-instar larval organs of the blowfly, *Calliphora vicina* (Insecta: Diptera), *J. Biochem.*, 111, 662, 1992.
61. **Spates, G. E., Bull, D. L., and Chen, A. C.,** Hydrolysis of sphingomyelin and phosphatidylcholine by midgut homogenates of the stable fly, *Arch. Insect Biochem. Physiol.*, 14, 1, 1990.
62. **Miller, R. R., Jr., Yates, J. W., and Geer, B. W.,** Dietary ethanol stimulates the activity of phosphatidylcholine-specific phospholipase D and the formation of phosphatidylethanol in *Drosophila melanogaster* larvae, *Insect Biochem. Molec. Biol.*, 23, 749, 1993.
63. **Bounias, M., Debevec, M., and Popeskovic, D.,** A comparison of hemolymph lipid classes at different stages of honeybee development, *Acta Veterinaria*, 35, 273, 1985.

64. **Judge, D. N.,** Flight activity and hemolymph diacylglyceride concentrations in *Heliothis zea* (Boddie) (Lepidoptera: Noctuidae), M. S. Thesis, Virginia Polytechnic Institute and State University, 1988.
65. **Judge, D. N., Mullins, D. E., and Eaton, J. L.,** Microquantity analysis of insect hemolymph lipids by high performance thin layer chromatography, *J. Planar Chromatogr.*, 2, 442, 1989.
66. **Stein, K. J. and Fell, R. D.,** Egg weights, energy reserves, and internal nest temperatures in embryo nests of *Dolichovespula maculata* (Hymenoptera: Vespidae), *Ann. Entomol. Soc. Am.*, 87, 554, 1994.
67. **Katagiri, C.,** Structure of lipophorin in insect blood: location of phospholipid, *Biochim. Biophys. Acta*, 834, 139, 1985.
68. **Singh, T. K. A. and Ryan, R. O.,** Lipid transfer particle-catalyzed transfer of lipoprotein-associated diacylglycerol and long chain aliphatic hydrocarbons, *Arch. Biochem. Biophys.*, 286, 376, 1991.
69. **Petzel, D. H., Parrish, A. K., Ogg, C. L., Witters, N. A., Howard, R. W., and Stanely-Samuelson, D. W.,** Arachidonic acid and prostaglandin E_2 in malpighian tubules of female yellow fever mosquitoes, *Insect Biochem. Molec. Biol.*, 23, 431, 1993.
70. **Nelson, D. R. and Fatland, C. L.,** Novel esters on long-chain methyl-branched alcohols in pupae of the tobacco hornworm, *Manduca sexta*: Propionate, butyrate, isobutyrate, pentanoate and isovalerate, *Insect Biochem. Molec. Biol.*, 22, 99, 1992.
71. **Nelson, D. R. and Fatland, C. L.,** Verification of structure and direction of biosynthesis of long-chain methyl-branched alcohols and very long-chain methyl-branched alcohols in pupae of the tobacco hornworm, *Manduca sexta*, *Insect Biochem. Molec. Biol.*, 22, 111, 1992.
72. **Guo, L., Nelson, D. R., Fatland, C. L., and Bloomquist, G. J.,** Very long-chain methyl-branched alcohols and their acetate esters in pupae of the southern armyworm, *Spodoptera eridania*: identification and biosynthesis, *Insect Biochem. Molec. Biol.*, 22, 277, 1992.
73. **Juárez, P., Brenner, R. R., and Gros, E. G.,** The epicuticular lipids of *Triatoma infestans* – 1. Glycerides, *Comp. Biochem. Physiol.*, 78B, 427, 1984.
74. **Juárez, P. and Brenner, R. R.,** Hydrocarbons of *Triatoma pallidipennis*, *Comp. Biochem. Physiol.*, 87B, 233, 1987.
75. **Juárez, P. and Brenner, R. R.,** Fatty acid biosynthesis in the integument tissue of *Triatoma infestans*, *Comp. Biochem. Physiol.*, 93B, 763, 1989.
76. **Juárez, P.,** Inhibition of cuticular lipid synthesis and its effect on insect survival, *Arch. Insect Biochem. Physiol.*, 25, 177, 1994.
77. **Hamilton, J. G. C., Sonenshine, D. E., and Lusby, W. R.,** Cholesteryl oleate: mounting sex pheromone of the hard tick *Dermacenter variabilis* (Say) (Acari: Ixodidae), *J. Insect Physiol.*, 35, 873, 1989.
78. **Morse, D. and Meighen, E. A.,** Differences in oxidase and esterase activities involved in pheromone biosynthesis in two species of *Choristoneura*, *J. Chem. Ecol.*, 16, 1485, 1990.
79. **Ragab, A., Bitsch, C., Thomas, J. M. F., Bitsch, J., and Chap, H.,** Lipoxygenase conversion of arachidonic acid in males and inseminated females of the firebrat, *Thermobia domestica* (Thysanura), *Insect Biochem.*, 17, 863, 1987.
80. **Ragab, A., Bitsch, C., Ragab-Thomas, J. M. F., Gassama-Diagne, A., and Chap, H.,** Phospholipase A2 activity in reproductive tissues of the firebrat *Thermobia domestica* (Insecta: Thysanura), *Insect Biochem. Molec. Biol.*, 22, 379, 1992.
81. **Downing, D. T.,** Lipids, in *Densitometry in Thin Layer Chromatography: Practice and Applications*, Touchstone, J. C. and Sherma, J., Eds., John Wiley and Sons, New York, 1979, 367–391.
82. **Pernes, J. F., Nurit, Y., and Detteaulme, M.,** Lipids: thin-layer chromatographic separation in twelve fractions by three successive unidirectional developments on the same plate, *J. Chromatogr.*, 181, 254, 1980.
83. **Fried, B. and Sherma, J.,** *Thin-Layer Chromatography; Techniques and Applications*, 2nd ed., Marcel Dekker, New York, 1986.
84. **Spates, G. E., Elissalde, M., and Williams, H.,** Lipid composition of housefly microsomes, *Arch. Insect Biochem. Physiol.*, 7, 47, 1988.
85. **Junker-Buchheit, A. and Jork, H.,** Mondansyl cadaverine as a fluorescent marker for carboxylic acids – in situ prechromatogaphic derivatization, *J. Planar Chromatogr.*, 2, 65, 1989.
86. **Christie, W. W.,** *Lipid Analysis; Isolation, Separation, Identification, and Structural Analysis of Lipids*, Pergamon Press, Oxford, 1973.
87. **Morgan, D. W. and Poole, C. F.,** The extraction and determination of ecdysones in arthropods, *Adv. Insect Physiol.*, 12, 17, 1976.
88. **Wilson, I. D., Scalia, S., and Morgan, E. D.,** Reversed-phase thin-layer chromatography for the separation and analysis of ecdysteroids, *J. Chromatogr.*, 212, 211, 1981.
89. **Wilson, I. D., Bielby, C. R., and Morgan, E. D.,** Studies on the reversed-phase thin-layer chromatography of ecdysteroids on C_{12} bonded and paraffin-coated silica, *J. Chromatogr.*, 242, 202, 1982.
90. **Wilson, I. D.,** Thin-layer chromatography of ecdysteroids, *J. Chromatogr.*, 318, 373, 1985.
91. **Wilps, H. and Zöller, T.,** Origin of ecdysteroids in females of the blowfly *Phormia terraenovae* and their relation to reproduction and metabolism, *J. Insect Physiol.*, 35, 709, 1989.
92. **O'Reilly, D. R, Brown, M. R., and Miller, L. K.,** Alteration of ecdysteroid metabolism due to baculovirus infection of the fall armyworm *Spodoptera frugiperda*: host ecdysteroids are conjugated with galactose, *Insect Biochem. Molec. Biol.*, 22, 313, 1992.

93. **Wilson, I. D. and Lafont, R.,** Thin-layer chromatography and high-performance thin-layer chromatography of [³H]metabolites of 20-hydroxyecdysone, *Insect Biochem.*, 16, 33, 1986.
94. **Wilson, I. D., Morgan, E. D., Robinson, P., Lafont, R., and Blais, C.,** A comparison of radio-thin-layer and radio-high-performance liquid chromatography for ecdysteroid metabolism studies, *J. Insect Physiol.*, 34, 707, 1988.
95. **Dinan, L.,** Thin-layer chromatography of ecdysone acyl esters and their 2,3-acetonide derivatives, *J. Chromatogr.*, 411, 379, 1987.
96. **Whiting, P. and Dinan, L.,** The occurrence of apolar ecdysteroid conjugates in newly-laid eggs of the house cricket, *Acheta domesticus*, *J. Insect Physiol.*, 34, 625, 1988.
97. **Whiting, P. and Dinan, L.,** Identification of the endogenous apolar ecdysteroid conjugates present in newly-laid eggs of the house cricket (*Acheta domesticus*) as 22-long-chain fatty acyl esters of ecdyson, *Insect Biochem.*, 19, 759, 1989.
98. **Horn, D. H. S. and Bergamasco, R.,** Chemistry of ecdysteroids, in *Comprehensive Insect Physiology Biochemistry and Pharmacology*, Kerkut, G. A. and Gilbert, L. I., Eds., Pergamon Press, Oxford, England, 1985, 185–248.
99. **O'Reilly, D. R., Howarth, O. W., Rees, H. H., and Miller, L. K.,** Structure of the ecdysone glucoside formed by a baculovirus ecdysteroid UDP-glucosyltransferase, *Insect Biochem.*, 21, 795, 1991.
100. **Coscia, C. J.,** *CRC Handbook of Chromatography, Terpenoids*, Volume I, CRC Press, Boca Raton, FL, 1984.
101. **Duve, H., Thorpe, A., Yagi, K. J., Yu, C. G., and Tobe, S. S.,** Factors affecting the biosynthesis and release of juvenile hormone bisepoxide in the adult blowfly Calliphora vomitoria, *J. Insect Physiol.*, 38, 575, 1992.
102. **Nijhout, H. F.,** *Insect Hormones*, Princeton Univ. Press, Princeton, NJ, 1994.
103. **Free, J. B.,** *Pheromones of Social Bees*, Comstock Publishing, Ithaca, NY, 1987.
104. **Wood, D. L.,** The role of pheromones, kairomones, and allomones in the host selection and colonization behavior of bark beetles, *Annu. Rev. Entomol.*, 27, 411, 1982.
105. **Pasteels, J., Grégoire, C., and Rowell-Rahier, M.,** The chemical ecology of defense in arthropods, *Annu. Rev. Entomol.*, 28, 263, 1983.
106. **Lefevere, K. S., Lacey, M. J., Smith, P. H., and Roberts, B.,** Identification and quantification of juvenile hormone biosynthesized by larval and adult Australian sheep blowfly *Lucilia cuprina* (Diptera: Calliphoridae), *Insect Biochem. Molec. Biol.*, 23, 713, 1993.
107. **Casas, J., Harshman, L. G., Messeguer, A., Kuwano, E., and Hammock, B. D.,** In vitro metabolism of juvenile hormone III and juvenile hormone III bisepoxide by *Drosophila melanogaster* and mammalian cytosolic epoxide hydrolase, *Arch. Biochem. Biophys.*, 286, 153, 1991.
108. **Campbell, P. M., Healy, M. J., and Oakeshott, J. G.,** Characterisation of juvenile hormone esterase in *Drosophila melanogaster*, *Insect Biochem. Molec. Biol.*, 22, 665, 1992.
109. **Casas, J., Harshman, L. G., and Hammock, B. D.,** Epoxide hydrolase activity on juvenile hormone in *Manduca sexta*, *Insect Biochem.*, 21, 17, 1991.
110. **Ottea, J. A., Harshman, L. G., and Hammock, B. D.,** Novel assay for determining the metabolic fate of juvenile hormone III: a study with *Drosophila melanogaster*, *Arch. Insect Biochem. Physiol.*, 8, 25, 1988.
111. **Meyer, W. R. and Lanzrein, B.,** Degradation of juvenile hormone and methylation of juvenile hormone acid by corpora cardiaca-corpora allata of the cockroach, *Nauphoeta cinerea*: I. Biochemical aspects, *Arch. Insect Biochem. Physiol.*, 10, 303, 1989.
112. **Meyer, W. R. and Lanzrein, B.,** Degradation of juvenile hormone and methylation of juvenile hormone acid by corpora cardiaca-corpora allata of the cockroach, *Nauphoeta cinerea*: II. Physiological aspects, *Arch. Insect Biochem. Physiol.*, 10, 317, 1989.
113. **Hammock, B., Nowock, J., Goodman, W., Stamoudis, V., and Gilbert, L. I.,** The influence of hemolymph binding protein on juvenile hormone stability and distribution in *Manduca sexta* fat body and imaginal discs in vitro, *Molec. Cellular Endocrin.*, 3, 167, 1975.
114. **Hammock, B. D. and Roe, R. M.,** Analysis of juvenile hormone esterase activity, *Meth. Enzymol.*, 111, 487, 1985.
115. **Crouteau, R. and Ronald, R. C.,** Terpenoids, in *Chromatography: Fundamentals and Applications of Chromatographic and Electrophoretic Methods, Part B: Applications*, Heftmann, E., Ed., Elsevier, Amsterdam, 1983, B149–B1189.
116. **Eaton, J. L. and Mullins, D. E.,** Quantitative high-performance thin-layer chromatography of dansyl derivatives of biogenic amines, *Anal. Biochem.*, 172, 484, 1988.
117. **Kayser, H.,** Pigments, in *Comprehensive Insect Physiology, Biochemistry and Pharmacology*, Vol. 10, Kerkut, G. A. and Gilbert, L. I., Eds., Pergamon Press, Oxford, 1985, 367–415.
118. **Wilson, T. G. and Jacobson, K. B.,** Isolation and characterization of pteridines from heads of *Drosophila melanogaster* by a modified thin-layer chromatography procedure, *Biochem. Genet.*, 15, 307, 1977.
119. **Ferré, J.,** Accumulation of kynurenic acid in the "cinnabar" mutant of *Drosophila melanogaster* as revealed by thin-layer chromatography, *Insect Biochem.*, 13, 289, 1983.
120. **Ferré, J., Silva, F. J., Real, M. D., and Méensura, J. L.,** Pigment patterns in mutants affecting the biosynthesis of pteridines and xanthommatin in *Drosophila melanogaster*, *Biochem. Genetics*, 24, 545, 1986.
121. **Reaume, A. G., Knecht, D. A., and Chovnick, A.,** The *rosy* locus in *Drosophila melanogaster*: Xanthine dehydrogenase and eye pigments, *Genetics*, 129, 1099, 1991.
122. **Summers, K. M., Howells, A. J., and Pyliotis, N. A.,** Biology of eye pigmentation, *Adv. Insect Physiol.*, 16, 119, 1982.

123. **Pepe, I. M. and Cugnoli, C.,** Isolation and characterization of a water-soluble photopigment from honeybee compound eye, *Insect Biochem.*, 30, 97, 1980.
124. **Shukolyukov, S. A. and Denisova, N. A.,** Opsin biosynthesis and *trans-cis* isomerization of aldehyde form chromophore in the blowfly *Calliphora erythrocephala* eye, *Insect Biochem. Molec. Biol.*, 22, 925, 1992.
125. **White, R. H.,** Insect visual pigments and color vision, in *Comprehensive Insect Physiology Biochemistry and Pharmacology*, Vol. 6, Kerkut, G. A. and Gilbert, L. I., Eds., Pergamon Press, Oxford, 1985, 431–493.
126. **Hori, M. and Riddiford, L. M.,** Isolation of ommochromes and 3-hydroxykynurenine from the tobacco hornworm, *Manduca sexta*, *Insect Biochem.*, 11, 507, 1981.
127. **Koch, P. B.,** Precursors of pattern specific ommatin in red wing scales of the polyphenic butterfly *Araschnia levana* L.: haemolymph tryptophan and 3-hydroxykynurenine, *Insect Biochem.*, 21, 785, 1991.
128. **Melber, Ch. and Schmidt, G. H.,** Identification of fluorescent compounds in certain species of *Dysdercus* and some of their mutants (Heteroptera: Pyrrhocoridae), *Comp. Biochem. Physiol.*, 101B, 115, 1992.
129. **Harborne, J. B.,** *Introduction to Ecological Biochemistry*, 3rd ed., Academic Press, London, 1988.
130. **Duffey, S. S.,** Sequestration of plant natural products by insects, *Annu. Rev. Entomol.*, 25, 447, 1980.
131. **Roeske, C. M., Seiber, J. N., Brower, L. P., and Moffitt, C. M.,** Milkweed cardenolides and their comparative processing by monarch butterflies (*Danaus plexippus* L.), *Recent Adv. Phytochem.*, 10, 93, 1976.
132. **Brower, L. P., Seiber, J. N., Nelson, C. J., Lynch, S. P., and Tuskes, P. M.,** Plant-determined variation in the cardenolide content, thin-layer chromatography profiles, and emetic potency of monarch butterflies *Danaus plexippus* reared on the milkweed, *Asclepias eriocarpa*, in California, *J. Chem. Ecol.*, 8, 570, 1982.
133. **Brower, L. P., Seiber, J. N., Nelson, C. J., Lynch, S. P., Hoggard, M. P., and Cohen, J. A.,** Plant-determined variation in the cardenolide content, thin-layer chromatography profiles, and emetic potency of monarch butterflies *Danaus plexippus* reared on the milkweed, *Asclepias californica*, in California, *J. Chem. Ecol.*, 10, 1823, 1984.
134. **Brower, L. P., Seiber, J. N., Nelson, C. J., Lynch, S. P., and Holland, M. M.,** Plant-determined variation in the cardenolide content, thin-layer chromatography profiles, and emetic potency of monarch butterflies *Danaus plexippus* reared on the milkweed, *Asclepias speciosa*, in California, *J. Chem. Ecol.*, 10, 601, 1984.
135. **Brower, L. P., Nelson, C. J., Seiber, J. N., Fink, L. S., Bond, C.,** Exaptation as an alternative to coevolution in the cardenolide-based chemical defense of monarch butterflies (*Danaus plexippus* L.) against avian predators, in *Chemical Mediation of Coevolution*, Spencer, K. C., Ed., Academic Press, Inc., San Diego, 1988, 447–475.
136. **Lynch, S. P. and Martin, R. A.,** Cardenolide content and thin-layer chromatography profiles of monarch butterflies, *Danaus plexippus* L., and their larval host-plant milkweed, *Asclepias viridis* Walt., in northwestern Louisiana, *J. Chem. Ecol.*, 13, 47, 1987.
137. **Cohen, J. A. and Brower, L. P.,** Cardenolide sequestration by the dogbane tiger moth (*Cycnia tenera*; Arctiidae), *J. Chem. Ecol.*, 9, 521, 1983.
138. **Moore, L. V. and Scudder, G. G. E.,** Selective sequestration of milkweed (*Asclepias* spp.) cardenolides in *Oncopeltus fasciatus* (Dallas) (Hemiptera: Lygaeidae), *J. Chem. Ecol.*, 11, 667, 1985.
139. **Barnby, M. A., Kioche, J. A., Darlington, M. V., and Yamasaki, R. B.,** Uptake, metabolism, and excretion of injected tritiated 22,33-dihydroazadirachtin in last instar larvae of *Heliothis virescens*, *Entomol. Exp. Appl.*, 52, 1, 1989.
140. **Cates, R. G. and Redak, R. A.,** Variation in the terpene chemistry of Douglas-fir and its relationship to western spruce budworm success, in *Chemical Mediation of Coevolution*, Spencer, K. C., Ed., Academic Press, San Diego, 1988, 317–344.
141. **Chou, J. C. and Mullin, C. A.,** Distribution and antifeedant associations of sesquiterpene lactones in cultivated sunflower (*Helianthus annuus* L.) on western corn rootworm (*Diabrotica virgifera virgifera* LeConte), *J. Chem. Ecol.*, 19, 1439, 1993.
142. **Center, T. D. and Wright, A. D.,** Age and phytochemical composition of waterhyacinth (*Pontederiaceae*) leaves determine their acceptability to *Neochetina eichhorniae* (*Coleoptera: Curculionidae*), *Environ. Entomol.*, 20, 323, 1991.
143. **Shapiro, J. P.,** Xenobiotic absorption and binding by proteins in hemolymph of the weevil *Diaprepes abbreviatus*, *Arch. Insect Biochem. Physiol.*, 11, 65, 1989.
144. **Bogner, F. and Boppré, M.,** Single cell recordings reveal hydroxydanaidal as the volatile compound attracting insects to pyrrolizidine alkaloids, *Entomol. Exp. Appl.*, 50, 171, 1989.
145. **Barker, R. J.,** Poisoning by plants, in *Honey Bee Pests, Predators, and Diseases*, 2md ed., Morse, R. A. and Nowogrodzki, R., Eds., Cornell Univ. Press, Ithaca, NY, 1990, 306–328.
146. **Scott, P. M., Coldwell, B. B., and Wiberg, G. S.,** Grayanotoxins. Occurrence and analysis in honey and a comparison of toxicities in mice, *Fd. Cosmet. Toxicol.*, 9, 179, 1971.
147. **Fell, R. D.,** Unpublished data, 1992.
148. **Sütlüpinar, N., Mat, A., and Satganoglu, Y.,** Poisoning by toxic honey in Turkey, *Arch. Toxicol.*, 67, 148, 1993.
149. **Picman A. K., Ranieri, R. L., Towers, G. H. N., and Lam, J.,** Visualization reagents for sesquiterpene lactones and polyacetylenes on thin-layer chromatograms, *J. Chromatogr.*, 189, 187, 1980.
150. **Bowers, M. D. and Larin, Z.,** Acquired chemical defense in the lycaenid butterfly, *Eumaeus atala*, *J. Chem. Ecol.*, 15, 1133, 1989.
151. **Godin, P.,** A new spray reagent for paper chromatography of polyols and cetoses, *Nature*, 174, 134, 1954.
152. **Mattocks, A. R.,** *Chemistry and Toxicology of Pyrrolizidine Alkaloids*, Academic Press, London, 1986.

Chapter 6

THIN-LAYER CHROMATOGRAPHY OF THE SKIN SECRETIONS OF VERTEBRATES

Paul J. Weldon

CONTENTS

I. Introduction 106

II. Obtaining Skin Material for TLC 106
    A. Epidermis and Epidermal Derivatives 106
    B. Exocrine Gland Secretions 110
    C. Storage and Preservation of Skin Material 111

III. Sample Preparation 111
    A. General Considerations 111
    B. Dissolving and Simple Extraction Procedures 111
    C. Complex Extraction Procedures — Partition of Extracts 111
    D. Column Chromatography 112

IV. Chromatographic Systems and Detection of Compounds 113
    A. General Considerations 113
    B. Nonpolar Lipids 113
    C. Polar Lipids..114
        1. Phosphatides..114
        2. Ceramides and Glucosylceramides 115
        3. Sphingosines 116
        4. Steryl Glycosides 116
        5. Sulfolipids 117
    D. Alkaloids 117

V. Quantification 118

VI. TLC Analyses in Vertebrate Systematics 121

VII. Validation of Results 121

VIII. Selected Protocols 124
    A. General Considerations 124
    B. Polar and Nonpolar Lipids from the Epidermis of Lizards and Snakes
       (Squamata, Reptilia) 124
    C. Ceramides and Acylglucosylceramides
       from the Epidermis of the Pig (*Sus scrofa*) 125
    D. Tetrodotoxin from Newts (*Taricha* spp.) and other Amphibians 126

References 127

I. INTRODUCTION

Secretions from the epidermis and exocrine skin glands of vertebrates contribute in various ways to survivorship and reproductive success. Epidermal secretions thwart the growth of microorganisms and prevent desiccation by retarding transepidermal water loss. Glandular exudates repel predators and ectoparasites; mark territories and signal alarm, sexual receptivity, and other states to conspecifics; and generally condition and maintain the skin. The multifunctionality of vertebrate skin secretions is reflected in the vast array of compounds they contain.

Thin-layer chromatography (TLC), used either as an analytical or preparative technique, has contributed substantially to the study of vertebrate skin chemistry. Analytical TLC has provided information on the complexity of mixtures by indicating the minimum number of components in skin extracts (and their purity prior to spectral analyses), permitted comparisons of the chemical profiles of different organs and organisms, and provided evidence for the presence in the integument of general chemical classes or specific compounds. Interspecific[1-18] (Figures 6.1 and 6.2), sexual,[18-26] individual,[16,26-29] ontogenetic,[21,23,30-32] mutational,[33] populational,[7,34,35] anatomical,[9,16,18,26,36-39] seasonal,[16,30,40,41] (Figure 6.3) dietary,[42] and hormonally mediated[20,23] variation in the composition of skin secretions, and the involvement of microorganisms in generating skin products,[43-47] are suggested by analytical TLC. In addition, this technique coupled with autoradiography has elucidated mechanisms of skin chemical biosynthesis by permitting the detection of labeled compounds recovered in surface extracts.[48-52]

Preparative TLC has been used in preliminary isolations of skin-derived compounds for subsequent analyses by gas chromatography–mass spectrometry (GC-MS) or other methods. This is a simple, inexpensive, and valuable technique of isolating physiologically and behaviorally active compounds, including some volatile components.[53]

This chapter describes the key methods and salient results of TLC analyses of the skin products of vertebrates, specifically tetrapods (amphibians, reptiles, birds, and mammals). The focus here is on comparative studies of lipids and alkaloids in the skin secretions of nonhuman species; TLC studies of human skin lipids are described where they involve comparisons with other species, or where their methods or results are exemplary.

II. OBTAINING SKIN MATERIAL FOR TLC

The methods by which epidermal and skin gland products are obtained for TLC vary with the taxon and organ sampled (see Downing,[54] Downing and Stewart,[55] and Nicolaides and Kellum[56] for techniques applied primarily to mammals). The degree of manipulation and invasiveness to which the secretion donors may be subjected will determine also the sampling methods used.

A. EPIDERMIS AND EPIDERMAL DERIVATIVES

An array of skin tissues from a diversity of vertebrates can be obtained from zoos. The naturally sloughed epidermal sheets of lizards and snakes and the molted feathers of birds are routinely available and have abundant chemicals for TLC. Lipid residues of up to approximately 15% of the dry masses of shed snake skins, for example, have been obtained by extraction with organic solvents.[2,15,57] Snakes preparing to slough, as evidenced by opaque skin, and molting birds can be confined to separate, clean quarters for easier access to shed materials. Soiled or wet epidermal materials generally should be discarded to avoid contamination and decomposition during storage.

Where comparisons are made between organisms, samples should be obtained from homologous anatomical sites using uniform methods; consistency in sampling procedures is essential for meaningful comparisons between and within species. Materials from different individuals may be pooled to obtain greater quantities of extracts, although separate analyses according to sex, age-class, and other variables, where possible, may provide useful information on sources of variation in TLC profiles.

Chemicals from the skin surface of mammals can be obtained from hair cut with solvent-extracted scissors; electric clippers should be avoided or at least cleaned with solvents to prevent contamination by mechanical lubricants. Solvents either may be poured directly over the intact

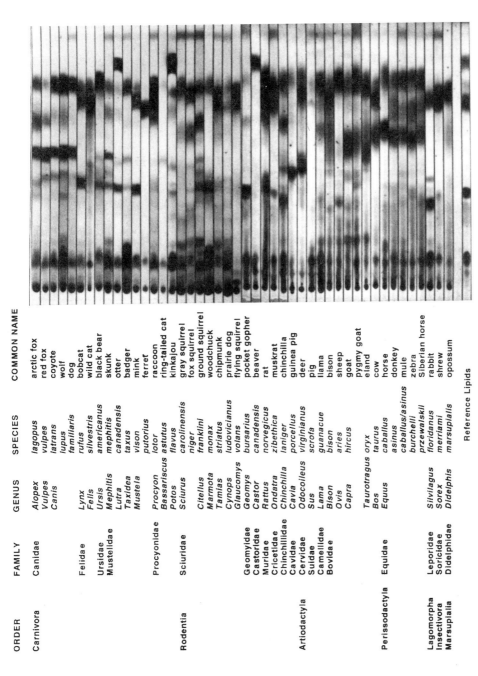

FIGURE 6.1 Chromatograms of skin surface lipids of 45 species of mammals, showing species variation in secretion composition. Extracts were applied to a plate developed successively with hexane, then toluene, and finally (half-way) with hexane–diethyl ether–acetic acid (70:30:1). The plate was sprayed with sulfuric acid and heated on a hot plate to char the lipids. Reference lipids (right lane) are (1) cholesterol, (2) triolein, (3) stearyl oleate, (4) cholesteryl oleate, and (5) squalene. (Reprinted from *Comp. Biochem. Physiol.*, 69B, 75, ©1981, Lindholm, J. S. et al., Variation of the skin surface lipid composition among mammals, with kind permission from Elsevier Science Ltd., The Boulevard, Langford Lane, Kidlington, OX5 1GB, UK.)

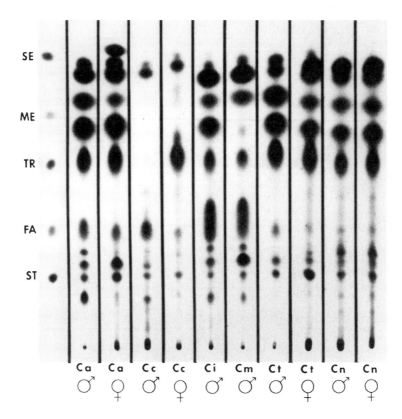

FIGURE 6.2 Chromatograms of the secretions of the paracloacal glands, paired exocrine organs embedded in the cloacal walls of crocodilians, showing species variation in secretion composition. Samples were obtained from the following crocodiles (*Crocodylus*): *C. acutus* (Ca), *C. cataphractus* (Cc), *C. intermedius* (Ci), *C. mindorensis* (Cm), *C. moreletii* (Ct), and *C. niloticus* (Cn); most are represented by samples from a male and a female. Extracts were applied to a plate developed successively with hexane, then toluene, and finally hexane–diethyl ether–acetic acid (40:10:1). The plate was sprayed with sulfuric acid and heated on a hot plate to char the lipids. Reference lipids are cholesterol (sterol, ST), oleic acid (free fatty acid, FA), triolein (triacylglycerol, TR), methyl oleate (methyl ester, ME), and cholesterol oleate (steryl ester, SE). (Reprinted from *Biochem. System. Ecol.*, 19, 133, ©1991, Weldon, P. J. and Tanner, M. J., Gular and paracloacal gland secretions of crocodilians: a comparative analysis by thin-layer chromatography, with kind permission from Elsevier Science Ltd., The Boulevard, Langford Lane, Kidlington, OX5 1GB, UK.)

pelage and recovered by scraping a glass beaker upward against the coat[58,59] or irrigated by pipet in an open glass cylinder pressed firmly against an area of exposed skin.[9,56] The irrigation method applied to humans yielded 50 to 350 μg of lipid per square centimeter of skin, depending upon the subject and anatomical site sampled.[36]

Extracts of the epidermis and epidermal derivatives may be obtained by immersing anesthetized or dead subjects in solvent, with care taken to avoid the mouth or cloacal/anal regions. The anus, feet, and other potential sources of contamination can be wiped with solvent-soaked cotton pads if it is not feasible to otherwise avoid exposing these areas during extraction.[56] Animals that lick themselves may be fitted with a conical collar to prevent contamination by saliva or food. Skin gland secretions ordinarily spread over the skin surface during grooming may be eliminated from epidermal extracts by excising the organs from which the secretions arise.[39]

Several materials have been applied to the skin of mammals, primarily humans, to absorb epidermal or glandular products. In one method, several layers of absorbent paper are tightly stacked onto the skin, removed after several hours, and extracted with diethyl ether.[60] Bentonite clay also has been used to collect skin secretions for TLC.[21,50,61,62] A film of an aqueous gel containing 15% bentonite clay and 0.2% carboxymethyl cellulose applied to the skin dries into an adherent, oil-absorbing layer within 10 min. After the required collecting time, the clay is removed and extracted with diethyl ether.

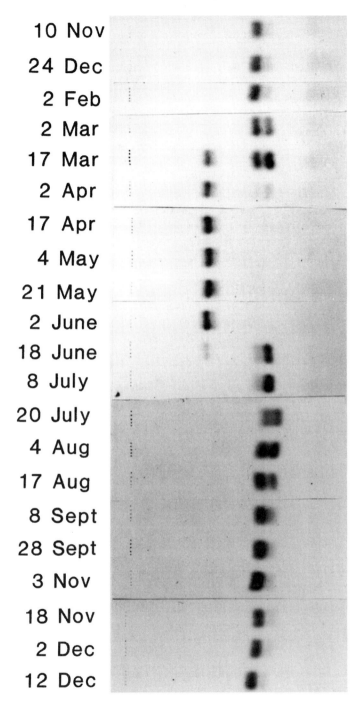

FIGURE 6.3 Chromatogram of the secretions of the uropygial gland secretions of female mallard ducks (*Anas platyrhynchos*), showing seasonal variation in secretion composition. Secretions were applied to a plate developed with hexane–diethyl ether–formic acid (90:10:1). The top doublet evident in samples obtained 10 November – 17 March and 18 June – 12 December corresponds to monoester waxes. The lower band evident in samples obtained 17 March – 2 June corresponds to diesters of 3-hydroxy acids. The origin is at the left. (From Kolattukudy, P. E., et al., Diesters of 3-hydroxy fatty acids produced by the uropygial glands of female mallards uniquely during the mating season, *J. Lipid Res.*, 28, 582, 1987. With permission.)

Epidermal samples can be obtained following necropsy by clipping away hair and placing a heated metal surface directly against the shaved skin[63] or by excising sections of skin and placing them epidermis-down for 1 to 2 min on a hot plate.[64] The epidermis is then removed in sheets by scraping with a spatula and drying by lyophilization before extraction; histological examination confirms that this procedure removes the epidermis without disturbing the underlying layers. Enzymes such as trypsin may be applied to sectioned pieces of skin to separate the stratum corneum from the unkeratinized portion of the epidermis.[4,63]

Hard keratinized tissues, such as hooves and horns, are washed and reduced to powder by filing with a coarse rasp before extraction.[1,65] Beaks and claws are freed of soft tissue and pulverized under liquid nitrogen.

The degradation of chemicals *in situ* resulting from the oxidation of native components and the activity of microorganisms is a persistent concern for investigators of vertebrate skin chemistry. The proportion of oxidative breakdown products can be reduced by prior removal of surface chemicals to ensure that only recently secreted substances are recovered.[54,55] Microorganisms cannot be eliminated entirely, but their contribution to skin extracts may be reduced (and assessed) by applying antibiotics to the organs sampled.[45-47]

B. EXOCRINE GLAND SECRETIONS

Secretions from macroscopic skin glands are obtained first by wiping the area around the gland duct openings with a towel or cotton swab, gently palpating the body of the gland, and directing the extruded fluids into vials placed near the duct opening. Where secretions occur in quantities too small to flow freely from glands, capillary tubes or solvent-extracted cotton swabs may be inserted directly into the glandular lumen or used to scrape residues from around duct openings. For example, secretions of the gular glands of crocodilians, paired organs located in skin folds on the lower jaw, were collected by manually everting each gland, compacting the dry residues surrounding the gland ducts into capillary tubes, and breaking off the tubes into vials with solvent.[18]

Secretions also can be collected postmortem by excising, homogenizing, and extracting glands. This method is not preferred because the entire gland, and not just the secretory products, are recovered. Nonetheless, some extracts have been prepared in this way. For example, the uropygial glands, bilobed organs of many birds situated on the rump at the base of the tail feathers, are excised (with care taken to avoid the surrounding, triacylglycerol-rich tissues) and homogenized to obtain samples for comparative studies.[66] These glands are voluminous in some birds, as in gulls and terns (Laridae), from which up to 600 mg of crude wax can be obtained by palpating the gland area.[67] In other taxa, such as songbirds (Passeriformes), maximum yields are in the range of 0.5 mg. The excision technique is adopted in preparing extracts of uropygial gland secretions because it can be applied to all birds that possess these organs, and thus it provides uniform samples for comparison.

Special methods are needed to collect materials from glands that are small, diffusely distributed, or otherwise difficult to access. Sebaceous glands of mammals can be isolated for secretion collection from sections of skin or punch biopsies by soaking the skin in 2 M $CaCl_2$ for several hours and then peeling off the loosened epidermis.[55,68] The sebaceous glands withdrawn along with pilosebaceous follicles are then harvested under a dissection scope. Otherwise, the glands may be forced to the surface by squeezing the surrounding tissue with forceps. Either method, however, may result in the contamination of skin samples by subcutaneous fat.

Secretions produced by amphibians in granular glands, small organs widely distributed over the integument, discharge their contents in response to a mild electrical stimulation of the skin[69] or following the injection of epinephrine.[70] Excised skins, fresh or dried, also can be extracted. Exudates of the parotoid glands, raised oblong organs in the nuchal region of toads (*Bufo* spp.), are obtained by firmly holding subjects in one hand and pressing a glass plate onto their dorsum with the other.[71] The glands are then pinched between the thumb and forefinger to extrude their milky secretions onto the plate, from which they are scraped after drying.

Most aquatic and semiaquatic turtles possess paired Rathke's or "musk" glands that open through pores on the shell bridge or on the margin of the shell in the axillary or inguinal areas. These glands are tightly applied to the inside of the plastron and thus they cannot be palpated manually. To collect Rathke's gland secretions, the shell bridge of a subject is dried and a mild electric stimulus is applied to its axillary or inguinal regions to elicit gland discharge.[72,73] The fluids exuded onto the shell surface are then taken up into capillary tubes or drawn by suction into a vial.

C. STORAGE AND PRESERVATION OF SKIN MATERIAL

Keratinaceous materials, including shed skins, feathers, and hair, are stored in glass tubes or wrapped in aluminum foil and kept frozen (–20°C). Samples from different species or individuals should be kept separate to avoid contamination.

Secretions from skin glands are routinely stored in vials with Teflon-lined screw caps. Acetone, hexane, chloroform–methanol (1:1 or 2:1), diethyl ether, or other solvents are added to prevent the loss of volatiles and impede the degradative activity of endogenous or microbial-derived enzymes. Alkaloid-bearing materials are stored in ethanol or methanol. Vials are flushed with nitrogen or other gases before being sealed and are kept frozen, particularly if they are stored without added solvents.

The addition of solvents and storage under nitrogen may forestall, but not entirely inhibit, sample degradation. Extracts of human hair, for example, were shown to contain one or more enzymes that generate free fatty acids from triacylglycerols.[74] This lipolytic activity is not inhibited by nitrogen, diethyl ether, or tetracycline.

III. SAMPLE PREPARATION

A. GENERAL CONSIDERATIONS

The preparation of samples of epidermal and skin gland secretions for TLC analyses entails methods widely used for other tissues. A primary consideration is the use of solvents appropriate to recover the compounds of interest.

B. DISSOLVING AND SIMPLE EXTRACTION PROCEDURES

A variety of solvents, including acetone, chloroform, chloroform–methanol (2:1), ethanol, and hexane have been used to extract lipids from the surface of live animals. As Nicolaides and Kellum[56] state, a major consideration for *in vivo* extraction involves selecting a solvent that recovers the array of substances present but does not damage the skin. A mixture of chloroform–methanol (2:1) recovers most nonpolar and polar compounds, but may irritate the skin. Nonpolar solvents, such as hexane, pentane, and petroleum ether, extract primarily nonpolar lipids, but fail to recover the bulk of the polar compounds. Moderately polar solvents, such as diethyl ether, extract most lipids without skin irritation and are widely used to recover human surface lipids. In recent studies of human skin lipids, 95% ethanol has been used.[28]

Hair, shed skins, and other keratinaceous materials for lipid extraction generally are immersed for 2 h each in three successive mixtures of chloroform–methanol (2:1, 1:1, and 1:2), the last of which is heated at 45°C. The extracts are combined and filtered, and the solvents are removed by rotary evaporation under reduced pressure. The concentrates are transferred in chloroform–methanol (1:1 or 2:1), evaporated to dryness under a stream of nitrogen, and desiccated *in vacuo*.

C. COMPLEX EXTRACTION PROCEDURES — PARTITION OF EXTRACTS

Solvent partitioning is performed to remove impurities and unwanted compounds from skin extracts prior to TLC analysis. Solvent partitioning to recover lipids is exemplified by the extraction of secretions from scent glands,[16] paired organs of snakes located in the base of the tail, using a modified method of Folch et al.[75]

Scent gland samples were obtained by manually compressing the base of the tail of snakes to induce the discharge of secretions from the two duct openings at the margin of the cloaca. The exudates were dissolved in a mixture of chloroform–methanol (2:1) at a concentration of 50 mg/ml. The mixture was homogenized and allowed to stand until the denatured proteins precipitated. The solution was then filtered into a separatory funnel. A volume of distilled water equal to one quarter that of the chloroform–methanol extract was acidified to pH 2.0 by adding concentrated HCl to the funnel. The funnel was shaken to mix the two phases and allowed to stand until the phases resolved. The aqueous upper layer was decanted and replaced with a fresh aliquot of Folch upper phase (chloroform–methanol–water, 36:576:564) that was acidified to pH 2.0. The funnel was shaken again and the phases were allowed to resolve. The upper layer was replaced, shaken, and resolved a third time, after which the bottom organic layer was drawn off, evaporated to near dryness under nitrogen, and transferred to a tared glass vial. The solvent was evaporated and the residue was dried *in vacuo* over anhydrous $CaCl_2$ before weighing. Lipid samples were then subjected to silicic acid column chromatography (see Section III.D).

Integral lipids, compounds that are covalently bound to epidermal derivatives such as hair, resist extensive extraction with solvents.[76,77] Integral lipid compounds, including cholesterol, cholesteryl sulfate, and fatty acids and alcohols, were obtained from 100 g of extracted, dried human hair saponified by heating for 2 h at 60°C in 200 ml of 1 M NaOH in 90% methanol.[76] The solution was cooled to room temperature and 88 ml of water and 360 ml of chloroform were added. The mixture was shaken and the phases resolved in a separatory funnel. The lower chloroform layer was transferred to a flask, and the upper phase and hair residue were acidified by adding 50 ml of 6 N HCl. The acidified upper phase was extracted with an additional 360 ml of chloroform, and the combined chloroform extracts were taken to dryness on a rotary evaporator. The dried residue was dissolved in 5 ml of chloroform–methanol (2:1) and washed with 1 ml of 2 M KCl containing 0.1 M HCl. The lower chloroform phase was filtered, and the filtrate was taken to dryness, weighed, and redissolved in chloroform–methanol before being applied to TLC plates.

A solvent partitioning procedure widely used to obtain alkaloids from the skins of amphibians has been refined over the years.[78] The method described here has been used primarily to obtain extracts for GC-MS analyses, but it is also recommended for TLC.[79] A skin sample is excised and weighed before being cut into smaller pieces and macerated in a mortar and pestle two or three times with separate portions of methanol (1 part skin/4–20 parts methanol). The combined methanol extracts are diluted with an equal volume of water. For large skin samples, the combined methanol extracts are first concentrated at 30°C *in vacuo*. The aqueous methanol is extracted three times each with 1 volume of chloroform. The combined chloroform layers are dried over Na_2SO_4 and concentrated to a volume of 1 to 10 ml. The concentrated chloroform is restored to the original volume with *n*-hexane, which is added to reduce the solubility of alkaloid hydrochlorides in the organic phase during partitioning with aqueous HCl. The *n*-hexane solution, which contains a small amount of chloroform, is extracted three times, each with a half volume of 0.1 N HCl. The combined 0.1 N HCl fractions are adjusted to pH 9.0 with 1 N NH_4OH, followed by three extractions each with a half volume of chloroform. The combined chloroform extracts are dried over anhydrous Na_2SO_4 and evaporated to dryness at 30°C *in vacuo* with a water aspirator. The resulting residue is dissolved in methanol, such that 1 µl of the alkaloid fraction corresponds to 1 mg of the original wet weight of the skin, and stored at –20°C in glass vials with Teflon-lined caps.

D. COLUMN CHROMATOGRAPHY

Column chromatography may be used to fractionate lipids into general compound classes prior to TLC separations. A simple column fractionation is one in which lipid extracts are separated into polar and nonpolar components. The epidermal extracts of lizards, for example, obtained by immersing shed skins in a series of chloroform–methanol solutions, were separated into polar and nonpolar fractions[17] using the method of Borgström.[80] The dried lipids were dissolved in chloroform and applied to a silicic acid column (approximately 3 g in a 30 ml cylindrical filter tube). The column was washed first with 75 ml of chloroform to elute the nonpolar lipids, and then with 75 ml of methanol to elute the polar lipids.

Lipids from the skin and scent gland secretions of snakes were fractionated by silicic acid column chromatography prior to TLC analysis[16] using the method of Hirsch and Ahrens.[81] Samples of 30 mg or more were separated on columns packed with silicic acid equal to 60 times the weight of extracted lipid and eluted with large volumes of solvent. Samples less than 20 mg were separated on columns constructed of clean pipets eluted with smaller eluate volumes. Hydrocarbons, wax esters, and cholesteryl esters (fraction I) were eluted with 100 ml of hexane, and other nonpolar lipids (fraction II) were eluted with 100 ml of diethyl ether; polar lipids (fraction III) were eluted with 100 ml of methanol. In subsequent fractionations, fraction I was eluted with 350 ml of hexane–diethyl ether (99:1) followed by 60 ml of diethyl ether; fraction II was eluted with 300 ml of diethyl ether; and fraction III was eluted with 200 ml of methanol. Each fraction was evaporated to dryness in a rotary evaporator. Fractions I and II were then transferred with two washes each of 1 ml of hexane; fraction III was transferred in 1 ml of methanol.

Polar lipids from cultures of human epidermal cells were fractionated by column chromatography prior to TLC analysis[4] using the method of Gray.[82] The column, solvent container, and fraction collecting tubes were wrapped in aluminum foil to protect solvents and lipids from light-catalyzed oxidation. The epidermal lipids were loaded onto a column of silica gel H in chloroform–methanol (49:1). A mixture of chloroform–methanol (49:1) was passed through the column to remove nonpolar lipids, after which it was washed from the column with tetrahydrofuran–benzene (1:4). A mixture of tetrahydrofuran–dimethoxymethane–methanol–water (10:6:4:1) was passed through the column to elute acidic phospholipids, glycosphingolipids, and finally phosphatidylethanolamine. The remaining phospholipids on the column, phosphatidylcholine and sphingomyelin, were eluted with chloroform–methanol (1:4).

IV. CHROMATOGRAPHIC SYSTEMS AND DETECTION OF COMPOUNDS

A. GENERAL CONSIDERATIONS

The purpose of this section is to describe the main chemical classes encountered in the skin secretions of vertebrates and the TLC sorbents and solvents frequently used to resolve them. Methods of visualizing TLC bands also are described (see Section VII). Plates for analytical TLC studies here generally were coated with a 0.25- or 0.50-mm layer of adsorbent; plates for preparative TLC were coated with a 0.50- or 1-mm layer. From 1 to 20 µl of lipid solutions, containing from 2 to 300 µg of lipids, have been developed on plates.

B. NONPOLAR LIPIDS

Nonpolar lipids in the skin secretions of vertebrates, including sterols, free fatty acids, alcohols, ketones, steryl esters, wax esters, triacylglycerols, and hydrocarbons, are separated routinely on silica gel G plates. Chromatograms may be developed with single solvents, such as benzene[83] or chloroform,[39] but usually solvent mixtures are used, including benzene–acetone (80:20),[53] petroleum ether–diethyl ether–acetic acid (70:30:1),[13] hexane–diethyl ether–ethyl acetate (125:5:3),[22] and hexane–diethyl ether–acetic acid (70:30:1 or 80:20:2).[16] Multiple development systems, with plates air-dried between solvent runs, provide high resolution of nonpolar compounds. Common multiple development systems involve successive full developments with hexane, then toluene or benzene, and finally full or partial development with hexane–diethyl ether–acetic acid (70:30:1 or 80:20:2).

For analytical TLC, plates are sprayed with 50% sulfuric acid or 0.5% phosphomolybdic acid in isopropanol and heated on a hot plate to 220°C. For preparative TLC, plates are sprayed with a 2% solution of 2′,7′-dichlorofluorescein in ethanol or with 8-hydroxy-1,3,6-pyrenetrisulfonic acid trisodium salt (10 mg/100 ml of 95% ethanol) and viewed under ultraviolet light.

The uropygial gland secretions of some birds have been separated by TLC with n-hexane–diethyl ether–formic acid (90:10:1),[20,40,41] but the following solvent systems are recommended to resolve particular components of these exudates[66]: isooctane to separate hydrocarbons and distinguish between paraffins, olefinic hydrocarbons, and aromatic hydrocarbons; carbon tetrachloride–chloroform (1:1) to separate cholesteryl esters, triacylglycerols, and mono-, di-, and triester waxes; chloroform to separate wax esters, triacylglycerols, alcohols, sterols, and diols; benzene–ethyl acetate (9:1) to separate free alcohols, free fatty acids, and more polar lipids; and light petroleum ether–diethyl ether (4:6) to separate *erythro*- and *threo*-diols resulting in the hydrolysis of diols, where silica gel plates impregnated with 5% boric acid (w/w) are used. Diols can be derivatized by treatment with acetone and toluene *p*-sulfonic acid for 20 min at room temperature and developed in less polar solvent systems, e.g., carbon tetrachloride–chloroform (1:1).

Bufogenins (bufadienolides) are C_{24} steroids from the parotoid glands of toads (*Bufo* spp.). These compounds have been separated by TLC with ethyl acetate, ethyl acetate–cyclohexane (80:20), ethyl acetate–acetone (90:10), and ethyl acetate saturated with water.[84] A variety of bufogenins and other sterols, including 3-hydroxy and 14-anhydro isomers, also were resolved by single or two-dimensional development with acetone–chloroform–cyclohexane (3:3:4), acetone–cyclohexane (3:7), *n*-hexane–ethyl acetate (1:9), methanol–ethyl acetate–cyclohexane (2:6:7), and diethyl ether–ethyl acetate (6:4).[14,85] Solvent systems for both column chromatography and TLC separations of bufogenins and other compounds from the parotoid glands include diethyl ether–chloroform (1:1 and 4:1), chloroform–methanol–water (80:20:2.5), and acetone–chloroform–methanol–water (6:2:1:1).[86] Bufogenins were visualized under ultraviolet light as bright fluorescing bands after plates were sprayed with antimony trichloride and heated; the colors of some bufogenins so treated are described.[14,85]

Equolides from the skin surface of horses (*Equus* spp.) are giant-ring lactones of ω-hydroxyacids ranging in carbon-chain length from 28 to 36.[58,59] Equolides from the donkey (*E. asinus*), zebra (*E. grevyi*), and onager (*E. hemionus*) consist exclusively of straight-chain ω-hydroxyacids, while those from the domestic horse (*E. caballus*) are entirely branched. For analytical TLC, equolides were applied to plates developed successively with hexane, toluene, and (half-way) with hexane–diethyl ether–acetic acid (70:30:1).[59] Plates were sprayed with sulfuric acid and heated on a hot plate. For preparative TLC, lipid extracts were applied to plates developed with hexane–benzene (1:1), and bands corresponding to lactones were visualized under ultraviolet light after spraying with 2′,7′-dichlorofluorescein. The bands were scraped from the plate, extracted with diethyl ether, and separated on a second TLC plate developed twice with carbon tetrachloride. Bands were again eluted with diethyl ether, and the lactone fraction was resolved into saturated and monounsaturated fractions on silica gel $H/AgNO_3$ plates before GLC analysis.

Squalene, a hydrocarbon precursor of cholesterol, has been isolated by TLC from the skin secretions of several mammals.[5,87] The hydrolyzed secretions of the anal glands of the sugar glider (*Petaurus breviceps*), a marsupial, were dissolved in petroleum ether and applied to silica gel G plates that were developed with hexane, or to silica gel F_{254} plates that were developed with petroleum ether–diethyl ether–acetic acid (90:10:1). Squalene appeared as an aqua blue spot after plates were sprayed with a mixture of vanillin–sulfuric acid and heated on a hot plate.

C. POLAR LIPIDS
1. Phosphatides

Phosphatides, including phosphatidylcholine, phosphatidylethanolamine, phosphatidylinositol, phosphatidylserine, and lysophosphatidylcholine, from the skin secretions of reptiles, birds, and mammals generally have been separated on silica gel H plates. Chloroform–methanol (2:1) extracts of various keratinized tissues of vertebrates, including the hair, hooves, and horns of mammals; the feathers and epidermal scales of birds; and the shed skins of snakes, were separated on plates developed first with diethyl ether–acetic acid (100:1) and then chloroform–methanol–water (40:10:1).[1] Bands consistent with phosphatidylcholine and phosphatidylethanolamine were visualized after plates were sprayed with sulfuric acid. Evidence of these and other phosphatides also has been observed in chromatograms of the epidermis or skin gland secretions of reptiles on silica gel H plates developed with chloroform–methanol–water (60:10:1 or 65:35:5)[2,17,26,72,73,88] or chloroform–2-propanol–triethylamine–methanol–0.25% KCl (30:20:18:9:6),[16] or on silica gel HR plates developed first with acetone–hexane (1:1) and then with chloroform–methanol–acetic acid–water (25:15:4:2 [89] or 50:25:7:3 [15]).

Following column chromatography (see Section III.D), phospholipids (and glycosphingolipids) from extracted human epidermal cell cultures were separated and characterized by two-dimensional TLC.[4] Approximately 3 µl samples were applied to a corner of a 20 × 20-cm plate, 1 cm from the edge. Plates were then developed with chloroform–methanol–water (65:25:4) in the first direction and tetrahydrofuran–dimethoxymethane–methanol–4 M NH$_4$OH (10:6:4:1) in the second direction. For analytical TLC, plates were sprayed with sulfuric acid and heated. For preparative TLC, bands were visualized with iodine vapor, or with rhodamine 6G for fatty acid analysis, and eluted with chloroform–methanol–water (10:10:1). The components were then deacylated with methanolic NaOH and subjected to paper chromatography or esterified for gas-liquid chromatography (GLC).

Preparative TLC of phospholipids from human sebaceous glands was performed on plates developed with chloroform–methanol–acetic acid–water (25:14:1:1).[90] Plates were then sprayed with 2′,7′-dichlorofluorescein and bands were eluted. The recovered phosphatidylcholine was saponified and its substituent fatty acids were analyzed by GC.

2. Ceramides and Glucosylceramides

Ceramides and glucosylceramides are major polar lipids in the epidermis of reptiles, birds, and mammals. Silica gel 60 H plates for the separation of ceramides from the skin of mammals were predeveloped with chloroform–methanol (2:1)[91-94] or chloroform–methanol–acetic acid (200:100:3).[95] For analytical TLC, human stratum corneum lipids in chloroform–methanol (2:1)[91] were applied to plates developed first with chloroform–methanol–water (40:10:1) to resolve cholesteryl sulfate from ceramides. Ceramides were then resolved by one or more developments with chloroform–methanol–acetic acid (190:9:1)[91,94] followed by one development with diethyl ether–acetic acid (100:1).[94] Plates were sprayed with 50% sulfuric acid and heated on a hot plate to visualize bands before ceramides were quantified by photodensitometry.[95]

For preparative TLC of ceramides, plates were developed first with diethyl ether–acetic acid (99:1) to move nonpolar lipids to the top of the plate,[91,94,95] and then double-developed with chloroform–methanol–acetic acid (190:9:1).[91,95] In an alternative method, the silica gel at and below R_f 0.5 of the plate initially developed with diethyl ether–acetic acid was scraped away, eluted with chloroform–methanol–water (50:50:1), and applied to another plate developed twice with chloroform–methanol (19:1).[94] Plates were sprayed with 1′,7′-dichlorofluorescein in ethanol, dried, and viewed under ultraviolet light. The fluorescent bands were scraped from the plates and the lipids were eluted with chloroform–methanol (2:1)[94,95] or chloroform–methanol–water (50:50:1).[91] The lipids were redissolved in chloroform–methanol and the 1′,7′-dichlorofluorescein was removed by passage of the solution through a column of magnesium hydroxide in a pipet. The eluates were then reexamined by TLC.

Ceramides containing phytosphingosine were analyzed on plates with an adsorbent containing sodium arsenite, as described by Karlsson and Pascher.[96] Aliquots of isolated ceramide fractions were applied to separate lanes of plates developed in chloroform–methanol (95:5).[95] To identify ester linkages, each ceramide fraction in chloroform–methanol (2:1) was subjected to mild alkaline hydrolysis by heating with an equal volume of 1 N NaOH in methanol–water (19:1) at 45°C for 1 h. An aliquot of the resulting solution was then analyzed by TLC. To identify amide-linked acids, each ceramide fraction was hydrolyzed at 80°C in 1 ml of a mixture prepared from 8.6 ml of concentrated HCl and 9.6 ml of water diluted to 100 ml with methanol. After 18 h, the solution was evaporated to dryness under a stream of nitrogen, and the residue was treated with 1 ml of 10% BCl$_3$ in methanol at 50°C for 10 min to convert the fatty acids into their methyl ester derivatives. The solution was again evaporated to dryness and the residue was dissolved in chloroform–methanol before analysis by TLC.

The ethylenic bonds of ceramides were hydrogenated by passing hydrogen through ceramide solutions in diethyl ether–methanol (2:1) at 20°C for 20 min, using Pt as a catalyst. Ceramides and sphingosine were acetylated in acetic anhydride–pyridine at 45°C for 2 h. The reagents were then evaporated under a stream of nitrogen, and the residual product was purified on TLC plates developed with chloroform–methanol (100:2).

Glucosylceramides were separated by TLC using sorbents and solvent systems similar to those used for ceramides. Extracts were applied to plates developed first with diethyl ether[97] or diethyl ether–acetic acid (99:1)[1] and then with chloroform–methanol–water (40:10:1). For analytical TLC, plates were sprayed with sulfuric acid and heated on a hot plate. Glucosylceramides (and other glycolipids) were indicated by the red-violet color they exhibit during charring.[97] For preparative TLC, plates were sprayed with 2′,7′-dichlorofluorescein and the visualized bands were scraped from plates and eluted with chloroform–methanol. The recovered acylglucosylceramides were saponified according to the methods of Gray et al.,[98] and the resulting free fatty acid and glucosylceramide products were separated by TLC with chloroform–methanol–water (40:10:1). The saponified glycolipid was treated with 10% BCl_3 in methanol at 55°C for 18 h to liberate the ω-hydroxyacid methyl esters, which were purified on plates developed with hexane–diethyl ether–acetic acid (30:70:1). The hydroxy methyl esters were acetylated by treatment with acetic anhydride–pyridine (1:1) at room temperature for 1 h before being subjected to GLC.

3. Sphingosines

Sphingosines and phytosphingosines, intermediates in the synthesis of ceramides and glycosylceramides, have been resolved by TLC from the epidermis of mammals. To quantify sphingoid bases from the epidermis of the pig (*Sus scrofa*),[63] chloroform–methanol (2:1) extracts were applied to silica gel G plates developed with chloroform–methanol–2 M NH_4OH (40:10:1). Plates were air-dried for approximately 30 min at room temperature, and then heated at 100°C for 1 h to remove ammonia. After cooling, plates were sprayed with 0.2% ninhydrin in water-saturated *n*-butanol and air-dried for 20 min. Plates were then placed silica-side down over a steaming hot water bath for 5 min, and then silica-side up for 30 min, after which they were cooled for 1 h and scanned with a photodensitometer (see Section V) at a wavelength of 520 nm.

2,4-Dinitrophenyl derivatives of sphingosines for analyses of long-chain bases were prepared from lipid extracts[63] or fractions of them isolated on silica gel H plates developed with chloroform–methanol–2 M NH_4OH (40:10:1).[100] Lipids were dissolved in chloroform–methanol (2:1) and treated with 2,4-dinitrofluorobenzene in methanol. Potassium borate buffer at pH 10.5 was added and the mixture was heated at 60°C for 30 min. After shaking and centrifuging, the lower chloroform phase containing the dinitrophenyl derivatives was drawn off, dried under nitrogen, and redissolved in chloroform–methanol (2:1). The dinitrosphingosine products were purified on TLC plates developed with chloroform–hexane–methanol (5:5:2). The purified material was oxidized with lead tetraacetate in benzene at 60°C for 90 min, and the resulting aldehyde products were extracted with hexane and analyzed by GLC.

4. Steryl Glycosides

Steryl glycosides, unusual glycolipids hypothesized to play a primitive role in the epidermal barrier, have been observed in the epidermis of reptiles and birds. Extracts of the epidermis of the chicken (*Gallus domesticus*) in chloroform–methanol were separated by analytical TLC on silica gel G plates developed twice with chloroform–methanol–acetic acid (190:9:1) to separate acylglucosylsterols and ceramides from other lipids,[39] and with chloroform–methanol–water (40:10:1) to separate glucosylsterol and glucosylceramides from other lipids. Plates were then sprayed with sulfuric acid and heated to visualize the lipids.

For preparative TLC, extracts were applied to silica gel H plates developed with the solvent systems described above for the analytical studies. The acylglucosylsterols and ceramides in the fraction obtained after the initial development with chloroform–methanol–acetic acid (190:9:1) were separated on plates with chloroform–methanol–water (40:10:1). Acylglucosylsterols were saponified by treatment with chloroform–methanol–10 M aqueous NaOH (2:7:1) at 60°C for 1 h. The reaction mixture was acidified to pH 4 with 2 M HCl and the products were extracted with chloroform. Analytical TLC revealed two bands corresponding to free fatty acids and glucosylsterols. These bands were separated by preparative TLC with chloroform–methanol–water (40:10:1), and the recovered fatty acids were converted to methyl esters and analyzed by GLC.

The glucosylsterols were hydrolyzed with 1 M HCl in methanol containing 20 M water. After 18 h at 65°C, the reaction mixture was partitioned between chloroform and water. The chloroform was transferred to a clean tube and dried under nitrogen. Analytical TLC revealed one product that had a mobility that matched that of cholesterol. The free sterols were acetylated by treatment with acetic anhydride–pyridine (1:1) for 1 h at room temperature, and the products were analyzed by GLC and electron impact MS.

A TLC survey of snake skin lipids developed on silica gel G plates with chloroform–methanol–water (60:10:1) revealed, in all species, appreciable quantities of compounds that migrate similarly to ceramides.[2] These components were hypothesized to be glycolipids because they exhibited a red-violet hue during charring. A glucosylsterol and an acylglucosylsterol subsequently were isolated by TLC from the shed skins of the bull snake (*Pituophis melanoleucus*).[99] Extracts in chloroform–methanol (2:1) were applied to plates coated with silica gel 60 H and developed in chloroform–methanol–acetic acid (190:9:1). Plates were sprayed with an ethanolic solution of 8-hydroxy-1,3,6-pyrenetrisulfonic acid trisodium salt and viewed under ultraviolet light. Two glycolipid bands (R_f values 0.2 and 0.05) were scraped from the plate and the lipids were eluted with chloroform–methanol–water (50:50:1). These glycolipid fractions were acetylated and rechromatographed using hexane–ether–acetic acid (30:70:1) to examine their purity.

Each glycolipid was hydrolyzed, dried under a stream of nitrogen, and partitioned between chloroform and water, as described above for the preparation of components from the chicken. Sterols, free fatty acids, and fatty acid methyl esters were isolated by TLC and analyzed by GLC following derivatization. The structures of a glucosylsterol, containing glucose and cholesterol, and an acylglucosylsterol, containing glucose and C_{14}–C_{28} esterified fatty acids, were determined by GLC and nuclear magnetic resonance.

5. Sulfolipids

Cholesteryl sulfate has been detected in the epidermis and/or epidermal derivatives of the chicken[39] and several mammals.[4,65,76,77,91] Chloroform–methanol (1:1) extracts of the hoof of the horse (*E. caballus*) were dissolved in chloroform–methanol (1:1) and applied to silica gel 60 H plates developed with chloroform–methanol–water (40:10:1).[65] A band with a mobility matching that of cholesteryl sulfate was observed under ultraviolet light after spraying with 8-hydroxy-1,3,6-pyrenetrisulfonic acid sodium salt. This component was eluted from the silica gel with chloroform–methanol (1:1).

Cholesteryl sulfate was tentatively identified as a component of human epidermis by cochromatography of the skin-derived component with an authentic standard on silica gel H plates developed separately with the following solvent systems: chloroform–methanol–water (24:7:1 and 40:10:1), chloroform–methanol–water–acetic acid (40:10:1:1 and 40:20:1:1), and chloroform–methanol–3.75 M NH_3 (50:25:4).[91] The TLC band suspected of containing cholesteryl sulfate and that containing the authentic compound exhibited the same red color after plates were sprayed with sulfuric acid and heated.

D. ALKALOIDS

A diversity of alkaloids occurs in the skin of amphibians, and at least one such compound — homobatrachotoxin — has been isolated from the feathers and skin of birds (*Pitohui* spp.).[101] Following the solvent partitioning procedure described above (see Section III.C), alkaloids were separated by TLC by applying 10 μl samples in methanol, equivalent to the amount in 10 mg of the wet weight of skin, to silica gel GF plates developed in chloroform–methanol (9:1).[7,35] After drying, the resolved alkaloids were visualized as yellow-brown spots by exposing plates to iodine vapors. Alkaloids also may be visualized by spraying plates with a modified Ehrlich's reagent (0.1%

p-dimethylaminocinnamaldehyde in 0.1 N HCl).[101] The R_f values of some compounds isolated from the skins of frogs are given by Daly et al.[3,78]

For preparative TLC, plates are streaked with 100 to 200 μl of methanol extracts, developed in chloroform–methanol (9:1), and dried. Most of the adsorbent layer is covered tightly with a glass plate before being exposed to iodine vapor; thus alkaloid bands are visualized only on the uncovered portion of the layer. The silica gel corresponding to iodine-positive regions is scraped from the covered (unreacted) portion of the plate and placed into small columns eluted with a few milliliters of ethyl acetate or chloroform–methanol (1:1). Where samples are analyzed by MS, the solvent is concentrated and part of the isolated alkaloid fraction is checked for purity by analytical TLC.

Tetrodotoxin is a neurotoxic alkaloid found in the skin of newts (*Taricha* spp.), among other vertebrates. Methanol extracts of newt tissues were applied to silica gel G plates and developed separately in the following solvent systems: ethanol, ethanol–acetic acid (96:4), *n*-butanol–acetic acid–water (50:3:10), collidine–water (50:20 and 70:30), and phenol–ammonium–water (13:1.8:22).[102,103] Plates were sprayed with alcoholic KOH, heated at 130°C for 10 min, and viewed under ultraviolet light.

V. QUANTIFICATION

The results of TLC analyses can be quantified by a variety of methods (see Touchstone[104]). *In situ* photodensitometry of charred TLC plates is the most popular quantitative technique in studies of vertebrate skin chemistry.[55,105–108] In this procedure, a chromatogram is passed under a light beam of rectangular cross-section, and the proportion of light transmitted by each visualized spot is detected by a photocell and recorded on a strip chart. Photodensitometry in comparative analyses of vertebrate skin lipids has been used to measure the relative or absolute amounts of native components resolved from reptiles, including a lizard (*Iguana iguana*)[15] and some snakes,[15,109] and from mammals, including a mole (*Scalopus aquaticus*),[110] rodents,[5,111,112] ungulates,[58,59,113] carnivores,[5,6,114] and a monkey (*Macaca fascicularis*).[115] Photodensitometry allows for quantitative comparisons between chromatograms, and thus it has been used to demonstrate interspecific differences in the abundances of some skin components (Figure 6.4).

The materials and methods for the photodensitometry described here are adapted primarily from Downing[105–108] and Downing and Stewart.[55] TLC plates are coated with a 0.25-mm layer of silica gel G or H and dried by heating at 120°C for 2 h immediately after setting. Plates are cooled and then predeveloped in diethyl ether or chloroform–methanol (2:1) to wash contaminants to the top. After drying, vertical 5- to 7-mm lanes are scored on the plates with the tip of a dissecting needle; the development of chromatograms on scored lanes prevents spots from expanding beyond the width of the densitometer light beam. A line also is scored at the top of the plate to restrict the ascent of the solvent. The need to score plates disqualifies from use some commercially prepared plates carrying a hard layer of adsorbent because they may crack. Plates impregnated with a binding material or packaged in some paper or plastics also may be unusable because organic compounds incorporated or transferred onto the adsorbent may interfere with visualization.

Plates are activated by heating at 130°C for 30 min, and 1 to 5 μl of extract samples, containing 2 to 10 μg of lipids, are applied to them; the amount of material per zone should not exceed 20 μg of material to avoid distortion of spots. Accuracy in sample preparation is not required when determining the percent composition of resolved spots or the absolute concentrations of components where an internal standard of known concentration is used. In the absence of an internal standard, absolute quantification requires that sample solutions be prepared in precise concentrations, and that steps be taken to prevent solvent evaporation, e.g., the use of high-boiling solvents such as toluene.[21] Any compound that can be resolved by TLC from the constituents of a sample mixture can be used as an internal standard.

Ideally, an extract should be resolved on one chromatogram, with each band of interest sufficiently distinct for quantification by photodensitometry. For most skin extracts, however, this is not feasible and separate chromatograms are required for nonpolar lipids, phospholipids, and lipids of

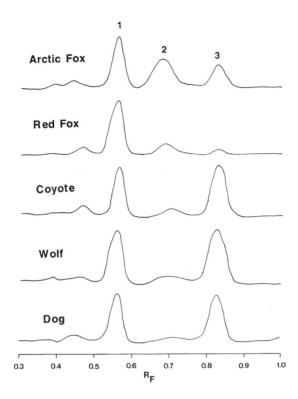

FIGURE 6.4 Photodensitometic recordings of surface lipid chromatograms shown in Figure 6.1 of five canids, showing species variation in (1) diesters of long-chain 1,2-diols containing one long-chain fatty acid and one short-chain acid, (2) diesters of long-chain 1,2-diols containing two long-chain acids, and (3) cholesteryl esters. (Reprinted from *Comp. Biochem. Physiol.*, 698, 75, ©1981, Lindholm, J. S. et al., Variation of the skin surface lipid composition among mammals, with kind permission from Elsevier Science Ltd., The Boulevard, Langford Lane, Kidlington, OX5 1GB, UK.)

polarities between those of cholesterol and phospholipids. Two-dimensional development of chromatograms is useful in examining the complexity of mixtures, particularly those containing phospholipids (see Section IV.C), but it is usually unsuitable for photodensitometric quantification.

To prevent the distortion of spots and to permit accurate quantification, filter paper soaked in solvent should line the developing tank and extend above the height of the TLC plate. In addition, the temperature of the plate and developing tank should be equal at the outset of and during development. Plates should be removed and allowed to air-dry for at least 10 min before being developed in a successive solvent system or sprayed evenly with a fine mist of visualizing agent, usually 50% sulfuric acid. Plates are then placed flush on a hot plate and heated gradually to a maximum temperature of 220°C to char the lipids.

The charred plate is then scanned with a photodensitometer with a 1 × 4-mm incident light beam. The dimensions of the scored lanes allow them to be aligned on the stage so that the beam of light passes symmetrically through each charred spot and scans each lane in a single uninterrupted run. Peaks on the strip chart are measured and the amounts of different components are then calculated with correction formula to account for the varying proportions of carbon in different lipids[55,105] and for the deviation of the optical density from the Lambert-Beer relation.[55,107,108]

Measurements of the rates at which lipids are secreted by the skin rely upon accurate methods of quantification by photodensitometry. In these investigations, the time course for ^{14}C from labeled substrates to appear in skin compounds is monitored by sampling skin secretions at intervals and analyzing them by a combination of TLC and autoradiography.[48–52]

The time course of sebum secretion in the guinea pig (*Cavia porcellus*) was measured following intradermal injections of [1–^{14}C]acetate in saline.[50] Lipids were collected daily over 12 days by immersing the injected sites of the body into solvents, absorbing surface lipids onto bentonite clay, and extracting skin biopsies. Aliquots of skin extracts, known amounts of reference lipids, and standard lipids containing known amounts of ^{14}C, e.g., methyl [1–^{14}C] palmitate, were applied to separate lanes of silica gel G plates developed successively with hexane, hexane–toluene (1:1), and hexane–diethyl ether–acetic acid (70:30:1). X-ray film was placed into a film holder with the TLC plates for 14 days and then developed. The plates were sprayed with 50% sulfuric acid and heated to 220°C on a hot plate. Both the charred plates and autoradiogram were scanned with a recording photodensitometer to quantify the lipids and radioactivity in each band. The peak areas for the charred bands from the standard lipids and for the autoradiogram bands produced by ^{14}C-labeled reference compounds were used to determine the radioactivity associated with the resolved lipids.

Radioactivity of steryl esters and wax diesters on the skin surface of *C. porcellus* exhibited a peak at 5 to 6 days, thus indicating the time elapsed between the synthesis and the secretion of these lipids (Figure 6.5). The secretion curves for radioactive cholesterol, free fatty alcohols, and triacylglycerols exhibited a similar time course, but the radioactivity emitted from each of these compounds was only about 20% of that from the steryl esters and wax diesters. This suggests that large proportions of sterols, free alcohols, and triacylglycerols are derived from lipids that were not synthesized at the injection sites. Glycerol ether diesters accounted for 30% of the recovered lipids, but radioactivity was not detected from them, suggesting that they are not synthesized in the skin.

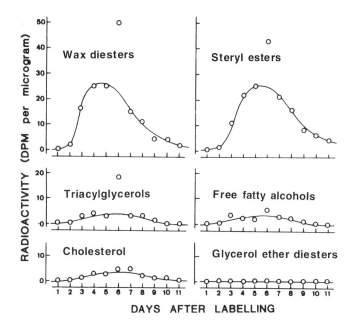

FIGURE 6.5 Secretion curves of skin lipid fractions from two guinea pigs (*Cavia porcellus*), expressed in decays per minute (DPM), showing the time course for the secretion of wax diesters, steryl esters, triacylglycerols, free fatty alcohols, cholesterol, and glycerol ether diesters. Secretions were collected daily over 11 days following intradermal injection of [1–^{14}C]acetate. The secretion curves were derived from photodensitometric scans of charred TLC plates and autoradiograms. (Reprinted from *Comp. Biochem. Physiol.*, 80B, 431, ©1981, Gaul, B. L. et al., The time course of sebum secretion in the guinea-pig, with kind permission from Elsevier Science Ltd., The Boulevard, Langford Lane, Kidlington, OX5 1GB, UK.)

VI. TLC ANALYSES IN VERTEBRATE SYSTEMATICS

TLC analyses often provide evidence of interspecific variation in skin products, but rigorous comparisons of chromatograms to investigate taxonomic relationships are seldom undertaken. Tolson's[16] phylogenetic study of boid snakes of the genus *Epicrates* (tree boas) from the Lesser

Antilles incorporated TLC analyses of epidermal and scent gland lipids, along with conventional morphological characters.

Polar and nonpolar lipids from the shed skins and scent gland secretions of *Epicrates* spp. were separated by column chromatography (see Section III.D) and applied to Whatman LQ6DF and LQ5DF plates, respectively. Hydrocarbons, wax esters, and steryl esters were developed with carbon tetrachloride–chloroform (96:4), and other nonpolar lipids were developed with hexane–ether–acetic acid (80:20:2). The polar lipids were developed with chloroform–2-propanol–triethylamine–methanol–0.25% KCl (30:20:18:9:6). Polar and nonpolar epidermal and scent gland lipids were compared among Antillean and continental congeners and to those of other boid genera, including a related genus, *Corallus*. Continental *Epicrates* and other boid genera displayed relatively few scent gland components, while some Antillean *Epicrates* displayed more than 20 components; this suggests the biosynthesis of new or different scent gland lipids as these snakes evolved on the islands.

Differences among *Epicates* spp. in lipid composition were reflected by the presence or absence of bands at particular R_f values in chromatograms developed with three solvent systems (Figure 6.6). Twenty-four lipid bands in the epidermal and scent gland secretions resolved with the total of the three developing solvent systems were used as character states. Bands were assigned a value of 0 if considered primitive, and 1 if derived. Highly variable characters were omitted from the analysis.

The assembled data, including both TLC and morphological characters, were analyzed with a computerized Wagner tree algorithm[116] to generate a cladogram. The method chosen accepts ordered character states, produces a best fit to all data, makes no assumptions about evolutionary rates, and allows for character state reversals. Seven of nine monophyletic groups of *Epicrates* were diagnosed by both lipid and morphological characters, with some scent gland components indicated as shared derived characters. Epidermal lipids, on the other hand, were similar in all species studied, a possible reflection of the common function of these compounds in retarding transepidermal water loss. Four lipid characters from among both epidermal and scent gland components appeared to have evolved by convergence or parallelism; similar results were not observed with respect to any morphological character. Two of the lipid characters (one epidermal and one scent gland component) were hypothesized reversals; two others (one epidermal and one scent gland component) were hypothesized parallelisms.

VII. VALIDATION OF RESULTS

Validation of TLC analyses of vertebrate skin secretions is needed primarily to characterize compounds indicated as visualized bands, and to distinguish between native constituents and contaminants.

General compound classes are tentatively identified by comparing the R_f values of skin components to those of authentic standards. These initial results should be interpreted cautiously due to the potential comigration of different compounds. For example, steryl esters, which are widely documented in the skin secretions of vertebrates, have been reported from TLC analyses of the shed skin extracts of snakes representing the Acrochordidae, Boidae, Colubridae, Elapidae, and Viperidae.[2,15,89] Tolson,[16] however, failed to confirm steryl esters in the epidermis of several genera and numerous species of boid snakes, primarily *Epicrates* spp., even after applying large quantities (800 μg) of skin extracts to plates. He observed that a wax ester standard (palmitoyl arachidate) comigrated with a steryl ester (cholesteryl palmitate) on plates developed with hexane–diethyl ether–acetic acid (80:20:2), thus raising the possibility that these compounds had been confused in other studies using the same or similar developing solvent systems.

Wax esters and steryl esters are notoriously difficult to separate by TLC. To distinguish between these compounds, specially devised methods and materials, such as TLC plates coated with an adsorbent layer containing magnesium hydroxide, are required.[11,117–119] The presence of steryl esters in the epidermal secretions of snakes may be explored further with these techniques. The identifi-

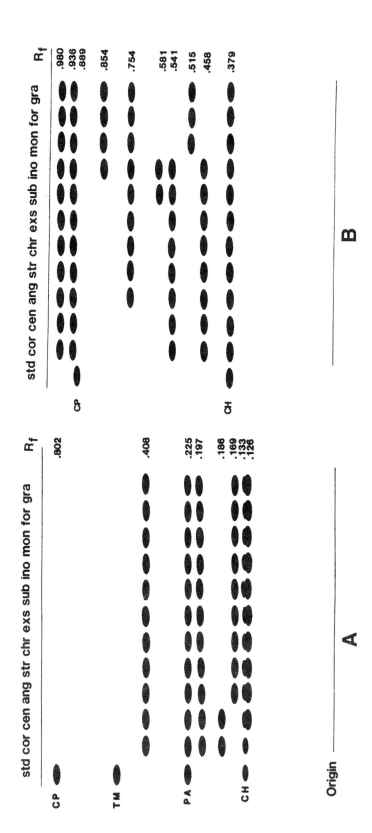

FIGURE 6.6 Schematic representation contrasting the chromatograms of the nonpolar lipids from (A) the epidermis and (B) scent gland secretions of the following boid snakes: *Corallus* sp., *Epicrates cenchria* (cen), *E. angulifer* (ang), *E. striatus* (str), *E. chrysogaster* (chr), *E. exsul* (exs), *E. subflavus* (sub), *E. inornatus* (ino), *E. monensis* (mon), *E. fordii* (for), and *E. gracilis* (gra). Extracts were applied to plates developed with hexane–diethyl ether–acetic acid (80:20:2). Plates were sprayed with phosphomolybdic acid in isopropanol to visualize the lipids. The standards (std) indicated in left lanes of the chromatograms are cholesterol (CH), palmitic acid (PA), trimyristin (TM), and cholesterol palmitate (CP). The R_f values of the observed bands are shown on the right of each chromatogram. (From Tolson, P. J., Phylogenetics of the boid snake genus *Epicrates* and Caribbean vicariance theory. *Occ. Pap. Mus. Zool. Univ. Michigan*, 715, 1, 1987. With permission.)

cation by GC-MS of mono- and diester waxes from skin extracts of the Indian python (*Python molurus*) corroborates TLC evidence for these compounds, at least, in the Boidae.[57]

Inference from TLC on the identity of compounds is strengthened when skin components and standards comigrate in more than one solvent system, and additionally so when they exhibit the same color during charring. A substance previously known as tarichatoxin from the skin of newts (*Taricha* spp.), for example, was demonstrated to be identical to a substance now known as tetrodotoxin from marine puffer fishes (Tetraodontoidae), in part, as a result of the comigration of TLC components from these taxa in several solvent systems.[103] Cholesteryl sulfate also was characterized as a component of the stratum corneum of humans due to the comigration of skin-derived and authentic compounds in five polar solvent systems and the display of the same color during charring.[91]

A variety of spray reagents may be used to confirm the identity of compounds resolved by TLC. Civetone, a component of the anal gland secretions of civet cats (*Civetta* spp.), appears as a yellow-green spot in visible light after spraying with 1 ml of 10% methanolic solution of phosphomolybdic acid and 49 ml of *p*-anisaldehyde–acetic acid–sulfuric acid–methanol (1:10:20:200) and heating at 105°C for 5 min.[120] Bufogenins exhibit an array of colors during heating after being sprayed with antimony trichloride in chloroform (20 g /100 ml).[14,85] A summary of reagents used in a TLC analysis of snake epidermal and scent gland lipids,[16] and the compounds indicated, are shown in Table 6.1.

TABLE 6.1
Detection Reagents and Lipids Indicated in a TLC Analysis of Epidermal and Scent Gland Secretions of Boid Snakes, Primarily *Epicrates* spp.[16]

Detection Reagent	Lipids Detected
20% Sulfuric acid in acetic anhydride	Cholesterol and cholesteryl esters
0.5% Phosphomolybdic acid in isopropanol	Neutral lipids
0.1% Bromcresol green in 99.5% sulfuric acid	Free fatty acids
Saturated potassium dichromate in 50% sulfuric acid	Polar lipids
0.2% Ninhydrin in *n*-butanol–10% acetic acid (95:5)	Aminophosphatides
2 mg/ml Orcinol in sulfuric acid–water (3:1)	Glycolipids
Iodine vapors	Unsaturated lipids
0.05% Rhodamine B in 96% ethanol	Glycerides

Note: Lipids were extracted, subjected to solvent partitioning (see Sections 2.III.C and 2.III.D), and chromatographed (see Section 2.VI) as described.

TLC components also may be characterized by derivatizing both eluants and authentic compounds and then comparing the migratory behavior of their products in follow-up TLC analyses. This method was used to demonstrate giant-ring lactones (equolides) in the skin surface extracts of the domestic horse (*E. caballus*).[58] A TLC band from the horse chromatogram exhibiting a mobility similar to that of wax diesters was hydrolyzed with 5% KOH in methanol–water–toluene (95:5:50) at 50°C. The recovered lipid band, which was presumed to contain ω-hydroxy acids, and authentic ω-hydroxy acids derived from carnauba wax exhibited identical TLC mobilities following (i) methylation with BCl_3–methanol, (ii) oxidation with chromic acid in acetone and conversion to dimethyl esters with BCl_3–methanol, and (iii) cyclization by refluxing in benzene with *p*-toluene sulfonic acid as catalyst.

The results of GC-MS and other spectral analyses may provide stronger evidence for interpretations of TLC results, particularly where novel skin compounds are involved. The skin surface extracts of a number of mammals, for example, displayed bands consistent with two types of diester waxes.[11,121] Type I diester waxes are hydroxy acids esterified with one fatty acid and one fatty alcohol; type II diester waxes are alkane-1,2-diols esterified with two fatty acids. The mouse (*Mus musculus*) and several other mammals displayed two bands, each thought to represent different

Type II waxes; the slower mobility of one band was attributed to the greater abundance of substituent unsaturated fatty acids and branched-chain alkanediols. The isolation of this component from the skin surface of hairless mice by high-performance TLC and its analysis by GC-MS revealed alkyldiacylglycerols,[122] unusual compounds for vertebrate skin products.

The presence of certain components in TLC studies occasionally has suggested the contamination or degradation of skin extracts. Small amounts of methyl esters, for example, have been observed in some exudates, as in the Rathke's gland secretions of turtles,[72,73] but where these compounds abound in methanol extracts, they may be artifacts.[123]

Epidermal hydrocarbons observed by TLC in extracts of reptiles[17,109] and mammals[37,51,52,124-126] are regarded by some investigators as endogenous components of the skin, and by others as contaminants from the environment, diet, or skin gland secretions. Thus, in one study, odd- and even-chained C_{18}–C_{35} n-alkanes isolated by TLC from human skin were deemed bona fide epidermal products since they were detected in lower concentrations in the outer regions of the stratum corneum than in preparations of the entire stratum corneum, and they consisted of different compounds than would be expected from the environment or than would likely be appropriated from the diet.[125] Another study, however, demonstrated that alkanes comprise only 0.5 to 1.7% of the total human skin lipids, and that not more than 2% of these compounds could have been produced by recent biosynthesis, as determined by ^{14}C-dating.[126] Moreover, alkanes were not detected in most samples of cerumen (ear wax), suggesting that such compounds present on the skin are acquired by direct contamination rather than by a systemic route.

VIII. SELECTED PROTOCOLS

A. GENERAL CONSIDERATIONS

The following protocols for the collection and TLC separation of epidermal and skin gland compounds are described for selected vertebrate taxa and skin compounds.

B. POLAR AND NONPOLAR LIPIDS FROM THE EPIDERMIS OF LIZARDS AND SNAKES (SQUAMATA, REPTILIA)*

In these studies, analytical TLC was used to examine polar and nonpolar lipids from the epidermis of lizards and snakes.

Sample preparation. Naturally shed skins of 23 species of lizards and 24 species of snakes were wrapped in aluminum foil and kept frozen. Skins were extracted by immersion in three successive mixtures of chloroform–methanol (2:1, 1:1, and 1:2) for 2 h each, the last of which was heated to 45°C. The extracts were combined and concentrated by rotary evaporation at 40°C and evaporated to dryness under a stream of nitrogen and then *in vacuo* before weighing. The residues of the snake skin extracts were redissolved in chloroform–methanol (1:1) to a concentration of 40 µg/ml and applied to TLC plates. Lizard skin residues were redissolved in chloroform and applied to a silicic acid column (approximately 3 g in a 30 ml cylindrical tube) to separate polar from nonpolar lipids, as described above (see Section III.D), and the separate eluents were redissolved in chloroform.

Chromatography. Aliquots of skin extracts 40 µg (1 µl for nonpolar lipids and 4 µl for polar lipids) were applied to 6 mm lanes on 20 × 20 cm glass plates coated with a 0.25-mm layer of silica gel G. Mixtures of lipid standards were applied to separate lanes of each plate. Nonpolar lipids were separated by successive development with hexane to 20 cm above the origin, toluene to 20 cm above the origin, and hexane–diethyl ether–acetic acid (80:20:1) developed twice to 11 cm. Lipids of high polarity from lizards were separated by successive development with hexane, toluene, and finally chloroform–methanol–water (60:10:1); lipids of high polarity from snakes were separated with only the last of these solvent systems. In snakes, moderately polar lipids also were separated by development with chloroform–methanol–acetic acid (190:9:1) to 20 cm. The chromatograms were sprayed with 50% sulfuric acid and heated on a hot plate to char the lipids.

---

* Adapted from Burken et al.[2] and Weldon and Bagnall.[17]

Results and Discussion. The chromatograms of the nonpolar lipids of most species displayed bands consistent with sterols, free fatty acids, and steryl esters. Bands were numerous and widespread in the regions to which triacylglycerols, wax esters, and/or methyl esters migrated. A number of lizards exhibited a band consistent with hydrocarbons; these were presumed to be endogenous components. Phosphatidylcholine, phosphatidylethanolamine, and sphingomyelin were observed among the polar lipids. The chromatograms of all snakes contained bands that displayed mobilities similar to those of ceramides, but these components became red-violet during charring, suggesting glycolipids.

C. CERAMIDES AND ACYLGLUCOSYLCERAMIDES FROM THE EPIDERMIS OF THE PIG (*SUS SCROFA*)*

In these studies, preparative TLC was used to isolate ceramides and acylglucosylceramides from the epidermis of the pig for subsequent structural determinations.

Sample preparation. The skin of a freshly killed pig was rinsed under tap water. Hair was removed with electric clippers, and sections of skin approximately 10 × 10 cm were excised and placed epidermis-down on a 95°C hot plate for 1 to 2 min. The epidermis was removed by scraping with a spatula and was stored at −20°C.

The epidermal material was dried *in vacuo* and extracted according to Gray.[127] Ten ml/g of tissue were minced and homogenized with chloroform–methanol (1:1). The mixture was filtered and the residue was reextracted twice with chloroform–methanol (2:1) at a concentration of 5 ml/g tissue. The extracts were combined and washed twice with 0.2 volumes of 0.1 M KCl. The aqueous upper phase was discarded and the chloroform phase was evaporated to dryness *in vacuo*. The lipids were redissolved in chloroform and stored at −20°C. The extracted lipid (530 mg from 7.3 g of dry epidermis) represented 7 to 8% of the total weight of the dry epidermis.

Chromatography. To isolate acylglucosylceramides, 53 mg of crude epidermal lipid were applied to a plate coated with a 0.5-mm layer of silica gel 60 H and predeveloped in chloroform–methanol (2:1). The plate was developed with diethyl ether–acetic acid (99:1). The silica gel below R_f 0.55 was scraped from the plate and lipids were eluted with chloroform–methanol–acetic acid–water (49:49:1:1). This material was reapplied to a plate coated with a 0.5-mm layer of silica gel H and developed with diethyl ether followed by chloroform–methanol–water (40:10:1). The plate was sprayed with 2′,7′-dichlorofluorescein and viewed under ultraviolet light to reveal four bands corresponding to acylglucosylceramides. These bands were scraped from the plate and eluted with chloroform–methanol–water (49:49:2). The least polar of these bands, porcine glucosylceramide A (PGLA), comprised 56% of the series and amounted to 6.2% of the total epidermal lipid.

PGLA was saponified by treatment with chloroform–methanol–10 M aqueous NaOH (2:7:1) at 37°C for 1 h. The free fatty acid and glucosylceramide products were separated by preparative TLC with chloroform–methanol–water (40:10:1). The fatty acids were converted to methyl esters by treatment with 10% BCl_3 in methanol at 70°C for 0.5 h, after which saturated, monounsaturated, and diunsaturated methyl esters were fractionated by argentation TLC; these compounds also were analyzed by GLC.

To liberate the long-chain bases, the glucosylceramides recovered after saponification of PGLA were treated with 1 *N* methanolic HCl containing 15 M water at 70°C for 18 h; this method released the amide-linked fatty acids as a mixture of free fatty acids and methyl esters, which were subsequently analyzed by GLC.

To isolate ceramides, the polar material was applied to silica gel plates and developed with chloroform–methanol–water (40:10:1), as described above. Four ceramide bands were observed under ultraviolet light after spraying with 2′,7′-dichlorofluorescein; however, the bands were poorly resolved, thus necessitating that the ceramide region be scraped from the plate and eluted from the silica gel with chloroform–methanol–water (25:25:1). The extract was washed with aqueous potassium carbonate to remove the 2′,7′-dichlorofluorescein and dried under nitrogen and finally *in vacuo*. The ceramide fraction so prepared accounted for 9.3% of the total epidermal lipid.

---

* Adapted from Wertz and Downing.[64,92]

The total ceramide fraction (27 mg) from several preparations was redissolved in chloroform–methanol (1:1) and streaked across a 20 × 20-cm plate coated with 0.5-mm of silica gel H and developed twice with chloroform–methanol (19:1). Six bands, visualized as described above, were individually scraped from the plate and eluted with chloroform–methanol–water (25:25:1). Cross-contaminated fractions were rechromatographed three or four times to achieve complete separation. The relative amounts of each of the chromatographically pure ceramide fractions were determined by quantitative TLC.

Results and Discussion. Four bands corresponding to acylglucosylceramides contained a predominance of C_{30}, C_{32}, and C_{34} ω-hydroxy acids consisting of 58% saturates, 35% monoenes, and 7% dienes. The structure for the acylglucosylceramides is formulated as 1-(3′-O-acyl)-β-D-glucosyl-N-(ω-hydroxy)acylsphingosines. Seven groups of ceramides were identified, including five sphingosines (designated ceramides 1, 2, 4, 5, and 6a) and two phytosphingosines (designated ceramides 3 and 6b). The predominant hydroxy acid moieties associated with these compounds range from C_{16} (in ceramide 5) to C_{32} (in ceramides 1 and 6a).

D. TETRODOTOXIN FROM NEWTS (*TARICHA* SPP.) AND OTHER AMPHIBIANS*

This study used analytical and preparative TLC to determine the distribution of tetrodotoxin among newts (*Taricha* spp.) and other amphibians. Different organs of *T. torosa* also were examined for the presence of this compound.

Sample preparation. Specimens were decapitated and whole carcasses were homogenized in a blender with an equal volume of water. The various organs of *T. torosa* were homogenized separately. To obtain purified tetrodotoxin, the homogenate was adjusted to pH 6.0 with acetic acid, dialyzed against distilled water, and concentrated by freeze-drying. The dialysate was dissolved in acetic acid–methanol–water (2:2:1) and applied to a column of Amberlite IRC-50, a weak cation exchange resin, slurried with 300 ml of 10% NaOH, and washed with water until the pH of the effluent was less than 10. The column was washed with water and tetrodotoxin eluted with 1 *N* acetic acid.

Chromatography. Crude extracts and purified tetrodotoxin were applied to silica gel G and GF plates developed separately with ethyl methyl ketone–pyridine–water–acetic acid (6:2:2:1), *n*-butanol–acetic acid–water (2:1:1), phenol–ammonium hydroxide–water (18:30:4), and 95% ethanol–acetic acid (50:1). Plates coated with cellulose powder (MN 300) were developed with acetic acid–*t*-butanol–water (2:2:1). Fluorescent spots corresponding to tetrodotoxin were observed on the silica gel plates after heating alone. Plates coated with cellulose powder were sprayed with 10% methanolic KOH and heated before being viewed under ultraviolet light (366 nm).

To determine the sensitivity of a chromatographic test for tetrodotoxin, silica gel plates spotted with 0.03 to 15 μg of pure toxin were developed with 95% ethanol–acetic acid (94:6). Plates were sprayed with 10% alcoholic KOH, heated at 130°C for 10 min, and then viewed under ultraviolet light.

Results and Discussion. Tetrodotoxin was detected in *Taricha* spp. and seven other salamanders (Salamandridae), but not in 14 other amphibians (salamanders, frogs, and toads). The highest concentrations of tetrodotoxin in male *T. torosa* were detected in the skin; in females, equally high concentrations were found in the skin and ovaries. The limit of detection of the chromatographic sensitivity test was 0.3 to 3.0 μg.

ACKNOWLEDGMENTS

J. Bogard, J. A. Campbell, L. Collins, J. W. Daly, D. T. Downing, B. Fried, H. Heatwole, J. Galef, K. Kenyon, R. Kuhse, P. E. Kolatukuddy, R. P. Reynolds, M. E. Stewart, P. J. Tolson, A. O. Tucker, T. L. Leto, H. B. Lillywhite, S. Weinstock, J. W. Wheeler, and A. Wynn provided information, figures, or comments on the manuscript. Photo Communications, Inc., Rockville, MD, provided some figures. The preparation of this paper was supported by a grant from the Research Enhance-

---

* Adapted from Wakely et al.[128]

ment Program of the Vice President for Research and Associate Provost for Graduate Studies at Texas A&M University.

REFERENCES

1. **Birkby, C. W., Wertz, P. W., and Downing, D. T.,** The polar lipids from keratinized tissues of some vertebrates, *Comp. Biochem. Physiol.*, 73B, 239, 1982.
2. **Burken, R. R., Wertz, P. W., and Downing, D. T.,** A survey of polar and nonpolar lipids extracted from snake skin, *Comp. Biochem. Physiol.*, 81B, 315, 1985.
3. **Daly, J. W., Highet, R. J., and Meyers, C. W.,** Occurrence of skin alkaloids in non-dendrobatid frogs from Brazil (Bufonidae), Australia (Myobatrachidae) and Madagascar (Mantellinae), *Toxicon*, 22, 905, 1984.
4. **Gray, G. M. and Yardley, H. J.,** Lipid compositions of cells isolated from pig, human, and rat epidermis, *J. Lipid Res.*, 16, 434, 1975.
5. **Lindholm, J. S. and Downing, D. T.,** Occurrence of squalene in the skin surface lipids of the otter, the beaver and the kinkajou, *Lipids*, 15, 1062, 1980.
6. **Lindholm, J. S., McCormick, J. M., Colton, S. W., VI, and Downing, D. T.,** Variation of the skin surface lipid composition among mammals, *Comp. Biochem. Physiol.*, 69B, 75, 1981.
7. **Meyers, C.W. and Daly, J.W.,** Preliminary evaluation of skin toxins and vocalizations in taxonomic and evolutionary studies of poison-dart frogs (Dendrobatidae), *Bull. Am. Mus. Nat. Hist.*, 157, 173, 1976.
8. **Meyers, C. W., Daly, J. W., and Malkin, B.,** A dangerously toxic new frog (*Phyllobates*) used by Emberá Indians of western Colombia, with discussion of blowgun fabrication and dart poisoning, *Bull. Am. Mus. Nat. Hist.*, 161, 307, 1978.
9. **Nicolaides, N.,** Skin lipids. II. Lipid class composition of samples from various species and anatomical sites, *J. Am. Oil Chem. Soc.*, 42, 685, 1965.
10. **Nicolaides, N., Fu, H. C., and Rice, G. R.,** The skin surface lipids of man compared with those of eighteen species of mammals, *J. Invest. Dermatol.*, 51, 83, 1968.
11. **Nicolaides, N., Fu, H. C., and Ansari, M. N. A.,** Diester waxes in surface lipids of animal skin, *Lipids*, 5, 299, 1970.
12. **Nikkari, T.,** The occurrence of diester waxes in human vernix caseosa and in hair lipids of common laboratory animals, *Comp. Biochem. Physiol.*, 29, 795, 1969.
13. **Oldak, P. D.,** Comparison of the scent gland secretion lipids of twenty-five snakes: implications for biochemical systematics, *Copeia*, 1976, 320, 1976.
14. **Omoto, T.,** Thin-layer chromatography of toad toxin, *Kagaku No Ryoiki, Zokan*, 64, 115, 1964 (in Japanese).
15. **Roberts, J. B.,** The role of epidermal lipids in limiting cutaneous evaporative water loss in squamate reptiles, Ph.D. dissertation, University of Kansas, Lawrence, 1980.
16. **Tolson, P. J.,** Phylogenetics of the boid snake genus *Epicrates* and Caribbean vicariance theory, *Occ. Pap. Mus. Zool. Univ. Michigan*, 715, 1, 1987.
17. **Weldon, P. J. and Bagnall, D.,** A survey of polar and nonpolar skin lipids from lizards by thin-layer chromatography, *Comp. Biochem. Physiol.*, 87B, 345, 1987.
18. **Weldon, P. J. and Tanner, M. J.,** Gular and paracloacal gland secretions of crocodilians: a comparative analysis by thin-layer chromatography, *Biochem. System. Ecol.*, 19, 133, 1991.
19. **Dunn, B. S., Jr., Weldon, P. J., Howard, R. W., and McDaniel, C. A.,** Lipids from the paracloacal glands of the Chinese alligator (*Alligator sinensis*), *Lipids*, 28, 75, 1993.
20. **Bohnet, S., Rogers, L., Sasaki, G., and Kolattukudy, P. E.,** Estradiol induces proliferation of peroxisome-like microbodies and the production of 3-hydroxy fatty acid diesters, the female pheromones, in the uropygial glands of male and female mallards, *J. Biol. Chem.*, 266, 9795, 1991.
21. **Jacobsen, E., Billings, J. K., Frantz, R. A., Kinney, C. K., Stewart, M. E., and Downing, D. T.,** Age-related changes in sebaceous wax ester secretion rates in men and women, *J. Invest. Dermatol.*, 85, 483, 1985.
22. **Mason, R. T., Chinn, J. W., and Crews, D.,** Sex and seasonal differences in the skin lipids of garter snakes, *Comp. Biochem. Physiol.*, 87B, 999, 1987.
23. **Nikkari, T. and Valavaara, M.,** The influence of age, sex, hypophysectomy and various hormones on the composition of the skin surface lipids of the rat, *Br. J. Derm.*, 83, 459, 1970.
24. **Simpson, J. T., Sharp, T. R., Wood, W. F., and Weldon, P. J.,** Further analysis of lipids from the scent gland secretions of Dumeril's ground boa (*Acrantophis dumerili* Jan), *Z. Naturforsch.*, 48c, 953, 1993.
25. **Weldon, P. J. and Sampson, H. W.,** The gular glands of *Alligator mississippiensis*: histology and preliminary analysis of lipoidal secretions, *Copeia*, 1980, 80, 1980.
26. **Weldon, P. J., Scott, T. P., and Tanner, M. J.,** Analysis of gular and paracloacal gland secretions of the American alligator (*Alligator mississippiensis*) by thin-layer chromatography, *J. Chem. Ecol.*, 16, 3, 1990.

27. **Downing, D. T., Strauss, J. S., and Pochi, P. E.,** Variability in the chemical composition of human skin surface lipids, *J. Invest. Dermatol.*, 53, 322, 1969.
28. **Green, S. C., Stewart, M.E., and Downing, D. T.,** Variation in sebum fatty acid composition among adult humans, *J. Invest. Dermatol.*, 83, 114, 1984.
29. **Wilkinson, D. I.,** Variability in composition of surface lipids: the problem of the epidermal composition, *J. Invest. Dermatol.*, 52, 339, 1969.
30. **Neiminen, E., Leikola, E., Koljonen, M., Kiistala, U., and Mustakallio, K. K.,** Quantitative analysis of epidermal lipids by thin-layer chromatography with special reference to seasonal and age variation, *Acta Derm.–Venereol.*, 47, 327, 1967.
31. **Oku, H., Urahashi, A., Yagi, N., Nagata, J., and Chinen, I.,** Fatty acid and lipid composition *in vitro* and *in vivo* of rat epidermis, *Comp. Biochem. Physiol.*, 105B, 293, 1993.
32. **Ramasastry, P., Downing, D. T., Pochi, P. E., and Strauss, J. S.,** Chemical composition of human skin surface lipids from birth to puberty, *J. Invest. Dermatol.*, 54, 139, 1970.
33. **Wilkinson, D. I. and Karasek, M. A.,** Skin lipids of a normal and a mutant (asebic) mouse strain, *J. Invest. Dermatol.*, 47, 449, 1966.
34. **Daly, J. W. and Meyers, C. W.,** Toxicity of Panamanian poison frogs (*Dendrobates*): some biological and chemical aspects, *Science*, 156, 970, 1967.
35. **Meyers, C. W. and Daly, J. W.,** Taxonomy and ecology of *Dendrobates bombetes*, a new Andean poison frog with new skin toxins, *Am. Mus. Nov.*, 2692, 1, 1980.
36. **Greene, R. S., Downing, D. T., Pochi, P. E., and Strauss, J. S.,** Anatomical variation in the amount and composition of human surface lipid, *J. Invest. Dermatol.*, 54, 240, 1970.
37. **Schoephoerster, R. T., Wertz, P. W., Madison, K. C., and Downing, D. T.,** A survey of polar and non-polar lipids of mouse organs, *Comp. Biochem. Physiol.*, 82B, 229, 1985.
38. **Weldon, P.J., Sampson, H. W., Wong, L., and Lloyd, H. A.,** Histology and biochemistry of the scent glands of the yellow-bellied sea snake (*Pelamis platurus*: Hydrophiidae), *J. Herpetol.*, 25, 367, 1991.
39. **Wertz, P. W., Stover, P. M., Abraham, W., and Downing, D. T.,** Lipids of chicken epidermis, *J. Lipid Res.*, 27, 427, 1986.
40. **Kolattukudy, P. E., Bohnet, S., and Rogers, L.,** Diesters of 3-hydroxy fatty acids produced by the uropygial glands of female mallards uniquely during the mating season, *J. Lipid Res.*, 28, 582, 1987.
41. **Kolattukudy, P. E., Rogers, L., and Flurkey, W.,** Suppression of a thioesterase gene expression and the disappearance of short chain fatty acids in the preen gland of the mallard duck during eclipse, the period following postnuptial molt, *J. Biol. Chem.*, 260, 10789, 1985.
42. **Downing, D. T., Strauss, J. S., and Pochi, P. E.,** Changes in skin surface composition induced by severe caloric restriction in man, *Am. J. Clin. Nutr.*, 15, 161, 1964.
43. **Gower, D. B., Nixon, A., and Mallet, A. I.,** The significance of odorous steroids in axillary odor, in *Perfumery: The Psychology and Biology of Fragrance*, Van Toller, S. and Dodd, G. H., Eds., Chapman and Hall, New York, 1988, 49.
44. **Nixon, A., Mallet, A. I., Jackman, P. J. H., and Gower, D. B.,** Testosterone metabolism by isolated human axillary *Corynebacterium* spp.: a gas chromatographic-mass-spectrometric study, *J. Steroid Biochem.*, 24, 887, 1986.
45. **Marples, R. R., Downing, D. T., and Kligman, A. M.,** Control of free fatty acids in human surface lipids by *Corynebacterium acnes*, *J. Invest. Dermatol.*, 56, 127, 1971.
46. **Marples, R. R., Downing, D. T., and Kligman, A. M.,** Influence of *Pityrosporum* species in the generation of free fatty acids in human surface lipids, *J. Invest. Dermatol.*, 58, 155, 1972.
47. **Marples, R. R., Kligman, A. M., Lantis, L. R., and Downing, D. T.,** The role of aerobic microflora in the genesis of fatty acids in human surface lipids, *J. Invest. Dermatol.*, 55, 173, 1970.
48. **Colton, S. W., III and Downing, D. T.,** Comparison of acetate and glucose incorporation into rat and horse skin lipids, *Biochem. Biophys. Acta*, 837, 190, 1985.
49. **Downing, D. T., Strauss, J. S., Ramasastry, P., Abel, M., Lees, C. W., and Pochi, P. E.,** Measurement of the time between synthesis and surface excretion of sebaceous lipids in sheep and man, *J. Invest. Dermatol.*, 64, 215, 1975.
50. **Gaul, B. L., Stewart, M. E., and Downing, D. T.,** The time course of sebum secretion in the guinea pig, *Comp. Biochem. Physiol.*, 80B, 431, 1985.
51. **Hedberg, C. L., Wertz, P. W., and Downing, D. T.,** The nonpolar lipids of pig epidermis, *J. Invest. Dermatol.*, 90, 225, 1988.
52. **Hedberg, C. L., Wertz, P. W., and Downing, D. T.,** The time course of lipid biosynthesis in pig epidermis, *J. Invest. Dermatol.*, 91, 169, 1988.
53. **Thiessen, D. D., Regnier, F. E., Rice, M., Goodwin, M., Isaaks, N., and Lawson, N.,** Identification of a ventral scent marking pheromone in the male Mongolian gerbil (*Meriones unguiculatus*), *Science*, 184, 83, 1974.
54. **Downing, D. T.,** Mammalian waxes, in *Chemistry and Biochemistry of Natural Waxes*, Kolattukudy, P. E., Ed., Elsevier, New York, 1976, 17.
55. **Downing, D. T. and Stewart, M. E.,** Analysis of sebaceous lipids, in *Methods in Skin Research*, Skerrow, D. and Skerrow, C. J., Eds., John Wiley & Sons, New York, 1985, 349.
56. **Nicolaides, N. and Kellum, R. E.,** Skin lipids. I. Sampling problems of the skin and its appendages, *J. Am. Oil Chem. Soc.*, 42, 685, 1965.

57. **Jacob, J., Ziemsen, B., and Hoppe, U.,** Cast skin lipids of the Indian python (*Python molurus bivittatus*, Kuhl, 1820), *Z. Naturforsch.*, 48c, 80, 1993.
58. **Downing, D. T. and Colton, S. W., VI,** The skin surface lipids of the horse, *Lipids*, 15, 323, 1980.
59. **Colton, S. W., VI and Downing, D. T.,** Variation in skin surface lipid composition among the Equidae, *Comp. Biochem. Physiol.*, 75B, 429, 1983.
60. **Strauss, J. S. and Pochi, P. E.,** The quantitative gravimetric determination of sebum production, *J. Invest. Dermatol.*, 36, 293, 1961.
61. **Downing, D. T., Stranieri, A. M., and Strauss, J. S.,** The effect of accumulated lipids on measurements of sebum secretion in human skin, *J. Invest. Dermatol.*, 79, 226, 1982.
62. **Stewart, M. E. and Downing, D. T.,** Measurement of sebum secretion rates in young children, *J. Invest. Dermatol.*, 84, 59, 1985.
63. **Wertz, P. W. and Downing, D. T.,** Free sphingosines in porcine epidermis, *Biochim. Biophys. Acta*, 1002, 213, 1989.
64. **Wertz, P. W. and Downing, D. T.,** Acylglucosylceramides of pig epidermis: structure determination, *J. Lipid Res.*, 24, 753, 1983.
65. **Wertz, P. W. and Downing, D. T.,** Cholesteryl sulfate: the major polar lipid of horse hoof, *J. Lipid Res.*, 25, 1320, 1984.
66. **Jacob, J.,** TLC, GLC and MS of complex lipid mixtures from uropygial secretions, *J. Chromatogr. Sci.*, 13, 415, 1975.
67. **Jacob, J.,** Bird waxes, in *Chemistry and Biochemistry of Natural Waxes*, Kolattukudy, P. E., Ed., Elsevier, New York, 1976, 93.
68. **Kellum, R. E.,** Isolation of human sebaceous gland lipids, *Arch. Dermatol.*, 93, 610, 1966.
69. **Willams, T. A. and Anthony, C. D.,** Technique to isolate salamander granular gland products with a comment on the evolution of adhesiveness, *Copeia*, 1994, 540, 1994.
70. **Barthalmus, G. T.,** Biological roles of amphibian skin secretions, in *Amphibian Biology, Vol. 1, The Integument*, Heatwole, H. and Barthalmus, G. T., Eds., Surrey Beatty & Sons, Chipping Norton, U.K., 1993, 382.
71. **Meyer, K. and Linde, H.,** Collection of toad venoms and chemistry of the toad venom steroids, in *Venomous Animals and Their Venoms*, Vol. 2, Bucherl, W. and Buckley, E. E., Eds., Academic Press, New York, 1971, 521.
72. **Weldon, P. J. and Tanner, M. J.,** Lipids in the Rathke's gland secretions of hatchling loggerhead sea turtles (*Caretta caretta*), *Copeia*, 1990, 570, 1990.
73. **Weldon, P. J., Mason, R. T., Tanner, M. J., and Eisner, T.,** Lipids in the Rathke's gland secretions of hatchling Kemp's ridley sea turtles (*Lepidochelys kempi*), *Comp. Biochem. Physiol.*, 96B, 705, 1990.
74. **Downing, D. T.,** Lipolysis by human skin surface debris in organic solvents, *J. Invest. Dermatol.*, 54, 395, 1970.
75. **Folch, J., Lees, M., and Sloane-Stanley, G. H.,** A simple method for the isolation and purification of total lipids from animal tissues, *J. Biol. Chem.*, 226, 497, 1957.
76. **Wertz, P. W. and Downing, D. T.,** Integral lipids of human hair, *Lipids*, 23, 878, 1988.
77. **Wertz, P. W. and Downing, D. T.,** Integral lipids of mammalian hair, *Comp. Biochem. Physiol.*, 92B, 759, 1989.
78. **Daly, J. W., Garraffo, H. M., and Spande, T. F.,** Amphibian alkaloids, in *The Alkaloids*, Vol. 43, Cordell, G. A., Ed., Academic Press, San Diego, 1993, 185.
79. **Daly, J. W.,** personal communication.
80. **Borgström, B.,** Investigation on lipid separation methods, *Acta Physiol. Scand.*, 25, 101.
81. **Hirsch, J. and Ahrens, E. H.,** The separation of complex lipid mixtures by the use of silicic acid chromatography, *J. Biochem.*, 233, 311, 1958.
82. **Gray, G. M.,** Chromatography of lipids. I. An improved chromatographic procedure for the quantitative isolation of the neutral ceramide-containing glycolipids from mammalian tissues, *Biochem. Biophys. Acta*, 84, 35, 1964.
83. **Sharaf, D. M., Clark, S. J., and Downing, D. T.,** Skin surface lipids of the dog, *Lipids*, 12, 786, 1977.
84. **Zelnik, R. and Ziti, L. M.,** Thin-layer chromatography: chromatoplate analysis of the bufadienolides isolated from toad venoms, *J. Chromatogr.*, 9, 371, 1962.
85. **Komatsu, M., Kamano, Y., and Suzuki, M.,** Thin-layer chromatography of *Bufo* steroids on silica gel, *Bunseki Kagaku*, 14, 1049, 1965 (in Japanese).
86. **Linde-Tempel, H. O.,** Konstitution der bufotoxine, *Helv. Chim. Acta*, 53, 2188, 1970.
87. **Autrum, H., Fillies, K., and Wagner, H.,** Squalen im Sekret der Analbeutel von *Petaurus breviceps papuanus* Th. (Phalangeridae, Marsupialia), *Biol. Zbl.*, 89, 681, 1970.
88. **Weldon, P. J., Dunn, B. S., Jr., McDaniel, C. A., and Werner, D. I.,** Lipids in the femoral gland secretions of the green iguana (*Iguana iguana*), *Comp. Biochem. Physiol.*, 95B, 541, 1990.
89. **Roberts, J. B. and Lillywhite, H. B.,** Lipid barrier to water exchange in reptile epidermis, *Science*, 207, 1077, 1985.
90. **Stewart, M. E., Downing, D. T., Pochi, P. E., and Strauss, J. S.,** The fatty acids of human sebaceous gland phosphatidylcholine, *Biochim. Biophys. Acta*, 529, 380, 1978.
91. **Long, S. A., Wertz, P. W., Strauss, J. S., and Downing, D. T.,** Human stratum corneum polar lipids and desquamation, *Arch. Dermatol. Res.*, 277, 284, 1985.
92. **Wertz, P. W. and Downing, D. T.,** Ceramides of pig epidermis: structure determination, *J. Lipid Res.*, 24, 759, 1983.
93. **Wertz, P. W., Downing, D. T., Freinkel, R. K., and Traczyk, T. N.,** Sphingolipids of the stratum corneum and lamellar granules of fetal rat epidermis, *J. Invest. Dermatol.*, 83, 193, 1984.
94. **Wertz, P. W., Miethke, M. C., Long, S. A., Strauss, J. S., and Downing, D. T.,** The composition of the ceramides from human stratum corneum and from comedones, *J. Invest. Dermatol.*, 84, 410, 1985.

95. **Robson, K. J., Stewart, M. E., Michelsen, S., Lazo, N. D., and Downing, D. T.,** 6-Hydroxy-4-sphingenine in human epidermal ceramides, *J. Lipid Res.*, 35, 2060, 1994.
96. **Karlsson, K-.A. and Pascher, I.,** Thin-layer chromatography of ceramides, *J. Lipid Res.*, 12, 466, 1971.
97. **Wertz, P. W., Colton, S. W., VI, and Downing, D. T.,** Comparison of the hydroxyacids from the epidermis and from the sebaceous glands of the horse, *Comp. Biochem. Physiol.*, 75B, 217, 1983.
98. **Gray, G. M., White, R. J., and Majer, J. R.,** 1-(3'-O-acyl)-β-glucosyl-N-dihydroxypentatriacontadienoylsphingosine, a major component of the glucosylceramides of pig and human epidermis, *Biochim. Biophys. Acta*, 528, 127, 1978.
99. **Abraham, W., Wertz, P. W., Burken, R. R., and Downing, D. T.,** Glucosylsterol and acylglucosylsterol of snake epidermis: structure determination, *J. Lipid Res.*, 28, 446, 1987.
100. **Wertz, P. W. and Downing, D. T.,** Free sphingosines in human epidermis, *J. Invest. Dermatol.*, 94, 159, 1990.
101. **Dumbacher, J. P., Beehler, B. M., Spande, T. F., Garraffo, H. M., and Daly, J. W.,** Homobatrachotoxin in the genus *Pitohui*: chemical defense in birds?, *Science*, 258, 799, 1992.
102. **Buchwald, H. D., Durham, L., Fischer, H. G., Harada, R., Mosher, H. S., Kao, C. Y., and Fuhrman, F. A.,** Identity of tarichatoxin and tetrodotoxin, *Science*, 143, 474, 1964.
103. **Mosher, H. S., Fuhrman, F. A., Buchwald, H. D., and Fischer, H. G.,** Tarichatoxin-tetrodotoxin: a potent neurotoxin, *Science*, 144, 1100, 1964.
104. **Touchstone, J. C.,** *Practice of Thin Layer Chromatography*, John Wiley & Sons, New York, 1992.
105. **Downing, D. T.,** Photodensitometry in the thin-layer chromatographic analysis of neutral lipids, *J. Chromatogr.*, 38, 91, 1968.
106. **Downing, D. T.,** Lipids, in *Densitometry in Thin Layer Chromatography*, Touchstone, J. C. and Sherma, J., Eds., John Wiley, New York, 1979, 367.
107. **Downing, D. T. and Stranieri, A. M.,** Correction for the deviation from the Lambert-Beer Law in the quantitation of thin-layer chromatograms by photodensitometry, *J. Chromatogr.*, 192, 208, 1980.
108. **Downing, D. T.,** Secretions of the sebaceous glands, in *Thin Layer Chromatography: Quantitative Environmental and Clinical Applications*, Touchstone, J. C. and Rogers, D., Eds., John Wiley & Sons, New York, 1980, 495.
109. **Ahern, D. G. and Downing, D. T.,** Skin lipids of the Florida indigo snake, *Lipids*, 9, 8, 1974.
110. **Downing, D. T. and Stewart, M. E.,** Skin surface lipids of the mole, *Scalopus aquaticus, Comp. Biochem. Physiol.*, 86B, 667, 1987.
111. **Yeung, D., Nacht, S., and Cover, R. E.,** The composition of the skin surface lipids of the gerbil, *Biochim. Biophys. Acta*, 663, 524, 1981.
112. **Downing, D. T. and Sharaf, D. M.,** Skin surface lipids of the guinea pig, *Biochim. Biophys. Acta*, 431, 378, 1976.
113. **Downing, D. T. and Lindholm, J. S.,** Skin surface lipids of the cow, *Comp. Biochem. Physiol.*, 73B, 327, 1982.
114. **Colton, S. W., VI, Lindholm, J. S., Abraham, W., and Downing, D. T.,** Skin surface lipids of the mink, *Comp. Biochem. Physiol.*, 84B, 369, 1986.
115. **Nishimaki-Mogami, T., Minegishi, K., Takahashi, A., Kawasaki, Y., Kurokawa, Y., and Uchiyama, M.,** Characterization of skin-surface lipids from the monkey (*Macaca fascicularis*), *Lipids*, 23, 869, 1988.
116. **Farris, J. S.,** Methods for computing Wagner trees, *System. Zool.*, 19, 83, 1970.
117. **Haahti, E. and Nikkari, T.,** Separation and isolation of waxes and sterol esters of skin surface fat with thin layer chromatography, *Acta Chem. Scand.*, 17, 536, 1963.
118. **Nicolaides, N., Fu, H. C., Ansari, M. N. A., and Rice, G. R.,** The fatty acids of wax esters and sterol esters from human skin surface, *Lipids*, 7, 506, 1972.
119. **Stewart, M. E. and Downing, D. T.,** Separation of wax esters from steryl esters by chromatography on magnesium hydroxide, *Lipids*, 16, 355, 1981.
120. **Ohno, Y. and Tanaka, S.,** Detection and identification of civetone in natural civets by thin-layer chromatography and GC-MS method, *Bunseki Kagaku*, 26, 232, 1977 (in Japanese).
121. **Nikkari, T. and Haahti, E.,** Isolation and analysis of two types of diester waxes from the skin surface lipids of the rat, *Biochim. Biophys. Acta*, 164, 294, 1968.
122. **Oku, H., Shudo, J., Mimura, K., Haratake, A., Nagata, J., and Chinen, I.,** 1-O-Alkyl-2,3-diacylglycerols in the skin surface lipids of the hairless mouse, *Lipids*, 30, 169, 1995.
123. **Lough, A. K., Felinski, L., and Garton, G. A.,** The production of methyl esters of fatty acids as artifacts during the extraction or storage of tissue lipids in the presence of methanol, *J. Lipid Res.*, 3, 478, 1962.
124. **Elias, P. M., Goerke, J., and Friend, D. S.,** Mammalian epidermal barrier layer lipids: composition and influence on structure, *J. Invest. Dermatol.*, 69, 535, 1977.
125. **Lampe, M. A., Williams, M. L., and Elias, P. M.,** Human epidermal lipids: characterization and modulations during differentiation, *J. Lipid Res.*, 24, 131, 1983.
126. **Bortz, J. T., Wertz, P. W., and Downing, D. T.,** The origin of alkanes found in human skin surface lipids, *J. Invest. Dermatol.*, 93, 723, 1989.
127. **Gray, G. M.,** Phosphatidyl-(N-acyl)-ethanolamine, a lipid component of mammalian epidermis, *Biochim. Biophys. Acta*, 431, 1, 1967.
128. **Wakely, J. F., Fuhrman, G. J., Fuhrman, F. A., Fischer, H. G., and Mosher, H. S.,** The occurrence of tetrodotoxin (tarichatoxin) in Amphibia and the distribution of the toxin in the organs of newts (*Taricha*), *Toxicon*, 3, 195, 1966.

Chapter 7

THIN-LAYER CHROMATOGRAPHY IN CLINICAL CHEMISTRY

Raka Jain

CONTENTS

I. Introduction 132

II. Historical Aspects of TLC 132

III. Principles of TLC 132

IV. Experimental Variables 132
    A. Backing Sheets 133
    B. Stationary Phase 133
    C. Preparation of TLC Plates 134
    D. Bioanalysis 134
        1. Specimen Collection and Storage..134
        2. Preparation of Biological Samples for Analysis 135
            a. Dilution 135
            b. Precipitation and Deproteinization 135
            c. Ultrafiltration..135
            d. Dialysis 136
            e. Hydrolysis 136
            f. Liquid-Liquid Extraction 136
            g. Liquid-Solid Extraction 137
    E. Sample Application..137
    F. Equilibration 137
    G. Development 137
    H. Detection and Identification of Separated Spots...138
    I. Instrumentation — Densitometry for Quantification in TLC......................138
    J. Validation of TLC Results..138
        1. Calibration Linearity 139
        2. Specificity..139
        3. Sensitivity..139
        4. Precision and Accuracy..139
        5. Extraction Recovery 140

V. Recent Development in TLC Plates 140

VI. Comparison of Instrumental TLC with Other Analytical Techniques 140

VII. Application of TLC in Clinical Chemistry...141
    A. Amino Acids 141
    B. Bile Acids 142
    C. Carbohydrates 142
    D. Drugs 143

  E. Lipids 144
  F. Phospholipids 145
  G. Prostaglandins 146
  H. Porphyrins 147
  I. Steroid Hormones 147
  J. Other Compounds of Clinical Interest 148

VIII. Conclusions 148

References 148

I. INTRODUCTION

In the field of analytical chemistry, thin-layer chromatography (TLC) is considered to be the most simple, robust, economical, and rapid technique. Since its advent in 1938, tremendous improvements have been made in stationary phases and in equipment for sample application, chromatogram development, and *in situ* quantification. In great measure, these notable advances can be attributed to technological advances in laboratory equipment and instrumentation. TLC is considered to be an ideal chromatographic system for clinical laboratories. It is used clinically as an aid in diagnosis. The relative simplicity of the method and inexpensive cost of operation make it the method of choice in clinical chemistry.

II. HISTORICAL ASPECTS OF TLC

Chromatography is a generic term for a technique that allows separation of a mixture of compounds. Chromatography received its name from the work of a Russian botanist named Michael Teswett in 1906. The basic principles of thin-layer chromatography were first described by Izmailov and Shraiber in 1938. Although Consden et al., Kirchner, and Kelley had experimented with a variety of layers and surfaces, the TLC methodology using fine-grained silica gel particles was developed by Stahl et al. and Randerath. Several comprehensive texts[1-7] are now available on TLC techniques. The effort here is not to replace the existing reviews and books available on TLC technique, but to outline recent developments of this technique and its application in clinical chemistry.

III. PRINCIPLES OF TLC

Thin-layer chromatography has been in use for more than the past 40 years. Today, it is one of the most widely used separation methods. It is a physicochemical separation method, in which the components to be separated are distributed between two phases. One of these phases constitutes a stationary phase of large surface area and the other is a mobile phase that traverses along the stationary phase. The components to be separated are differentially attracted to the stationary phase because of variation of their physicochemical properties. This distinctive influence is manifested by different rates and distances migrated. Substances that are less strongly held on the stationary phase would tend to move faster in the mobile phase and vice-versa. In other words, the speed of migration of a particular component would depend on its sorptive affinity to the other component present.

IV. EXPERIMENTAL VARIABLES

Operationally, TLC is very simple. In this technique the basic steps followed for analysis of biological samples are shown in Figure 7.1. These are discussed briefly as follows:

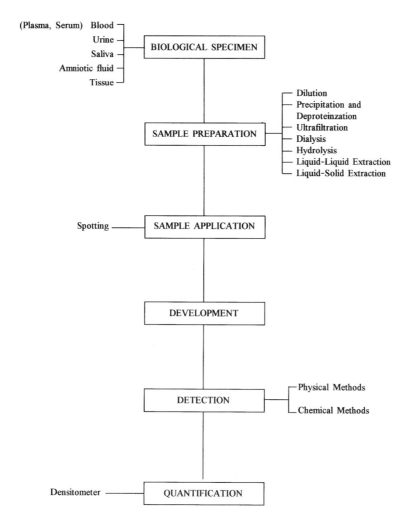

FIGURE 7.1 Basic steps in TLC for separation of biological compounds.

A. BACKING SHEETS

In thin-layer chromatography, glass plates, plastic sheets, or aluminum foils are used as backings. The sole function of the backing sheet is to provide a flat surface to which the stationary phase can adhere.

B. STATIONARY PHASE

The stationary phase in thin-layer chromatography consists of a finely powdered solid. An important feature of a stationary phase is the surface area. The adsorbents used as stationary phases in TLC techniques can be polar or nonpolar. Polar adsorbents can be classified as either acidic or basic. Silica gel and magnesium silicate are acidic adsorbents, whereas alumina is a basic adsorbent. Acidic adsorbents are used for the separation of acidic compounds, as they preferentially retain basic solutes. Similarly, basic adsorbents are used for the separation of basic compounds, as they preferentially retain acidic solutes. Nonpolar adsorbents tend to retain compounds of high molecular weight and compounds with aromatic properties. Charcoal is a classic example of a nonpolar adsorbent.

Relatively few layer materials have found widespread used in TLC. Silica gel is by far the most widely used adsorbent. Other TLC phases used include alumina, cellulose, diatomaceous earth (kieselguhr), ion exchange resins (Dowex 50-X8), Sephadex, and polyamide powder. The particle size of most commercially available sorbents is about 20 to 50 μm for TLC and 5 to 15 μm for high performance TLC (HPTLC). An inorganic binder such as gypsum ($CaSO_4$) is usually added to the sorbents to adhere the stationary phase to the backing. Fluorescent additives may be useful if detection of separated spots by examination in ultraviolet light is intended.

In the last decade, the introduction of bonded silica has led to a renewed interest in reversed phase (RP)-TLC.[8] These include C_{18}, C_{12}, C_8, C_2, aminopropyl, diphenyl, and cyanopropyl bonded stationary phases. Further, complex-forming stationary phases or chiral phases have also been introduced for separation of amino acid enantiomers and similar compounds. RP-TLC has proved useful for the chromatography of polar compounds. Advantages of RP-TLC over normal phase (NP)-TLC are higher recoveries of materials from the plate, ease of optimization of solvent system, and minimum decomposition of sensitive compounds. Although there are many similarities between RP-TLC and RP-HPLC, it is unlikely that reverse phase will achieve the dominant position in TLC that it occupies in HPLC. However, RP-TLC represents a useful extension of the technique of TLC, especially for polar compounds.

C. PREPARATION OF TLC PLATES

The glass plates can be of various sizes (20 × 20 cm, 20 × 10 cm, 10 × 5 cm, or micro slides). These plates can be prepared easily in a laboratory. There are many ways to coat these plates in the laboratory. The most common techniques used are dipping and spreading. The dipping technique is mainly used for preparation of microslides. In general, a homogeneous, uniform, air bubble-free adsorbent slurry is made by mixing with distilled water (about 50 g in 100 ml of water). The slurry is spread on the TLC plate by use of a manual or mechanical spreading device. The two types of commercial spreaders used are stationary trough (or Kirchner) type or the movable trough (or Stahl) type. These spreaders provide excellent uniform thickness of about 0.25 mm. The coated plates are air dried for 10 to 20 min to allow the binder to set. Commercially available precoated TLC and HPTLC plates are being used in most laboratories these days.

D. BIOANALYSIS

1. Specimen Collection and Storage

The biological samples most often collected for determination of various analytes are plasma, serum, and urine. Other biological specimens (e.g., whole blood, saliva, cerebrospinal fluid, gastric fluid, and body tissues) are also analyzed, but less frequently. Quality control in the clinical laboratory begins before the sample is collected from the patient. The collection process is quite important.[9,10] In all instances, it is essential to ensure that the proper and appropriate specimen is collected, as the validity of the test results will depend upon the specimen being collected. The collection containers must be clean and leak-proof. All specimens should also be clearly labeled with the date and time of sample collection. The laboratory must be concerned with the transportation of the samples, as it can affect results. During transportation to the laboratory, sample containers should be properly stoppered or sealed and should be kept in the stopper-up position. This will provide clot formation, reduce agitation, and minimize the possibility of hemolysis in blood samples.

For forensic purposes and for substance abuse testing programs, the biological specimen, preferably urine, must be collected under close supervision and should be transported under custody to avoid pilferage.[11-14] The time between sample collection and receipt by the laboratories should be fast. It, however, depends on the analyte and the type of specimen being analyzed. Ideally, quick transport and short period of storage can significantly contribute to the quality of results.

Proper storage and preservation of the sample is also important in clinical chemistry.[10] Analyte loss, leading to erroneous results, will occur if the sample is not properly stored. Certain samples will require stabilization of the analyte if the analysis is not performed immediately or at least on the same day as the samples are collected. Depending on the analyte, the sample may need refrigeration, freezing, or deep freezing until it is analyzed. In addition, some blood and urine samples may need acid, alkaline, or solvent stabilization. The proper choice depends on the substance to be analyzed.

2. Preparation of Biological Samples for Analysis

Biological samples are very complex multicomponent mixtures. Often the substance of interest is present at very low concentration in the biological matrix. Selective isolation of the analytes from the samples and preconcentration procedures are often necessary to remove extraneous material prior to separation and measurement. Removal of these interfering substances enhances the reliability and sensitivity of the chromatographic system. The chief physiochemical methods[15-17] used to isolate and enrich analytes include dilution, protein precipitation, ultrafiltration, dialysis, hydrolysis, liquid-liquid extraction (LLE), and liquid-solid extraction (LSE). Less commonly used techniques include lyophilization, saponification, microwave processing, and supercritical fluid extraction.

a. Dilution

Dilution is a very simple method of sample preparation, if an analyte is present in a sufficiently high concentration. The removal of particulate matter is accomplished by addition of diluting fluid such as water or buffer to the sample, followed by analysis. The diluting fluid disrupts bonding between the analyte and plasma protein. It also reduces the viscosity or ionic strength of the sample. As this is a nonspecific method to detect free drugs and their metabolites, care may need to be taken in the interpretation of the results.

b. Precipitation and Deproteinization

Removal of proteins by precipitation or denaturation is an effective method of sample preparation. A number of agents are commonly used to reduce the solubility of proteins and to facilitate their removal from the biological samples. These are either acidic or organic solvents or inorganic salts or metallic ions. The commonly used acids for the precipitation of proteins are tricholoroacetic, perchloric, tungstic, and metaphosphoric acids. These acids act by forming insoluble salts with the positively charged amino group of the protein molecules at a pH below their isoelectric point. The control of pH is essentially important in the case of tungstic and metaphophoric acids. These agents give optimum results when a cold solution of 5 to 20% is used. Addition of organic solvents that are miscible with water, such as methanol, ethanol, acetone, and acetonitrile, have been very popular as protein precipitants. Their relative order of effectiveness in precipitating protein is acetonitrile > acetone > ethanol > methanol, which is approximately inversely related to their polarities. Sometimes the use of two solvents may be essential for quantitative recovery of analytes as shown by Li et al.[18] to extract porphyrins from liver.

Ammonium sulfate is a classic protein precipitant. It should be noted that protein denaturation with ammonium sulfate is reversible. The efficiency of this precipitant could probably be improved by controlling the pH so that the plasma proteins are at or near their isoelectric points.

Proteins can also be removed by heavy metal cations such as zinc and copper. They act by forming insoluble salts with negatively charged carboxyl group of the protein at a pH value above the isoelectric point.

c. Ultrafiltration

The ultrafiltration technique is used for rapid removal of low-molecular-weight dissolved solutes from the matrix by passing through an extremely fine membrane filter. The procedure uses cone-shaped membranes that fit into the centrifuge tubes, and centrifugal force is used to drive water and molecules of molecular weight < 50,000 through the anisotropic membrane filter. This approach is widely used to concentrate proteins, as membranes permit passage of smaller dissolved solute molecules. After centrifugation, the filtrate is evaporated to dryness, reconstituted in a small volume of the solvent, and preserved for further analysis.

d. Dialysis

Removal of analytes from the matrix can be achieved by dialysis through a semipermeable membrane. Dialysis membranes are prepared from cellophane or collodion and contain pores that allow passage of solute molecules whose molecular weight is less than 5000 and hold back the macromolecules. Thus proteins of high molecular weight are retained within a dialysis bag, while low-molecular-weight solutes diffuse through the pores into the dialysis medium. Diffusion is a slow process and is driven by a concentration gradient until equilibrium is established. Complete removal of low-molecular-weight solutes requires repeated changes of dialysate, and the sample becomes diluted. Thus dialysis is a time-consuming technique and it is not normally preferred for sample preparation.

e. Hydrolysis

Another aspect of sample preparation is hydrolysis. This step is essential where analytes are present in their conjugated form in biological samples. The main aim of the technique is to clean the conjugate and release the analyte in free form. Hydrolysis can be accomplished by the use of either enzymes, such as β-glucoronidase and aryl sulfatase, or nonspecific acidic or basic hydrolysis methods. Acidic or basic hydrolysis are rapid and efficient methods as compared to enzymatic hydrolysis. However, the hydrolysis conditions generally involve strong acids or bases and, therefore, precautions should be taken to ensure that the resulting analyte is stable under the conditions used.

f. Liquid-Liquid Extraction

Liquid-liquid extraction (LLE) is a versatile technique that is commonly used for sample preparation. In this extraction procedure, the pH of the aqueous phase is adjusted so that the sample is in its unionized form. This enables preferential partitioning of the analyte(s) into the organic solvent and leaves behind unwanted endogenous material. In general, the less polar the solvent, the more selective it is. The solvent is then evaporated to dryness, and the dried residue is reconstituted in a small volume of solvent and reserved for further chromatographic analysis.

A major advantage of LLE is selectivity. The compound of interest can be extracted from the biological matrix by an appropriate choice of solvents and pH adjustment of the aqueous phase. Selectivity can be enhanced by multiple extraction procedures or back extraction whenever required, i.e., by dissolving the analyte back and forth from solvent to aqueous phase with pH adjustment and subsequently reextracting into an organic solvent.

One of the major problems in liquid-liquid extraction is formation of emulsions that are extremely difficult to break. Emulsion formation causes loss of analyte by occlusion within the emulsion. Emulsion formation can be minimized by the use of large volumes of the extracting solvent, by less vigorous mixing, centrifugation, and ultrasonication. Other drawbacks of this procedure are requirement of large solvent volumes, long analysis time, and low analyte recovery, besides being tedious and labor intensive. The lower recoveries of analyte are due to degradation or volatilization by heat and inability to complete transfer of the extracting solvent. LLE is also not suitable for highly polar compounds. However, the use of a suitable ion pairing reagent can markedly influence the extraction efficiency.

In recent years, the use of liquid-liquid extraction columns has gained popularity for isolation of compounds from biological samples. These are disposable columns that contain diatomaceous earth as an adsorbent. They are commercially available with the trade names Tox Elut, Chem Elut, and Extrelut. The diluted sample to be extracted is poured through the column. The analyte is retained on the porous support material as a very thin film. The column is then eluted with the extracting solvent that is not miscible with water. The liquid surface area of the film allows very efficient extraction of the analytes of interest. By the use of these columns, the difficulties arising with the conventional liquid-liquid extraction, i.e., emulsification, phase separation, and consumption of high volume of solvent, are overcome.

g. Liquid-Solid Extraction

An important addition to sample preparation techniques has been the introduction of liquid-solid extraction (LSE). This consists of mixing the solvent with an adsorbent, separating it, and then eluting the analyte with an appropriate solvent. This extraction scheme works on the principles of modern liquid chromatography. The approach typically entails either retaining the analyte on a solid phase allowing isolation and clean up, or eluting it rapidly in the minimum possible volume, prior to analysis. The capacity factor (k') should be > 1000 for retention and < 0.001 for the elution step.[17]

The conventional adsorbents, such as carbon, celite, alumina, nonionic resins, ion-exchange resins, and Sephadex gel, are still widely used for sample preparation. In recent years, the development of bonded silica chemistry has given an impetus for advances in sample preparation using LSE. Many different types of cartridges for solid-phase extraction (SPE) are commercially available from an increasing number of vendors, among which Analytichem International, Bond Elut, Water Assoc, Sepak, Baker Bond, Supelco, and Alltech are internationally known. The SPE cartridges are packed in a luer-tipped polypropylene column into which 100 to 500 mg of sorbent is dry packed between two 20 µ polyethylene frits. In this procedure, the column is first conditioned with an appropriate solvent (e.g., methanol, hexane, and chloroform) to solvate the functional groups of the sorbent. The sorbent is further conditioned with the sample matrix solvent (usually aqueous), and then the test sample solution is forced through the sorbent by aspiration or positive pressure. The analyte is retained on the column and it is subsequently washed with an appropriate solvent that selectively elutes the impurities but leaves the analyte on the column. The purified analyte is finally eluted with a solvent strong enough to displace the analyte from the sorbent. Further details of this technique have been reviewed by Mcdowall et al.[16,17]

E. SAMPLE APPLICATION

The air-dried TLC plates are activated in a hot air oven at about 110 to 120°C for 1 to 2 h. During this activation step, water that could interfere with the action of the solvent is driven off from the plate. The dried and activated chromatoplate should be kept in a storage cabinet to protect them from moisture, dust, and laboratory fumes. After activation, the TLC plates are spotted with test material, and the spot is allowed to air dry. The samples are spotted 1.5 cm up from the lower edge of the plate and 1.0 cm in from the edge of the plates. For qualitative analysis, sample is applied mainly using a hypodermic syringe, disposable fine glass capillary, or lambda pipette, whereas for quantitative purposes, a wide range of autospotters are used that are capable of applying solution spots or streaks. These devices are fully microprocessor-controlled. The amount of the sample to be applied in one spot depends not only upon the thickness of the layer but also upon the composition of the sample and the principle of chromatography employed. Kaiser[19] has reexamined the handling of sample working volumes.

F. EQUILIBRATION

Prior to spotting TLC plates, it is necessary to prepare the developing chamber into which they are to be placed. This is done by lining the three sides of the developing tank with the filter paper and filling it to a height of 0.5 cm with the developing solvent and allowing the solvent within the chamber to equilibrate for 1 to 2 h. The chamber is closed with the glass lid and thus saturates the chamber with solvent vapors. Saturated chambers give reproducible and consistent R_f values.

G. DEVELOPMENT

After spotting, the TLC plates are then placed in the developer, and the solvent is allowed to migrate the required distance needed to obtain the desired separation. There are currently in existence a large variety of development strategies[7,8] that improve the performance and speed of thin-layer separation. The methods include continuous development, multiple development and its instrumentalized variants of programmed multiple development (PMD) and automated multiple development (AMD), continuous multiple development, circular (or radial) and anticircular development, over-pressed TLC, two-dimensional TLC, two-phase triangular development, and centrifugal development. Recently, a computerized statistical method has also been published for the selection of the solvent system.[20]

H. DETECTION AND IDENTIFICATION OF SEPARATED SPOTS

After development, the plates are allowed to air dry, and the separated spots are visualized by nondestructive physical or destructive chemical methods. Physical methods are fluoresced in ultraviolet light or a counter is used for the detection of radioactive components. Such methods have the advantage that the chemical substances are not altered and thus can be recovered intact. In chemical methods, staining is used. This is done either by dipping the plate into the developing reagent or by spraying, which produces characteristic color reactions. Identification is achieved by co-chromatographing reference standards with the unknown, followed by comparison of their R_f values. The R_f value is a characteristic of a particular substance and is described as the ratio of the distance traveled by the constituent to the distance traveled by the solvent. Postchromatographic derivatization for TLC has been extensively reviewed.[21] Measurement of the *in situ* fluorescence and *in situ* UV absorbance has increased the detection sensitivity of the separated spots by a factor of 100 to 1000.

I. INSTRUMENTATION — DENSITOMETRY FOR QUANTITATION IN TLC

Instrumentation for *in situ* measurement of TLC chromatograms first appeared in the mid 1960s. It is best carried out using flying spot densitometers. Scanning densitometers offer a fast and simple means of directly quantitating the chromatographic spots with great sensitivity. *In situ* measurements of substances on TLC plates can be performed by a variety of methods: reflectance, transmission, simultaneous reflectance and transmission, fluorescence quenching, and fluorescence. The methods are based on measuring the difference in optical response between blank portions of the medium and regions where a separated substance is present. When monochromatic light strikes the opaque medium, some light may be reflected from the surface, some may be absorbed by medium, and the remainder will be diffusely reflected or transmitted by the medium. The measurement of the signal diminution between the light transmitted or reflected by blank zone of the plate and a zone containing sample provide the mechanism for quantitative measurements by absorption. For fluorescence measurements, the sample beam serves to excite the sample to emit secondary radiation of longer wavelength. Blank zones of the plates are optically dark and the sample zone can, therefore, be considered as a point source of illumination superimposed on the "dark background." The most sensitive *in situ* method is fluorometry. Hence, it should always be used whenever substances exhibit an intrinsic fluorescence or where it is possible to convert substances into a fluorescent compound.

Significant technology advances have been made in the densitometric technique in recent years. Modern densitometers have a high degree of automation, both in the recording process and in the manipulation of the data obtained. They are equipped for automated peak location, multiple wavelength scanning, and spectral comparison of fractions and are capable of measurement in any mode. Scanning, calibration, and quantification are both time-consuming and tedious processes when performed manually. These procedures are also prone to variable errors, such as those arising from positioning of the spot in the measuring beam and estimating the correct shape of the calibration curve. These errors have been minimized by built-in computer control of the densitometer. For further details of densitometers, readers may go through the other available published literature.[22-25]

J. VALIDATION OF TLC RESULTS

Validation of results is an important step in the analysis of biological samples, as the analytical results are often used in making decisions in the diagnosis and treatment of diseases in patients. The validation of TLC methods requires the demonstration of calibration linearity, specificity, sensitivity, extraction recovery, precision, and accuracy.

1. Calibration Linearity

Calibration is one of the most important steps in quantitative chemical analysis. Without a good calibration procedure, precision and accuracy cannot be achieved. A preferred method of preparing the calibration standard of the test compound in a particular biological fluid is to supplement the fixed volume of the same blank biological fluid with accurately measured concentrations of the reference test standards in an increasing order. A fixed concentration of internal standard in each sample can also be added to ensure good and consistent recovery. The reference standard and unknown samples are then processed similarly as per the assay procedure. The residues of each of the processed samples are dissolved in a fixed volume of solvent, and then fixed volumes of a series of standard processed samples are spotted in an increasing concentration on the same plate as the unknown samples. After development, the plate is scanned with a densitometer, with or without spraying of a detection reagent, as required. A standard curve can then be prepared by measuring peak area to peak height from which concentrations of unknown samples can be determined. In order to check the reproducibility of the results, chromatography can be repeated three or four times on different plates, and the mean value of such experimentally obtained data can be plotted for each concentration spotted.

It is important to mention that the calibration curve for *in situ* reflectance and transmission absorption measurements are inherently nonlinear. The propagation of light within an opaque medium is very complex, because opaque media such as thin-layer plates diffusely scatter light to a significant degree, and adherence to the Beer-Lambert law cannot be expected. In recent years, extensive efforts have been made to linearize absorption calibration curves by mathematical transformation of raw data combined with regression analysis or the use of electronic transformation of the initial densitometric signal based on the explicit hyperbolic solution of the Kubelka-Munk model.[25] Currently, densitometers are specifically designed to gather data based on the explicit hyperbolic solution of the Kubelka-Munk model that allow the linearization.

2. Specificity

This term refers to the degree to which a TLC test can discriminate between closely related compounds. The specificity of a chromatographic assay depends largely on the chromatographic separation. Specificity in TLC methods is achieved by a combination of differential pH extraction, different mobilities of compounds on TLC plates (i.e., different R_f values), and different staining reactions with chromogenic detection sprays. The separability of compounds on a TLC plate is also influenced by the size of the plate, the relative solubility of the compounds of interest in the developing system, and the affinity of the compounds to the adsorbent layer. The detection systems commonly used in instrumental TLC, such as UV absorbance and fluorescence, also contribute to assay specificity.

3. Sensitivity

Sensitivity is defined as the minimal concentration of an analyte that can be detected in the original undiluted biological fluids. Because of the great variation among TLC methods and extraction procedures, it is difficult to determine single values for the sensitivity of TLC methods to detect different compounds. It also varies from laboratory to laboratory. In general, sensitivity in TLC procedures can be enhanced by using differential staining reactions with chromogenic detection sprays. Moreover, including a hydrolysis step further increases the sensitivity of the assay where compounds are excreted in biological fluids in the conjugated form. As has been mentioned earlier, various detection devices are also used for enhancing the sensitivity of the assay.

4. Precision and Accuracy

Precision and accuracy determine the error of analysis and are the most important criteria for assessing the performance of an analytical method. The precision of the analysis is estimated as the relative standard deviation (coefficient of variation) of the measured concentrations of replicate samples, and the accuracy is estimated as the percent differences (bias) between the mean values and the true or known concentrations. When the analysis of the spiked control samples (approximately N=20) of the same concentration are estimated on the same day, the precision data are reported as same-day precision (intra-assay coefficient of variation), and if they are estimated on different days, the precision data are reported as between-day precision (inter-assay coefficient of variation). In the author's opinion, assay results with < 10% bias may be considered acceptable in the analysis of biological specimens.

5. Extraction Recovery

The recovery of analyte by various analytical procedures from a biological matrix is a crucial step for all chromatographic techniques. Various extraction techniques have already been discussed in the section on sample preparation methods. Each step in the extraction process has to be performed carefully to avoid loss of analyte. Loss of analyte occurs due to incomplete extraction, adsorption, volume loss, or co-precipitation. Recovery of compounds greatly affects the sensitivity, precision, and accuracy of the test procedure. Recovery should also be reproducible. However, good precision and accuracy cannot be achieved if the recovery of the analytes is not reproducible. It is, therefore, essential that the recovery of the compound of interest from the matrix be checked and compared to that of a nonprocessed standard. The extraction recovery can be estimated by comparing the slope of the calibration curve based on the calibration standards prepared by spiking with accurately measured amounts of test compound of high purity (reference standard) and processed as per the assay procedure with the slope of the calibration curve of the pure unprocessed reference test compound of same concentration.

V. RECENT DEVELOPMENT IN TLC PLATES

A recent development in TLC is the introduction of high performance TLC plates (HPTLC). These plates have enhanced the capability of thin-layer chromatography. HPTLC layers are slightly thinner than conventional TLC materials (0.20 mm instead of 0.25 mm), with smaller particle size (7 µm instead of 12 µm), and they give better separation performance over a shorter migration distance (50 mm instead of 100–200 mm). On HPTLC layers, chromatography takes place in the fast capillary flow range of the respective developing solvent, giving faster separation and reduced diffusion, i.e., better efficiency. In addition, it has the advantage of lower detection limits by a factor of 10 to 15 and is more economical due to less solvent consumption and lower plate costs per sample.

VI. COMPARISION OF INSTRUMENTAL TLC WITH OTHER ANALYTICAL TECHNIQUES

In recent years, quantitative thin-layer chromatography (QTLC) has gained analysts' interest.[26] More and more, analysts accept the fact that both HPLC and quantitative TLC have their specific merits and are complementary techniques. Some typical features can make quantitative TLC the method of choice over liquid chromatography due to the following merits[27]:

1. Sample preparation is usually simple, because the stationary phase is used only once. On the other hand, HPLC requires relatively clean sample extracts. More complicated sample pretreatment is necessary as the stationary phase is intended for multiple reuse. If the sample is not properly cleaned, expensive columns would be damaged.
2. Due to the off-line character, sample application, chromatogram development, and evaluation (*in situ* scanning) can be done concurrently.
3. TLC offers a much higher sample throughput because of the possibility of performing separation of many samples simultaneously.

4. Mobile phase requirement per sample is much less and degassing is not required as for HPLC.
5. Detection takes place in the absence of the mobile phase and, therefore, UV-absorbing mobile phases can be used without interfering with detection.
6. As the chromatograms remain unaffected even after scanning, it is possible to reassess the chromatograms with the same or with different parameters without repeating. Moreover, the same plate can be used for recording the absorbance of the positively identified fractions and to recover the fractions for further analysis.
7. A major advantage of TLC is positive identification of compounds by pre- or postchromatographic derivatization, by spectroscopy, or by a combination of both.
8. Co-chromatography of standards and unknowns can be done under identical conditions. This is not possible in HPLC because samples have to be injected one after the other.
9. Running costs (solvents) and cost per analysis are relatively low in comparison to HPLC.
10. Any combinations of mobile and stationary phases can be used in QTLC. This is not true in HPLC.
11. Due to the off-line principle, QTLC is extremely flexible.

There are, however, certain limitations of QTLC for which HPLC does have the following advantages:

1. The chromatographic resolution power of HPLC is better than that of QTLC due to the short migration distance of the latter.
2. Quantitative precision of HPLC is somewhat better than that of QTLC.

VII. APPLICATIONS OF TLC IN CLINICAL CHEMISTRY

Thin-layer chromatography is widely used in clinical laboratories for detection of a number of substances like amino acids, bile acids, carbohydrates, cerebrosides, drugs and their metabolites, gangliosides, lipids, phospholipids, porphyrins, and prostaglandins in various biological specimens, which are of significance in medical diagnosis. In view of the complexity of the subject at hand, it is not possible to cover all of the applications extensively. Therefore, important selected applications of this technique will be discussed in the remainder of this chapter.

A. AMINO ACIDS

Chromatographic separation of amino acids has become of great importance in biomedicine. For clinical purposes, the estimation of free amino acids in biological fluids and tissues aids in making diagnosis of inborn errors of amino acid metabolism.[28,29] Early diagnosis and treatment may prevent neurological damage and mental retardation in young infants with inborn errors of amino acid metabolism. Of the chromatographic techniques, one- and two-dimensional thin-layer chromatography has proved useful to screen and quantitate free abnormal amino acids in urine and blood samples.[29]

For amino acid analysis, pretreatment of sample is necessary prior to chromatographic analysis to remove proteins, lipids, inorganic salts, or other substances that interfere with chromatographic separation. Of the possible choices of stationary phases, use of cellulose is popular in most laboratories for separation of amino acids. The main advantage of using cellulose is that it gives higher chromatographic resolution in a shorter time. A large number of solvent systems have been in use for separating amino acid mixtures. The most common solvent system for separating amino acids by one-dimensional TLC[30] is *n*-butanol–acetone–acetic acid–water (35:35:10:20, v/v), while pyridine–acetone–ammonia–water (80:60:10:35, v/v) and butan-1-ol–acetone–water–acetic acid–water (35:35:10:20, v/v) are used in two-dimensional TLC.[31] The most widely used staining reagent for qualitative and quantitative assessment of amino acids is ninhydrin-collidine, a polychromatic staining reagent.

B. BILE ACIDS

Bile acids are 24-carbon steroid derivatives. They are formed by the conversion of cholesterol to cholic and chenodeoxy cholic acids (primary bile acids). These are then conjugated with glycine or taurine via amide linkage in the liver. After conjugation they are eliminated together with the bile. Most of the bile acids entering the gut are reabsorbed in the terminal ileum, while some of them undergo bacterial deconjugation. In this process, secondary bile acids (deoxy cholic, lithocholic, and ursodeoxycholic acids) and a number of keto-bile acids occurring in feces are formed. The investigation of bile acids has clinical importance in the diagnosis of certain liver or intestinal disorders. In biomedicine, concentrations of bile acids in human feces are used for diagnosis of bile acid malabsorption and chologenic diarrhea. In liver disorders, serum levels of bile acids are elevated and their measurement is a sensitive index of liver disease. Bile acids are not found in urine owing to efficient uptake by the liver and excretion into the intestine. In hepatocellular disease and obstructive jaundice, however, their urinary excretion increases.

The determination of bile acids is done in various biological specimens,[32] e.g., bile, serum, duodenal contents, and crude fecal extracts. Thin-layer chromatography (TLC) has been used extensively for the separation of bile acids. Prior to TLC analysis of bile acid, purification of the specimen is essential. Sample cleanup steps would, however, depend upon the type of biological specimens being analyzed. Isolation of individual free bile acids in serum involves protein separation, alkaline hydrolysis, solvolysis of conjugated bile acids, and purification of lipids. Kindel et al.[33] published a very useful paper in 1989 for separation of bile acids by TLC. These authors separated five predominant bile acids (cholic, chenodeoxycholic acid, deoxycholic acid, lithocholic acid, and ursodeoxycholic acid) in human stool specimens on a silica gel plate using two different solvent systems. The plate was first developed in isooctane–2-propanol–acetic acid (30:10:1, v/v) for 40 min, dried, and developed again with isooctane–ethyl acetate–acetic acid (10:10:2, v/v) for 65 min. For quantitative analysis, the plate was dipped in 0.2% 2,7-dichlorofluorescein ethanolic solution, and bile acid fluorescence was measured by TLC with direct scanning fluorometry (TLC-DSF). In clinical practice, this TLC procedure constitutes a reliable, simple, and time-saving alternative to gas chromatography.

A binary mixture (methanol-chloroform, 2.5:97.5, v/v) was proposed by Ferreira et al.[34] for separation of pentachlorophenyl esters of bile acids. Derivatization to methyl esters and chromatographic mobilities of 26 theoretically possible hydroxylated derivatives substituted in one or more of the C_3, C_7, and C_{12} positions of methyl-5-beta-cholanate were thoroughly studied by Iida et al.[35]

In general, fluorimetry is the most sensitive detection technique for quantitative analysis of bile acids.[36] Bile acids are converted into fluorescent derivatives by reaction with 5% perchloric acid[37] in methanol. However, spraying with 10% sulfuric acid and subsequent heating at 90°C for 10 min yields fluorescent, light blue spots on a dark violet background, which has proved very successful in detection.[7]

C. CARBOHYDRATES

Carbohydrates are naturally occurring organic substances that contain mainly carbon, oxygen, and hydrogen. Several diseases are accompanied by increased elimination of various sugars in the urine and feces. In biomedical research and clinical laboratories, all chromatographic techniques are applied for separation of sugars in various biological samples. Paper chromatography (PC) and TLC have been used extensively in the analysis of sugars, but now PC has been superseded by TLC.[38] Although numerous sugars exist, only a few are present in biological samples. Various sorbents are used in the TLC of sugars. The most popular are magnesium silicate, alumina, kieselguhr, silica gel, or mixture of the last two, and amino propyl-bonded silica. Cellulose TLC is used for sugar analysis by some laboratories. It is based on earlier procedures developed for PC analysis of sugars. Separation of sugars (sucrose and fructose) in biological samples has been recommended on silica gel with ethyl acetate–isopropanol–water (60:30:10, v/v) and isopropanol–n-butanol–0.5% aqueous boric acid (50:30:30, v/v) followed by n-butanol-acetone-0.5% boric acid (40:50:10, v/v), respectively.[7]

In clinical laboratories, TLC has been used for determination of urinary excretion of lactulose of patients with cystic fibrosis using silica gel with propanol–borate buffer solvent and dilute sulfuric acid as a spraying reagent.[39] Menzies et al.[40] reported quantitative estimation of clinically important monosaccharides in plasma by rapid thin layer on silica gel plates with ethyl acetate–pyridine–glacial acetic acid–water (60:30:10:10, v/v) solvent system and n-amino benzoic acid as spraying agent. In various inherited lysosomal diseases, abnormal oligosaccharides are excreted in urine, and thin-layer chromatography has proved quite successful in their separation.[41–44]

D. DRUGS

Drug analysis has always been of interest to clinical chemists. Drug analysis in body fluids is usually carried out for two purposes: the identification and quantification of drugs taken for nontherapeutic purposes, usually analgesics, sedatives, tranquilizers, cannabinoids, and stimulants; and therapeutic monitoring of drug levels to ensure drug dosage. The former is usually confined to the emergency investigation of the unconscious or semiconscious patient, and drug screening is increasingly being sought by drug dependence treatment centers. Detection of drugs of abuse in biological fluids provides concurrent and independent validation of drug abuse. These tests supplement and confirm clinical history of drug intake, monitor treatment efficacy with regard to illicit drug use, contribute to the clarification of the medical-legal problems, and promote pharmacokinetic and metabolic research. Therapeutic drug monitoring has also become a routine part of many clinical laboratories. Plasma drug concentrations at any given point in time are used to adjust and optimize dosage on an individual basis.

Several types of techniques are used for drug analysis. These include chromatographic techniques, immunological methods, spectrophotometry, spectrofluorometry, gas chromatography–mass spectrometry, among others. The optimal choice of technique depends on many factors, which include cost, workload, and program needs. Of the chromatographic techniques, TLC is one of the oldest methods and is still used for large-scale multiple drug screening programs.[45–47]

A number of body fluids and tissues can be used for drug abuse testing. Urine, however, is the specimen of choice for the mass screening as it is readily accessible in adequate quantities, can be obtained noninvasively, and most drugs and drug metabolites are present in urine in relatively high concentrations. Urine is a simpler matrix than other biological specimens because of the usual absence of protein and cellular constituents, which simplifies its sample preparation and analysis. As for the analysis of other compounds, sample pretreatment is necessary prior to TLC. The sample pretreatment step requires pH-dependent extraction followed by purification and concentration of the drug from the biological sample.[15,48,49] In some laboratories, commercially available prepacked columns (Tox Elut, Extrelut, Chem Elut, and Bond Elut) are commonly used for the purification step in drug analysis.

Most TLC plates used for drug detection are coated with silica gel. For toxicological drug screening, silica gel G with a fluorescent indicator is most commonly used for separating abused drugs. A large variety of solvent mixtures have been proposed for development of TLC plates.[50] The most commonly described developing solvent system is ethyl acetate–methanol–ammonium hydroxide (85:10:5, v/v), first proposed by Davidow et al.[51] in 1966. Chloroform–acetone (9:1, v/v) is another common mixture, used especially for separation of barbiturates and other acidic drugs. The cannabinoids are usually separated with heptane–ethyl ether–acetic acid (80:10:4, v/v) or chloroform–methanol–ammonium hydroxide (85:15:2, v/v). Chromogenic sprays[50] used for detecting drugs of abuse include ninhydrin (amphetamine and primary amines), diphenyl carbazone and mercuric sulfate (barbiturate and glutethimide), iodoplatinate (narcotic analgesics and cocaine, phenothiazines, antidepressants, antihistamines), Dragendorff's reagent (narcotic analgesics), Fast blue BB salt, and Fast blue B salt (cannabinoids). Currently, a standardized TLC kit, the Toxi Lab system, is commercially available. It has acceptable sensitivity for abused drugs and is used for screening biological specimens in autopsy and emergency cases.[52,53] In recent years, detection and quantification of abused drugs, viz., cannabinoids,[54,55] buprenorphine,[56] benzodiazepine,[57,58] heroin, and cocaine,[59] by densitometric techniques on HPTLC and TLC plates have proved quite successful.

As was stated earlier, quantitative densitometric TLC methods are found to be quite suitable for estimation of drugs like chlorpromazine, tricyclic antidepressants, and anticonvulsant drugs at a therapeutic level in blood serum and plasma.[60,61] Analysis of these drugs by TLC is found to be rapid and reproducible, and it is well suited to other drugs, like sulfonamides,[62] steroids,[63] and antibiotics.[64]

E. LIPIDS

Lipids are ubiquitous in body tissues and play a vital role in virtually all aspects of biological life. They are critical components of the enzymatic conditions and transport systems of biological membranes. Lipids may be polar (phospholipids or sphingolipids), nonpolar, or slightly polar (glycerides and sterol esters). Analysis of lipids in serum, mainly cholesterol ester, has great diagnostic value in hepatic diseases and lipid metabolism disorders. In the field of cutaneous research, analysis of sebaceous lipids has an important part in defining the role of sebum in many skin disorders.

The chromatography of lipids is very well documented. Various chromatographic methods have been advocated for analysis of lipids and are extensively employed in biomedical research, but only a few of them are used in routine clinical chemistry. For biomedical purposes, TLC has proved suitable for quantitative lipid analysis and for obtaining a qualitative profile of total lipid. Several texts dealing with chromatographic analysis of lipids in biomedical research have been published.[65,66] A detailed description of various lipids is beyond the scope of this chapter.

Lipids are primarily extracted from biological samples before they are subjected to chromatography. Preliminary extraction is mainly done by the extraction procedure of Folch et al.[67] with chloroform–methanol (2:1, v/v). Extraction of lipids from total plasma or cerebrospinal fluid is performed as per the procedure of Nelson[68] or Mitchell.[69] Further details of extraction procedures have been covered in the *CRC Handbook of Chromatography: Lipids*.[70] After extraction, chromatography on a thin layer of silica gel is the method of choice for the fractionation of complex lipid mixtures into classes of compounds. Various single, binary, ternary, and multicomponent solvent systems are used for the separation of lipids. Two types of procedure are recommended in the literature for separation of lipids: starting with a solvent of strong eluting strength, with the use of a less polar solvent in a second or third run, or the reverse. Different classes of lipids are better resolved by successive developments to different developing distances. Excellent separation of lipid classes was obtained by Bounias,[71] who performed the first migration with benzene to 12 cm, followed by a 7-cm migration with trichloroethylene–acetonitrile.

Other developing systems used for the separation of lipids by various authors are (1) benzene, (2) chloroform, (3) trichloroethylene–acetonitrile (85:15, v/v), (4) heptane–diethyl ether–acetic acid (70:27:10, v/v), (5) benzene–chloroform–formic acid (35:15:1, v/v), chloroform–methanol–water (75:25:4, v/v), followed, after drying, by migration with chloroform to the middle of the plate and then hexane–chloroform (30:10, v/v) to the upper edge of the plate, methanol to 3 cm, followed by trimethylpentane–diethyl ether–acetic acid (75:25:2, v/v). Bhat et al.[72] reported an improved method for separation of ethyl and chloroethyl esters of fatty acids in tissue samples. These authors used silver-impregnated normal and reversed-phase HPTLC plates. The separation was achieved by circular phase TLC in two solvent systems, i.e., acetonitrile–methanol–water (10:2:2, v/v) or methanol–water (9:1, v/v). Normal phase HPTLC plates were sprayed with copper (II) sulfate reagent (10% w/v) in 8% (v/v) phosphoric acid and detected by heating at 100 to 200°C. Reversed-phase TLC plates were exposed to iodine vapors for location of spots. Yao and Rasteller[73] separated nonpolar lipids into two solvent systems: (1) benzene–diethyl ether–ethanol–acetic acid (60:40:1:0.05, v/v) followed, after drying with (2) hexane–diethyl ether (47:3, v/v). Sphingolipids and phospholipids were separated in methyl acetate–propanol–chloroform–water–0.25% potassium chloride (25:25:25:10:9, v/v) followed by benzene–diethyl ether–ethanol–acetic acid (60:40:1:0.05, v/v). Analysis and quantification of the individual components of nonpolar or slightly polar lipids is of great interest for the evaluation of assimilation diseases. Cholesteryl esters, triglycerides, and free cholesterol are well resolved on silica gel HPTLC plates with hexane–heptane–diethyl ether–acetic acid. (63:18.5:18.5:1, v/v) followed by further development with heptane–chloroform (60:40, v/v) for separation of cholesteryl esters.[74]

Thin-layer chromatography is also used for separation of lipids from patients with alcoholic disease, as alcoholic fatty liver is accompanied by a major increase in triglycerides and a smaller increase in cholesteryl esters.[75] Endonard et al.[76] reported separation and quantification of free cholesterol and cholesteryl esters in skin biopsies by TLC and recommended that this technique be used as a diagnostic tool to detect pathologies of skin lipid metabolism. A simple and accurate routine TLC procedure for qualitative analysis of skin surface lipids (cholesterol, free fatty acids, triglycerides, wax esters, and squalene) without sample pretreatment was developed by Saint Leger et al.[77] Separation of triglycerides, which constitute a sizeable percentage of sebum, was achieved by Jacobsen et al.[78] Separation of fatty acids or fatty acid esters by conventional TLC techniques is rather tedious. Such methods are either time consuming or the resolution is sometimes poor. In order to achieve separation by TLC, derivatization is required. Currently, Ohata et al.[79] have developed a rapid and quantitative procedure for determination of the fatty acid composition of plasma lipid using silica gel plates. The plates were partially developed in methanol, followed by chloroform–methanol (1:1, v/v) and then fully developed in hexane–diethyl ether–acetic acid (80:20:1, v/v). This method was recommended for screening large plasma samples as no extraction of lipids was involved.

Lipids are usually rendered visible and subsequently quantitated by either charring or by color-forming sprays. Iodine vapors are used for qualitative detection, whereas fluorescence measurements after derivatization are found to be most suitable for quantification. Manganese (II) chloride is mainly used as a derivatizing agent. Few lipids have significant UV chromophores, but post-chromatographic derivatization methods result in almost universal detection. In general, high performance thin-layer chromatographic techniques have resulted in better resolution between critical lipid groups.

F. PHOSPHOLIPIDS

This lipid class includes the phosphoglycerides and sphingolipids. These are amphipathic lipid constituents of membranes. The determination of phospholipid is important. Surface-active phospholipids play a vital role in the mechanical behavior of the lungs. Respiratory distress syndrome (RDS), or hyaline membrane disease, occurs in premature and full-term infants. This syndrome is due to immaturity of the lung with respect to its ability to synthesize and secrete a lipid protein lecithin material, known as a surfactant, into the alveolar space. Deficiency of surfactant causes collapse of the alveoli, which results in RDS. Thus determination of the ratio of phosphatidyl choline (lecithin) and sphingomyelin (L/S) in the amniotic fluid provides an index to fetal lung maturity.[80–84]

Presently TLC has become the most widely used technique for most of these analyses after the pioneering work of Gluck et al.[83,84] on the L/S ratio. Prior to chromatography, purification of phospholipids is essential. Purification involves the following steps: (1) centrifugation, (2) extraction, (3) acetone precipitation, (4) evaporation, and (5) reconstitution. The extraction of lipids is often carried out by the method of Gluck and Kulovich[84] or the modified procedure of Kolins et al.[85] A solid phase extraction procedure has also been proposed for extraction of lipids.[86] The TLC systems used most extensively employ a silica gel stationary phase and a mobile phase consisting of either chloroform–methanol–water or chloroform–methanol–ammonium hydroxide.[87] Improved resolution of lecithin and sphingomyelin occurs with addition of either acid or base to the water. A double development system using chloroform–methanol–acetic acid–water (70:30:4:3, v/v) and chloroform–methanol–acetic acid–water (60:30:4:3, v/v) has also been used for the separation of phosphatidyl choline and sphingomyelin by circular HPTLC.[7] Both one- and two-dimensional TLC methods have been reported for the determination of phospholipids.[7] Kolins utilized short bed continuous development and high performance thin-layer chromatography plates for resolution of phospholipids.[85]

Visualization of the separated phospholipids is done by color-forming sprays or charring. The intensity of the colored spots depends on the degree of unsaturation of the phospholipid constituent fatty acid. Iodine vapor and molybdate ions also react with the unsaturated bonds in the fatty acid moieties of the phospholipids. Other agents, such as cupric acetate and sulfuric acid with or without dichromate, react with the fatty acid double bond and give colored products. Staining with the periodic acid Schiff stain and charring agent, i.e., 10% copper sulfate in 8% orthophosphoric acid, have also been recommended for detection.[7] Fluorescent dyes reported in the literature for detection of phospholipids include rhodamine 6G, rhodamine B, 2′,7′-dichlorofluorescein (DCF), anilino-8-naphthalene sulfonate, and 8-anilino-1-naphthalene sulfonate (8-ANS).[88] The fluorometric TLC procedure is highly recommended for accurate and reproducible measurement of the L/S ratio.[88] Spraying with 2,5-bis (5-tert-butyl-2′-benzoxa-zolyl) thiophene(BBOT) has also been found very sensitive.[7] Laser densitometry is suitable for simple and efficient detection of phospholipids.[89] Quantitation at lower than 25 ng per sampling zone can be achieved by this technique. Colarow[89] reported the use of the lipophilic fluorescence agent, 4-(N, N-dihexadecyl) amino-7-nitrobenzene-2-oxo-1, 3-diazole (NBD dihexadecylamine), in the mobile phase. Using this agent, fluorescent phospholipid spots are quantified without any additional procedures by reflectance densitometry at 366 nm.[89]

Extremely high levels of ether phospholipids were detected in the phospholipid composition of human eosinophils, and these were separated by using a combination of TLC and gas chromatography.[90] These results suggest that alkyl and alkenyl ether phospholipids play an essential role in human eosinophils in various physiological and pathological conditions. Phospholipids of many cancer tumors also contain larger amounts of alkyl ethers than healthy tissues. Narayan et al.[91] studied alterations in sphingomyelin and fatty acids in human benign prostatic hyperplasia and prostatic cancer. These authors extracted lipids from tissues and separated them by TLC and GLC. Dembitsky[92] reported the separation and quantification of alkenyl ether phospholipids (plasmalogen). Recently, Cartwright[93] and Gregson[94] have also published papers on the separation of phospholipids by TLC.

G. PROSTAGLANDINS

Prostaglandins (PGs) are derivatives of polyunsaturated essential fatty acids. The parent compound is a 20-carbon unnatural fatty acid known as prostanoic acid that contains a five-membered cyclopentane ring. Derivatives that contain this structure (prostaglandins, prostacyclins, and thromboxanes) are known collectively as prostanoids. Although their full physiological role is not completely known, they have diverse biological actions. They are extremely potent, producing physiological actions at concentration of 1 ng/ml.

TLC is still widely used for the separation and identification of PGs and related compounds. Raajmakers[95] published a very useful paper on TLC using chloroform–methanol–acetic acid–water solvent system for separation of all prostaglandin groups from pure susbtances. Chloroform–isopropanol–ethanol–formic acid (45:50:0.5:0.3, v/v) and ethyl acetate–acetone–acetic acid (90:5:1, v/v) are claimed to be efficient for separating major PGs.[7] Tsunamoto et al.[96] used ethyl acetate–isooctane–acetic acid–water (11:5:2:10, v/v) and chloroform–methanol–acetic acid (90:8:6, v/v) for separation of PGs by two-dimensional TLC. RPTLC was used by Beneytout et al.[97] for separation of prostaglandins; PGF_2, PGE_1, PGE_2, PGA_1, and some arachidonic acid metabolites were separated on silica gel modified with phenyl methyl vinyl chlorosilane. PGs can also be separated by using a combination of TLC and HPTLC.[98]

Various spray reagents, e.g., concentrated sulfuric acid, phosphomolybdic acid, 2,4-dinitrophenylhydrazide, anisaldehyde-sulfuric acid, acidic ceric sulfate, and vanillin-phosphoric acid-ethanol, have been used for the detection of prostaglandins. These are, however, destructive reagents. Goswami et al.[99] proposed an extremely sensitive nondestructive spray reagent, 8-hydroxy-1,3,6-pyrene-trisulfonic acid trisodium salt (10 mg/100 ml methanol), for locating prostaglandins on thin-layer chromatograms. To improve detection limits of prostanoids, tritium scintillator radioautography was used to detect arachidonic acid metabolites with 2400 dpm activity per spot over a period of 24 h.[100]

H. PORPHYRINS

Porphyrins are metal-free cyclic tetrapyrrole derivatives. The main sites of porphyrin synthesis in humans are the bone marrow and the liver. The clinical-biochemical significance of the porphyrins is that disturbances in their synthesis can lead to erythoporietic or hepatic porphyrias with various clinical syndromes. Hepatic porphyrias are either congenital or acquired cirrhosis. Each of the porphyrias is characterized by a specific pattern of porphyrins and porphyrin precursors in body fluids, feces, and tissue as a result of overproduction and accumulation. Analysis of porphyrins in urine or feces involves lipid precipitation from acetone and subsequent extraction with chloroform prior to separation by chromatographic techniques.

Although HPLC is used to identify and quantitate porphyrins in biological samples,[101] TLC has also gained popularity for routine examination of porphyrins in clinical chemistry. Porphyrins can be separated by TLC in the free or esterified forms. Free porphyrins are rather rarely separated by TLC, but esterified porphyrins are normally chromatographed on silica gel. The developing solvents[102] that are commonly used for separation of porphyrins are binary mixtures of hexane with methylene chloride, chloroform, or ethyl acetate, and ternary mixtures of benzene–ethyl acetate–methanol, toluene–ethyl acetate–methanol, methylene chloride–carbon tetrachloride–ethyl acetate, and chloroform–kerosene–ethanol. Separation is according to the number of carboxylic groups. Free carboxylic porphyrins in urine were separated by reversed-phase TLC in a buffered mobile phase containing N-cetyl-N,N,N,-trimethyl ammonium bromide as an ion pairing reagent.[7] The porphyrins were detected by their fluorescence at 560 nm after excitation at 404 nm. The detection limit for porphyrins by this method was about 10 fmol and the calibration graph is linear between 0.15 to 3.0 fmol.

I. STEROID HORMONES

Steroid hormones are secreted by the adernal cortex, ovary, testis, corpus luteum, and placenta. They are derivatives of tetracyclic hydrocarbons and are based on the cyclopenta-[a]-phenanthrene skeleton. The steroid hormones are defined by their physiological functions. The main classes of steroid hormones include androgens, testrogens, progestogens, glucocorticoids, mineral corticoids, and vitamin D. The estrogens and androgens are female and male sex hormones. The progestogens are involved in the preparation and maintenance of pregnancy. The glucocorticoids have a distinct effect on carbohydrate metabolism. Excessive secretion of glucocorticoids causes Cushing's syndrome, and deficiency causes Addison's disease. They also have remarkable anti-inflammatory and anti-allergic action. Mineral corticoids promote the retention of Na^+ and the loss of K^+ by the kidneys. Through this action, mineralocorticoids maintain water and salt balance in the body. Vitamin D is involved in the regulation of calcium transport.

The chromatographic analysis of steroids is well documented. Some valuable books and reviews have been published on analysis and characterization of steroids.[103–105] The general practice of TLC is described in a recent handbook[106] that includes a chapter specially on the TLC of steroids. Thin-layer chromatography is widely used in steroid analysis either as a basic analytical method or as a cleanup procedure. Sample pretreatment is also essential in steroid analysis and is mainly done by enzymatic or acidic hydrolysis procedures followed by liquid-liquid or liquid-solid extraction. Silica gel is the most commonly used adsorbent in steroid analysis. Xu et al.[107] used silica gel plates impregnated with silver ion for separation of a large group of steroids. The availability of reversed-phase alkyl-bonded silica plates has given rise to new dimensions in the analysis of steroids in recent years. The two solvents systems chloroform–acetone (9:1, v/v) and cyclohexane–ethyl acetate–ethanol (77.5:20:2.5, v/v) are claimed to be efficient for separating major estrogens, androgens, and pregnones. Other mobile phases, like dichloromethane–methanol–water (225:15:1.5, v/v) and light petroleum b.p. 37–55°C–diethyl ether–acetic acid (48:50:2, v/v) are also used for androgen separations.[109] Watkins et al.[110] resolved estrone sulfate, estradiol-17β-glucuronide, estrone β-glucuronide, estriol 16-α-glucuronide, estriol-3-sulfate, and estriol-3-glucuronide on RP-18 and RP-8 layers with the ion-pairing mobile phase methanol–0.5% aqueous tetramethylammonium chloride

(45:55 or 50:50). Some workers have used two-dimensional developments and multidimensional development modes[108] for separation of steroids.

There are various spraying reagents that can be used for locating steroids *in situ*. Most of them have been adequately discussed in the *Handbook of Chromatography*.[70] Probably the most universal one involves spraying the plate with 10% sulfuric acid in methanol, heating for 10 min at 90°C, and fluorescence scanning (ex. 366 nm, em. 509 nm) for quantification. Other recommended reagents are a mixture of copper sulfate and *o*-phosphoric acid, and fast violet salt B. Spraying with 0.1 mM 2,3-dichloro-5,6-dicyano-1,4-benzoquinone in toluene–acetic acid and subsequent densitometry has been found to be very sensitive for detection of estriol 16β-glucuronide. Combinations of TLC-GC and TLC-MS have been advocated by Curticus et al.[111] and Kraft et al.[112] for detection and identification of steroids.

J. OTHER COMPOUNDS OF CLINICAL INTEREST

Apart from the above groups, a wide range of other compounds, e.g., gangliosides,[113-115] cerebrosides,[116] purines, pyrimines, derivatives of nucleic acids,[117] and urinary organic acids,[118] are also of some clinical interest. These have been analyzed by TLC in various laboratories.

VIII. CONCLUSIONS

In light of the above discussions, it is felt that TLC is a most economical and cost effective technique. It is ideally suited for performing qualitative and quantitative analysis of compounds of clinical interest from biological samples where the sample load is high. In recent years, considerable innovation has taken place in the area of instrumental TLC, HPTLC, and densitometric techniques, and much more is expected in the near future. The current trend is toward the use of hyphenated methods, such as TLCMS and TLC-FTIR. These will provide structural information as well as quantitation without an extraction procedure and will greatly extend the application of TLC in clinical chemistry.

REFERENCES

1. **Randerath, K.,** *Thin Layer Chromatography*, 2nd ed., Verlag Chemie, Weinheim, and Academic Press, New York, 1966.
2. **Stahl, E.,** *Thin Layer Chromatography, A Laboratory Handbook*, 2nd ed., Springer-Verlag, New York, 1969.
3. **Niederwieser, A. and Pataki, G., Eds.,** *Progress in Thin Layer Chromatography and Related Methods*, Vol. I, Ann Arbor–Humphrey Science Publishers, Ann Arbor and London, 1970.
4. **Kaiser, R. E.,** in *HPTLC – High Performance Thin Layer Chromatography*, Zlatkis, A., and Kaiser, R. E., Eds., Elsevier/North Holland Biomedical Press, Amsterdam, 1977, p. 73.
5. **Bertsch, W., Hara, S., Kaiser, R. E., and Zlatkis, A., Eds.,** *Instrumental HPTLC*, Dr. Alfred Heuthig Verlag, Heidelberg, 1980.
6. **Dallas, F. A. A., Read, H., Ruane, R. J., and Wilson, I. D., Eds.,** *Recent Advances in Thin Layer Chromatography*, Plenum Press, New York, 1988.
7. **Siouffi, A. M., Mincsovics, E., and Tyihak, E.,** Planar chromatographic techniques in biomedicine: current status, *J. Chromatogr.*, 492, 471, 1981.
8. **Wilson, I. D.,** Contemporary thin-layer chromatography: an introduction, in *Recent Advances in Thin-Layer Chromatography*, Dallas, F. A. A., Read, H., Ruane, R. J., and Wilson, I. D., Eds., Plenum Press, New York, 1988, Chap. 1.
9. **McMurray, J. R.,** Collection of specimens, in *Varley's Practical Clinical Biochemistry*, Gowenlock, A. H., McMurray, J. R., and McLauchlan, D. M., Eds., Heinemann Medical Books, London, 1988, Chap. 14.
10. **Pickard, M. A.,** Collection and handling of patient specimens, in *Clinical Chemistry, Theory Analysis and Correlation*, Kaplan, L. A., and Pesce, A. J., Eds., Mosby, St. Louis, 1989, Chap. 2.
11. Report of the Substance-Abuse Testing Committee, Critical Issues in Urinalysis of Abused Substances, *Clin. Chem.*, 34, 605, 1988.
12. **Jain, R.,** Interference of adulterants in thin layer chromatography method for drugs of abuse, *Ind. J. Pharmacol.*, 25, 240, 1993.

13. **Jain, R.,** Laboratory services for drug abuse screening in a clinical setting, *J. Forensic, Med. Toxicol.*, x, 26, 1994.
14. **Jain, R. and Ray, K.,** Detection of drugs of abuse and its relevance to clinical practice, *Ind. J. Pharmacol.*, 27, 1995 (in press).
15. **Harkey, M. R. and Stalowitz, M. L.,** Solid phase extraction techniques for biological specimens, in *Advances in Analytical Toxicology*, Baselt, R. C., Ed., Biomedical Publications, Foster City, CA, 1984, Chap. 9.
16. **Mcdowall, R. D., Pearce, J. C., and Murkitt, G. S.,** Liquid-solid sample preparation in drug analysis, *J. Pharm. Biomed. Anal.*, 4, 3, 1986.
17. **Mcdowall, R. D.,** Sample preparation for biomedical analysis, *J. Chromatogr.*, 492, 3, 1989.
18. **Li, F., Lim, C. K., and Peters, T. J.,** A high performance liquid chromatographic method for the assay of coproporphyrinogen oxidase activity in rat liver, *Biochem. J.*, 239, 481, 1986.
19. **Kaiser, R. E.,** *J. Planar Chromatogr.–Mod. TLC*, 1, 182, 1988, quoted in Siouffi, A. M., Mincsovics, E., and Tyihak, E., Planar chromatographic techniques in biomedicine: current status, *J. Chromatogr.*, 492, 471, 1989.
20. **Johnson, E. K. and Nurok, D.,** Computer simulation as an aid to optimizing continuous development two-dimensional thin layer chromatography, *J. Chromatogr.*, 302, 135, 1984.
21. **Treiber, L. R.,** *Quantitative Thin Layer Chromatography and Its Industrial Applications*, Marcel Dekker, New York, 1987.
22. **Poole, C. F., Poole, S. K., and Dean, T. A.,** Quantitative methods in thin-layer chromatography, in *Recent Advances in Thin Layer Chromatography*, Dallas, F. A. A., Read, H., Ruane, R. J. and Wilson, I. D., Eds., Plenum Press, New York, 1988, Chap. 2.
23. **Baeyens, W. R. G. and Lin Ling, B.,** Thin layer chromatography application with fluorodensitometric detection, *J. Planar Chromatogr.–Mod. TLC*, 1, 198, 1988.
24. **Poole, C. F., Poole, S. K., Dean, T. A., and Chirco, N. M.,** Sample requirements for quantitation in thin layer chromatography, *J. Planar Chromatogr.–Mod. TLC*, 2, 180, 1989.
25. **Poole, C. F. and Poole, S. K.,** Progress in densitometry for quantitation in planar chromatography, *J. Chromatogr.*, 492, 539, 1989.
26. **Jork, H.,** in *Proceedings IV Intl. Symp. Instrumental HPTLC*, Traitler, H., Studer, R., Kaiser R. E., Eds., Institute for Chromatography, Bad Durkheim, F. R. G., 1987, p. 193.
27. **Jaenchen, D. E.,** Instrument thin layer chromatography, state of the art and future trends, *Internat. Analyst*, 3, 36, 1987.
28. **Wu, J. T.,** Screening for inborn errors of amino acid metabolism., *Ann. Clin. Lab. Sci.*, 21, 123, 1991.
29. **Edwards, M. A., Grant, S., and Green, A.,** A practical approach to the investigation of amino acid disorders, *Ann. Clin. Biochem.*, 25, 129, 1988.
30. **Silverman, L. M. and Christenson, R. H.,** Amino acids and proteins, in *Tietz Textbook of Clinical Chemistry*, Burtis, C. A., and Ashwood, E. R., Eds., W. B. Saunders, Philadelphia, 1994, Chap. 19.
31. **Varley, H.,** Amino acids, in *Varley's Practical Clinical Biochemistry*, Gowenlocks, A. H., McMurray, J. R., and McLauchlan, D. M., Eds., Heinemann Medical Books, London, 1988, Chap. 18.
32. **Street, J. M., Trafford, D. J. H., and Makin, H. L. J.,** The quantitative estimation of bile acids and their conjugates in human biological fluids, *J. Lipid Res.*, 24, 491, 1983.
33. **Kindel, M., Ludwig-Koehn, H., and Lembcke, M.,** New and versatile method for the determination of faecal bile acids by thin layer chromatography with direct scanning fluorimetry, *J. Chromatogr.*, 497, 139, 1989.
34. **Ferreira, H. C. S. and Elliott, W. H.,** Pre-column derivatization of free bile acids for high performance liquid chromatographic and gas chromatographic–mass spectrometric analysis, *J. Chromatogr.*, 562, 697, 1991.
35. **Iida, T., Momose, T., Shinohara, T., Goto, J., Nambara, T., and Chang, F. C.,** Separation of allo bile acid stereoisomers by thin layer and high performance liquid chromatography, *J. Chromatogr.*, 366, 396, 1986.
36. **Vaden Ende, A., Radecker, C. E., and Mairuhu, W. M.,** Microanalysis of free and conjugated bile acids by thin-layer chromatography and *in situ* spectrofluorimetry, *Anal. Biochem.*, 134, 153, 1983.
37. **Klaus, R.,** Procedures for quantitative high pressure liquid chromatographic analysis of nonfluorescing compounds with bile acids as an example, *J. Chromatogr.*, 333, 276, 1985.
38. **Roberts, K. and Whitelaw, M.,** Chromatography of monosaccharides and disaccharides, *J. Chromatogr.*, 373, 81, 1986.
39. **Flick, J. A., Schnaar, R. L., and Perman, J. A.,** Thin layer chromatographic determination of urinary excretion of lectulose, simplified and applied to cystic fibrosis patients, *Clin. Chem.*, 3317, 1211, 1987.
40. **Menzies, I. S., Mount, J. N., and Wheeler, J. N.,** Quantitative estimation of clinically important monosaccharides in plasma by rapid thin layer chromatography, *Ann. Clin. Biochem.*, 15, 65, 1978.
41. **Blom, W., Luteyn, J. C., Hkelholt-Dijkman, H., Huismans, J. B., and Loonen, M. C. B.,** Thin layer chromatography of oligosaccharides in urine as a rapid indication for the diagnosis of lysosomal and maltose deficiency (Pompes' disease), *Clin. Chim. Acta.*, 134, 221, 1983.
42. **Humbel, R. and Collart, M.,** Oligosaccharides in urine of patients with glycoprotein storage diseases. I. Rapid detection by thin-layer chromatography, *Clin. Chim. Acta*, 60, 143, 1975.
43. **Sewell, A. C.,** An improved thin layer chromatographic method for urinary oligosaccharide screening, *Clin. Chim. Acta*, 92, 411, 1979.

44. **Friedman, R. B., Williams, M. A., Moser, H. W., and Kolodny, E. H.,** Improved thin layer chromatography method in the diagnosis of mannosidosis, *Clin. Chem.*, 24, 1576, 1978.
45. **Michaud, J. D. and Jones, D. W.,** Thin layer chromatographic screening procedure for broad spectrum drug detection, *Am. Lab.*, 12, 104, 1980.
46. **Lillsunde, P. and Korte, J.,** Comprehensive drug screen in urine using solid-phase extraction and combined TLC and GC/MS identification, *J. Anal. Toxicol.*, 15, 71, 1991.
47. **Singh, A. K., Granley, K., Ashraf, M., and Mishra, U.,** Drug screening and confirmation by thin layer chromatography, *J. Planar Chromatogr.–Mod. TLC.*, 3411, 1989.
48. **Jain, N. C.,** Forensic toxicology: general unknown, in *Developments in Analytical Methods in Pharmaceutical, Biomedical and Forensic Sciences*, Piemonte, G., Tagliaro, F., Marigo, M., and Frigerio, A., Eds., Plenum Press, New York, 1987, p. 51.
49. **Maickel, R. P.,** Procedures for the isolation of organic compounds from biological materials, in *Guidelines for Analytical Toxicology Programs*, Vol. I, Thoma, J. J., Bando, P. B., and Sunshine, I., Eds., CRC Press, Cleveland, Ohio, 1977, p. 189.
50. **Gorodetzky, C. H. W.,** Detection of drugs of abuse in biological fluids, in *Handbook of Experimental Pharmacology, Drug Addiction*, Vol. II, Martin, W. R., Ed., Springer-Verlag, New York, 1977, Chap. 6.
51. **Davidow, B., Petri, L. N., Quame, B., Searle, B., Fastlich, E., and Sovitzky, J.,** A thin layer chromatographic screening test for the detection of users of morphine or heroin, *Am. J. Clin. Path.*, 46, 5a8, 1966.
52. **Nishigami, J., Okshima, T., Takayasu, T., Kondo, T., Lin, Z., and Nagaro, T.,** Forensic toxicological application of Toxi-lab screening for biological specimens in autopsy and emergency cases., *Nippon-Hoigaku-Zasshi*, 47, 372, 1993.
53. **Wolf, K., Sanderson, M. J., Hay, A. W., and Barnes, I.,** An evaluation of the measurement of drugs of abuse by commercial and in-house horizontal layer chromatography, *Ann. Clin. Biochem.*, 30, 163, 1993.
54. **Vinson, J. A. and Patel, A. M.,** Detection and quantitation of tetrahydrocannabinol in serum using thin-layer chromatography and fluorometry, *J. Chromatogr.*, 307, 493, 1984.
55. **Vinson, J. A. and Lofatofsky, D. J.,** A semi-automated extraction and spotting system for drug analysis by TLC: procedure for analysis of the major metabolite of 9-tetrahydrocannabinol in urine, *J. Anal. Toxicol.*, 9, 6, 1985.
56. **Jain, R.,** Determination of buprenorphine and its N-dealkylated metabolite in urine by TLC densitometry, *Ind. J. Pharmacol.*, 26, 288, 1994.
57. **Jain, R.,** Simultaneous determination of diazepam and its metabolites in urine by thin-layer chromatography and direct densitometry, *J. Chromatogr.*, 615, 365, 1993.
58. **Inoue, T. and Ni Waguchi, T.,** Determination of nitrazepam and its main metabolites in urine by thin-layer chromatography and direct densitometry, *J. Chromatogr.*, 339, 163, 1985.
59. **Della Lasa, E. and Martone, G.,** A quantitative determination of heroin and cocaine samples by high performance thin layer chromatography, *Forensic Sci. Int.*, 32, 117, 1986.
60. **Fenimore, D. C., Meyer, C. J., Davis, C. M., Hsu, F., and Zlatkes, A.,** High performance thin layer chromatographic determination of psychopharmacologic agents in blood serum, *J. Chromatogr.*, 142, 399, 1977.
61. **Fenimore, D. C., Davis, C. M., and Meyer, C. J.,** Determination of drugs in plasma by high performance thin layer chromatography, *Clin. Chem.*, 24, 1386, 1978.
62. **Reimer, G. J. and Suraz, A.,** Development of a screening method for five sulfonamides in salmon muscle tissue using thin-layer chromatography, *J. Chromatogr.*, 555, 315, 1991.
63. **Chiong, D. M., Consegrao-Rodriguez, E., and Almirall, J. R.,** The analysis and identification of steroids, *J. Forensic Sci.*, 37, 488, 1992.
64. **Oka, A., Ikai, Y., Hayakawa, J., Masudo, K., Harada, K., and Suzaki, M.,** Improvement of chemical analysis of antibiotics. Part XIX : Determination of tetracycline antibiotics in milk by liquid chromatography and thin-layer chromatography/fast atom mass spectrometry, *J. AOAC. Int.*, 77, 891, 1994.
65. **Kuksis, A.,** *Chromatography in Biomedical Research and Clinical Diagnostics,* Journal of Chromatography Library, Vol. 37, Elsevier, Amsterdam, 1987.
66. **Kuksis, A. and Myher, J. J.,** Lipids and their constituents, *J. Chromatogr.*, 379, 57, 1986.
67. **Folch, J. M., Lees, M., and Sloane-Stanley, C. H.,** A simple method for the isolation and purification of total lipids from animal tissues, *J. Biol. Chem.*, 226, 497, 1957.
68. **Nelson, G. J.,** in Nelson, G. J., Ed., *Blood and Lipoproteins—Quantitation, Composition and Metabolism*, Wiley, New York, 1972, p. 3.
69. **Mitchell, R. H., Hawthorne, J. N., Coleman, R., and Karnosky, M. L.,** Extraction of polyphosphoinositides with neutral and acidified solvents – a comparison of guinea-pig brain and liver and measurement of rat liver inositol compounds which are resistant to extraction, *Biochem. Biophys. Acta*, 210, 86, 1970.
70. **Christie, W. W.,** Extraction and hydrolysis of lipids and some reactions of their fatty acid components, in *CRC Handbook of Chromatography, Lipids*, Vol. 1., Mangold, H. K., Zweig, G., and Sherma, J., Eds., CRC Press, Boca Raton, 1984, Chap. 2.
71. **Bounias, M.,** *Analyse Biochimique Quantitiative par Nanochromatographie en Couches Minces*, Masson, Paris, 1983, p. 83.

72. Bhat, H. K. and Ansari, G. A. S., Improved separation of lipid esters by thin-layer chromatography, *J. Chromatogr.*, 483, 369, 1989.
73. Yao, J. K. and Rasteller, G. M., Microanalysis of complex tissue lipids by high-performance thin layer chromatography, *Anal. Biochem.* 150, 111, 1985.
74. Schmitz, G., Assmann, G., and Bowyer, D. E., A quantitative densitometric method for the rapid separation and quantitation of the major tissue and lipoprotein lipids by high performance thin layer chromatography. I. Sample preparation, chromatography and densitometry, *J. Chromatrogr.*, 307, 65, 1984.
75. Peters, T. J. and Cairns, S. K., Analysis and subcellular localization of lipid in alcoholic liver disease, *Alcohol*, 2, 447, 1985.
76. Edouard, L., Doucel, F., Buxtof, J. C., and Beaumont, J. L., Quantitative determination of free cholesterol and cholesteryl esters in skin biopsies, *Clin. Physiol. Biochem.*, 3, 323, 1981.
77. Saint Leger, D. and Bague, A., A simple and accurate rapid procedure for qualitative analysis of skin surface lipids (SSL) in man, *Arch. Dermatol. Res.*, 271, 215, 1981.
78. Jacobsen, E., Stewart, M. E., and Dowing, D. T., Age related changes in sebaceous wax ester secretion rates in men and women, *J. Invest. Dermatol.*, 85, 483.
79. Ohta, A., Mayo, M. C., Kramer, N., and Lands, W. E., Rapid analysis of fatty acids in plasma lipids, *Lipids*, 25, 742, 1990.
80. Spillman, T. and Cotton, D. B., Current perspectives in assessment of fetal pulmonary surfactant status with amniotic fluid, *Crit. Rev. Clin. Lab. Sci*, 27, 345, 1989.
81. Serranto, de la., Cruz, D., Santillana, E., Mingo, A., Fuenmayor, G., Pantoja, A., and Fernandez, E., Improved thin layer chromatographic determination of phospholipids in gastric aspirate from newborns, for assessment of lung maturity, *Clin. Chem.*, 34, 736, 1988.
82. Toso F. H. and Zachman, R. D., Use of quantitative amniotic fluid phosphatidyl glycerol as a criterion for fetal lung maturation, *Am. J. Perinatol.*, 9, 34, 1992.
83. Gluck, L. and Kulovich, M. V., Lecithin–sphingomyelin ratio in amniotic fluid in normal and abnormal pregnancy, *Am. J. Obstet. Gynecol.*, 115, 539, 1973.
84. Gluck, L., Kulovich, M. V., and Borer, R. C., Jr., The diagnosis of the respiratory distress syndrome by amniocentesis, *Am. J. Obstet. Gynecol.*, 109, 440, 1972.
85. Kolins, M. D., Epstein, E., Civin, W. H., and Weiner, S., Amniotic fluid phospholipids measured by continuous development thin layer chromatography, *Clin. Chem.*, 26, 403, 1980.
86. Touchstone, J. C. and Alvarez, J. G., Separation of acidic and neutral glycolipids by diphasic thin-layer chromatography, *J. Chromatogr.*, 436, 515, 1988.
87. Freer, D. E. and Statland, B. E., Measurement of amniotic fluid surfactant, *Clin. Chem.*, 27, 1629, 1981.
88. Blass, K. G. and Ho, C. S., Sensitive and highly reproducible qualitative fluorescent thin layer chromatographic visualization technique for lecithin and sphingomyelin, *J. Chromatogr.*, 208, 170, 1981.
89. Colarow, L., Quantitation of phospholipid classes on thin layer plates with a fluorescence reagent in the mobile phase, *J. Planar Chromatogr.–Mod. TLC*, 2, 19, 1989.
90. Ojima-Uchiyama, A., Masuzawa, Y., Sugiura, T., Waka, K., Saito, H., Yui, Y., and Tomioka, H., Phospholipid analysis of human eosinophils: high levels of alkyl acyl glycerophosphocholine (PAF precursor), *Lipids*, 23, 815, 1988.
91. Narayan, P. and Dahiya, R., Alteration in sphingomyelin and fatty acids in human benign prostatic hyperplasia and prostatic cancer, *Biomed. Biochim. Acta*, 50, 1099, 1991.
92. Dembitsky, V. M., Quantification of plasmalogens, alkylacyl and diacyl glycerophospholipids by micro thin layer chromatography, *J. Chromatogr.*, 436, 467, 1988.
93. Cartwright, I. J., Separation and analysis of phospholipids by thin layer chromatography, *Methods Mol. Biol.*, 19, 153, 1993.
94. Gregson, N. A., The extraction and analysis of glycosphingolipids, *Methods. Mol. Biol.*, 19, 287, 1993.
95. Raajmakers, J. G. A. M., System for the identification for novel prostaglandins, *J. Chromatogr.*, 138, 355, 1977.
96. Tsunamoto, K., Todo, S., and Ismashuku, S., Separation of prostaglandins and thromboxane by two-dimensional thin layer chromatography, *J. Chromatogr.*, 417, 414, 1987.
97. Beneytout, J., Grenrt, D., and Tixer, M., *J. High Resolut. Chromatogr. Chromatogr. Commun.*, 7, 211, 1984, quoted in Siouffi, A. M., Mincsovics, E., and Tyihak, E., Planar chromatographic techniques in biomedicine: current status, *J. Chromatogr.*, 492, 471, 1989.
98. Herman, C. A., Hamberg, M., and Granstram, E., Quantitative determination of prostaglandins E1, E2, and E3 in frog tissue, *J. Chromatogr.*, 394, 353, 1987.
99. Goswami, S. K. and Kinsellas J. E., A nondestructive spray reagent for the detection of prostaglandins and other lipids on thin layer chromatography, *Lipids*, 16, 759, 1981.
100. Riutta, A., Seppala, E., and Vapaatalo, H., A rapid, sensitive method for detecting different arachidonic acid metabolites by thin layer chromatography: the use of autoradiography, *J. Chromatogr.*, 307, 185, 1985.
101. Jacob, K. and Lappa, P., Application of ion pair high performance liquid chromatography to the analysis of porphyrins in clinical samples, *Biomed. Chromatogr.*, 5, 122, 1991.

102. **Petryka, Z. J.,** Porphyrins in densitometry, in *Thin Layer Chromatography, Practice and Application*. Touchstone, J. C. and Sherma, J., Eds., Wiley–Interscience, New York, 1979, Chap. 24.
103. **Heftman, E.,** *Chromatography of Steroids*, Elsevier, Amsterdam, 1976.
104. **Touchstone, J. C.,** *CRC Handbook of Chromatography—Steroids*, Zweig, G. and Sherma, J., Eds., CRC Press, Boca Raton, FL, 1986.
105. **Lamparczyk, H.,** in *Handbook of Chromatography, Analysis and Characterization of Steroids*, Sherma, J., Ed., CRC Press, Boca Raton, FL, 1992.
106. **Sherma, J. and Fried, B., Eds.,** *Handbook of Thin Layer Chromatography*, Marcel Dekker, New York, 1991, p. 1047.
107. **Lamparczyk, H.,** Chromatographic techniques used in steroid analysis, in *Handbook of Chromatography—Analysis and Characterization of Steroids*, Sherma, J., Ed., CRC Press, Boca Raton, FL, 1992, Chap. 3.
108. **Vanlook, L., Deschuytere, Ph., and Van Peteghem, C.,** Thin layer chromatographic method for the detection of anabolics in fatty tissues, *J. Chromatogr.*, 489, 213, 1989.
109. **Brown, J. W., Carballeria, A., and Fishman, L. M.,** Rapid two-dimensional thin layer chromatographic system for the separation of multiple steroids of secretary and neuroendocrine interest, *J. Chromatogr.*, 439, 441, 1988.
110. **Watkins, T. R., Smith, A., and Touchstone, J. C.,** Reversed-phase thin layer chromatography of estrogen glucuronides, *J. Chromatogr.*, 374, 221, 1986.
111. **Curtius, H. C. and Muller, M.,** Combination of gas chromatography and thin layer chromatography in steroid investigations, *J. Chromatogr.*, 32, 222, 1968.
112. **Kraft, R., Otto, A., Makower, A., and Etzold, G.,** Combined thin-layer chromatography/mass spectrometry without substance elution: use for direct identification of phenols, steroids, nucleosides, biogenic amines and amino acids, *Anal. Biochem.*, 113, 193, 1981.
113. **Kono, K., Tschida, T., Kern, D. H., and Irie, R.,** Ganglioside composition of human melanoma and response to anti-tumor treatment, *Cancer Invest.*, 8, 161, 1990.
114. **Merritt, W. D., Der-Minassian, V., and Reaman, G. H.,** Increased GD3 gangliosides in plasma with children with T-cell and lyphoplastic leukemia, *Leukemia*, 85, 816, 1994.
115. **Nakakuma, H., Horikawa, H., Kawaguchi, T., Hidaka, M., Nagakura, S., Hirari, S., Kageshita, I., Ono, I., Kageshita, I., and Iwamori, M.,** Common phenotypic expression of gangliosides GH3 and GD3 in normal human tissues and neoplastic skin lesion, *Jpn. J. Clin. Oncol.*, 22, 308, 1992.
116. **Ogawa, K., Fujiwara, Y., Sugamata, K., and Abe, T.,** Thin layer chromatography of neutral glycosphingolipids an improved simple method for the resolution of GlcCer, GalCer, LaCer and Ga$_2$Cer, *J. Chromatogr.*, 426, 183, 1988.
117. **Alemany, G., Nicolav, M. C., Gamundi, A., and Rial, R.,** Thin layer chromatographic determination of brain catecholamine and 5-hydroxy tryptamine, *Biomed. Chromatogr.*, 7, 315, 1993.
118. **Lee, B. Y. and Thurmon, T. F.,** Compact gas chromatography used with thin layer chromatography for assessment of abnormalities of organic acids, *Clin. Chim. Acta*, 218, 215, 1993.

Chapter 8

THIN-LAYER CHROMATOGRAPHY IN ENVIRONMENTAL ANALYSIS

Jan Bladek

CONTENTS

I. Introduction 153

II. General Principles of Pollutant Analyses by TLC 154
    A. Sample Preparation 154
    B. Chromatogram Development Techniques 155
    C. Stationary and Mobile Phases 155
    D. Visualization and Quantitation 155

III. Applications 156
    A. Application of TLC to the Monitoring of the Environment 156
        1. Pesticides 156
        2. Polycyclic Aromatic Hydrocarbons 159
        3. Polycyclic Aromatic Heterocycles 160
        4. Phenols 161
        5. Miscellaneous Pollutants 162
            a. Polychlorinated Biphenyls 162
            b. Warfare Agents 162
            c. Trinitrotoluene 163
            d. Aldehydes and Ketones 164
        6. Inorganic Pollutants 164
    B. TLC as a Research Technique 164
        1. Measurement of Lipid-Water Partitioning 165
        2. Measurement of Pollutant Mobility in the Environment 165
        3. TLC as a Pilot Technique for HPLC 165

References 166

I. INTRODUCTION

The term *environmental pollution* means any physical, chemical, or biological change disturbing ecological equilibrium in the environment. It may be a result of random, accidental events (e.g., chemical disasters), emission of certain pollutants due to activity of nature itself (e.g., volcanic eruption), or human activities (e.g., industry, energy production, and use of automotive vehicles). As a result of the activity of nature, "natural pollutants" are emitted into the atmosphere. Human activity leads to the emission of pollutants called "anthropogenic pollutants." All the pollutants (natural and anthropogenic) emitted from a given source are called primary pollutants. A number of primary pollutants can undergo some changes due to reactions with other pollutants, as well as with some components of the environment. In this way, new compounds, often of higher toxicity, can be formed (secondary pollutants).

It is estimated that the number of chemical compounds (mainly organic) present in direct human surroundings amounts to 1,000,000. Only a very few of these compounds may be considered beneficial or neutral for health. The majority of them act harmfully on humans, even in minimal doses. They occur in all parts of the environment: atmosphere, hydrosphere, and soil; they may be biomagnified in the food chain. Thus, increasing attention is being given to environmental protection, which today is considered a global problem. One of more important ventures in this field is the analysis of pollutants. It may even be said that reliable analytical data are essential for environmental protection activities. Due to high sensitivity, specificity, and versatility, chromatographic methods are today the most widely used techniques of environmental studies. Thin-layer chromatography (TLC) plays a particular role in this field. It is used mainly for analytical goals; it may also be used as a cleanup or research technique.

II. GENERAL PRINCIPLES OF POLLUTANT ANALYSIS BY TLC

The use of TLC for the separation, identification, and quantitation of environmental pollutants has a long history, and there are innumerable references in the literature. It is used among other techniques in the analysis of pesticides, polycyclic aromatic hydrocarbons, phenols, and inorganic pollutants. Samples of air, water, soil, and sometimes plants and food are the object of studies. The methods of sampling, transport, and storage of samples are standardized and do not provide any serious analytical problem. Many more problems may occur due to the procedures of sample preparation, which is (in the case of the environmental studies) usually the most difficult stage of analysis. This is because of the necessity of determining very small amounts of examined compounds and considerable "overloading" of samples by interfering additives.

A. SAMPLE PREPARATION

A glass absorption tube packed with Carbowax, glass-fiber, or other filters containing air samples as well as samples of soil and sediments usually undergo extraction in a Soxhlet apparatus or ultrasonic bath with different solvents: acetone, benzene, toluene, and methylene chloride. The extracts obtained require a preliminary cleanup procedure before determination. The goal of cleanup is to remove as much interfering coextracted material and as little of the analyte as possible. The cleanup techniques most commonly employed for extracts are liquid-liquid partitioning or chromatographic cleanup, such as column chromatography, TLC, and solid phase extraction (SPE). Many methods require a combination of techniques. The possibility of applying different stationary phases is an advantage of the chromatographic cleanup techniques. Supercritical fluid extraction (SFE) is another method of sample preparation, mainly for the analysis of soils and sediments. It has been shown to be rapid and is less laborious, toxic, and hazardous than liquid extraction.

Liquid-liquid extraction (LLE) with the use of such solvents as chloroform or dichloromethane is applied most often in the case of water samples. In some analyses, salting-out or acidification of samples before extraction is recommended. The extracts obtained are dried in columns (filled, for example, with anhydrous sodium sulfate), purified, and then concentrated. Large consumption of solvents, lack of speed, and low recovery (particularly in the case of water-soluble compounds) have caused SPE to become more common than LLE in environmental studies. SPE techniques have numerous advantages in comparison with LLE techniques, including ease of automation, reduced cost, and higher sample throughput.

The first stage of preparation of plants and food samples is homogenization. Solvents of different polarity (from water to hexane) are used for the extraction. The most effective strategy is to choose a solvent, whenever possible, in which the recovery of analytes is maximized. For highly water-soluble compounds, the water content in the extract should be large, thus minimizing the amount of potentially interfering coextracted material. Similarly, relatively nonpolar solvents may efficiently extract less polar compounds without removing an excess of pigments. In this case, contaminated extracts should be cleaned up before the analysis.

Thin-Layer Chromatography in Environmental Analysis

B. CHROMATOGRAM DEVELOPMENT TECHNIQUES

In environmental analysis, the one-dimensional ascending technique has usually been used for the development of chromatograms in a closed chamber, allowing the mobile phase to migrate 10 to 15 cm from the starting line. Horizontal chambers, multiple, and two-dimensional development techniques have also been used. Recent development techniques requiring more complicated equipment such as forced-flow planar layer chromatography (FFPLC), automated multidevelopment (AMD), and gradient development techniques are becoming more frequently used. Multimodal TLC (e.g., TLC/GC = gas chromatography, TLC/MS, and OPLC = overpressured layer chromatography, HPLC = high performance liquid chromatography) have occasionally been used. Details regarding new TLC techniques, including gradient and forced-flow systems, were described in the review article by Jork.[1]

A tendency toward quantitative analyses and determinations of small amounts of pollutants, which in the case of TLC can be identified only with the application of relatively large volumes of analyzed solutions in the form of narrow zones or spots, has resulted in the development of apparatuses for sample application. In environmental analyses, the spray-on technique for applying bands is of particular interest. Narrow-band sample application provides the highest resolution attainable in any given TLC system and is favorable for densitometric quantitation.

C. STATIONARY AND MOBILE PHASES

The largest number of separations of environmental mixture samples is performed on commercial chromatographic plates coated with silica gel and bonded-phase silica gel (mainly with octadecyl). They allow separations in normal (NP) and reverse (RP) modes, respectively. Relatively often, mainly in the analyses of phenols and inorganic pollutants, modified silica gel is used. Modification is obtained through impregnation of silica gel with organic (e.g., aniline) or inorganic (e.g., salt solutions such as silver nitrate, ammonium molybdate, and iron (III) nitrate) substances. The use of impregnated layers leads to improvement of resolution and changes mobility as a result of complex interactions between the analytes and impregnates. The need for quantitation of very low pollutant concentrations suggests the necessity of using plates with concentration (preadsorbent) zones. They play a similar role to the spray-on techniques mentioned above. A description of properties and applications of TLC adsorbents was presented in the review paper by Hanh-Deinstrop.[2] In the scientific literature, publications dealing with "behavior" of certain groups of pollutants on investigated (usually commercially available) chromatographic plates are available.

The fact that industrial and academic laboratories frequently have several types of commercially available TLC plates on hand causes the majority of investigators to screen solvent systems much more than plates themselves. In separating environmental sample components, there are essentially no limitations connected with selection of solvents used to prepare mobile phases. Acetone, chloroform, hexane, and benzene in NP systems, or acetonitrile or methanol and water in RP systems, are most often used as components of mobile phases. In the case of analysis of compounds of distinctly different polarity (e.g., pesticides), multidevelopment or gradient elution may be necessary.

D. VISUALIZATION AND QUANTITATION

In environmental investigations direct visualization methods have crucial significance. Chemical, physical, and/or biological-physiological procedures have been used to locate the analytes on chromatographic plates. In certain analyses, one detection method can be combined with another to increase the selectivity of detection. Compounds that absorb ultraviolet light are detected through the measurement of fluorescence quenching. If the analyzed substances do not quench fluorescence or naturally fluoresce, or if it is necessary to increase selectivity of analyses, chromogenic or fluorogenic reagents capable of forming colored products with the separated species are sprayed onto chromatographic plates to detect the pollutants. Versatile reagents making possible visualization of many substances (e.g., arsenic trichloride–sulfuric acid for various pesticide classes) and specific reagents have the greatest significance. Phenols are usually detected by 7-chloro-4-

nitrobenzo-2-oxa-1,3-diazole (fluorescent 4-nitrobenzofuran derivatives are produced); phenols capable of coupling are detected by means of Pauly's reagent (intensively colored azo dyes are formed). Typical chromogenic reagents used to visualize inorganic pollutants include dithizone, alizarin, and aluminon. An enzymatic method has been used to detect pesticides, heavy metals, and toxic warfare agents. The majority of polycyclic aromatic hydrocarbons (PAHs) as well as their heterocyclic derivatives, and also certain pesticides, fluoresce and are readily detected. An exhaustive collection of information on reagents and detection methods is presented in the monograph by Jork and co-workers.[3]

Two of the above-mentioned direct methods draw particular attention in environmental investigations (fluorescence and enzymatic methods). They are distinguishable by high sensitivity and specificity and allow the analysis of certain pollutants in the presence of background impurities that do not interfere in these reactions. Fluorescence measurements concern not only compounds that fluoresce, e.g., PAHs or their heterocyclic derivatives, but also nonfluorescing compounds after their conversion into derivatives showing fluorescence. Methods for pretreatment of compounds to convert them into fluorescent species include hydrolysis, ligand-exchange reaction, pH effects, and heat treatment. Most of these have been applied to the analysis of pollutants. In enzymatic methods, chromatograms are first sprayed with an enzyme solution. Then, after appropriate incubation, the enzymatically altered components are detected by reaction with a suitable reagent. This method is characterized by very low detection limits; sometimes it allows analysis of enzyme inhibitors in samples without resorting to enrichment.

TLC applications in quantitative environmental analyses is increasing. Windhorst and De Klein[4] proved that results of quantitations obtained by use of TLC are comparable with those obtained by HPLC and GC. Quantitations are made primarily by densitometry (measurement of absorbance or fluorescence). Newer techniques of quantitation (photoacoustic measurement, fiber-optic remote sensor for fluorometric quantitations, and voltammetry measurement) until recently have been used in a very limited way. Detailed information on the topic of quantitative TLC is available in Zieloff's work.[5]

III. APPLICATIONS

TLC applications for environmental investigation can be classified based on the criteria of goals of experiments. Environmental monitoring is the main aim in such research in order to gain information on the type and quantity of pollutants in samples. On the other hand, assessment of physical and chemical properties of pollutants, e.g., their mobility, bioaccumulation, and biotransformation, is also an important area of TLC study. In these cases, TLC is used as a research tool in such experiments. TLC is also used as cleanup technique in various analyses.

A. APPLICATION OF TLC TO THE MONITORING OF THE ENVIRONMENT

Research papers connected with pollutant monitoring by TLC can be divided into two groups. The first concerns investigations of the method itself, usually done with standards. In these investigations, visualization methods and separation techniques are improved, and the principles of quantitative determination are studied. The other group involves investigation of applications. As a result of these studies, methods describing the treatment of the environmental sample from its collection until complete information on the type and amount of investigated group of pollutants or their reaction products is obtained (monitoring of environment). Such investigations are based on spiking the given matrix with the analyte, and then extraction and cleanup techniques, chromatographic systems, and selection of methods of detection are developed, and the efficiency of each stage of the analytical process is optimized. The developed method is then used to analyze actual environmental samples.

1. Pesticides
Pesticides (see Figure 8.1 for pesticide structures) comprise a group of compounds that are given great attention in environmental studies. In most cases, they are introduced into the environment

due to willful human activity. With respect to biological activity, there are three main groups: insecticides, herbicides, and fungicides. These pesticide classes belong to different groups of compounds (organophosphorous, organochlorine, carbamate, triazine, and chlorophenoxy acids). Pesticides are, above all, poisons. Some of them also demonstrate carcinogenic potential and teratogenic activity. In spite of their adverse effects on the environment, the total amount of pesticides used for agricultural and industrial purposes is still increasing. As a result, pesticides are currently present in all parts of the environment. Many of them undergo degradation, but others are persistent and may be biomagnified in the food chain. The range of TLC application in pesticide analysis is wide. Rathore and Begum[6] referred to about 300 papers in their review of thin-layer chromatographic methods for use in pesticide residue analysis. This list is supplemented by review articles written by Sherma.[7,8]

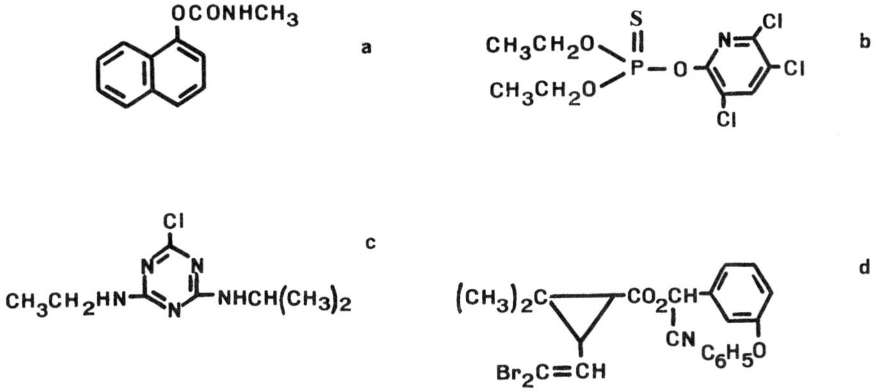

FIGURE 8.1 Representatives of pesticides: (a) carbaryl, (b) chlorpyrifos, (c) atrazine, (d) deltametrin.

Good examples of monitoring investigations include some historic papers by Lawrence, Frei, Mallet, and their co-workers,[9] which concerned visualization and quantitation of pesticides by fluorescence methods. They indicated that most fluorometric analyses of pesticides require pretreatment of the compounds to convert them to a fluorescent species. Certain organophosphorous pesticides fluoresce just after heating, others (e.g., guthion, carbaryl, or benzomyl) form fluorescent anions as a result of hydrolysis, and still others (e.g., organothiophosphorous compounds) can be quantitated in the presence of metal chelate compounds. Sulfur-containing pesticides were determined quantitatively by pH-sensitive fluorescent reagents. Significant attention was also devoted to the applications of fluorogenic labeling reagents; most advantageous were dansyl chloride and fluorescamine. The application of fluorogenic reagents allows detection of certain pesticides at the level of 10 ng and a linearity region in quantitations in the range 10 ng to 10 mg. Investigations by Judge et al.[10] can be treated as a continuation of this topic. The subject of this work was the study of fluorescence reagents. The authors indicated that F_{254} indicator present in the HPTLC plates proved to be the best detection method for atrazine, carbofuran, metachlor, and their by-products. These results confirmed earlier observations made by Sherma[11] that in some analyses of pesticides using effect of fluorescence quenching is advantageous. Chloropyrifos and its by-products were better detected by the fluorogenic reagent 1-pyrene carboxaldehyde. Several of the investigated fluorescent reagents were sensitive to the various pesticides and by-products tested. The most interesting result of these investigations was the observation that some fluorogenic substances can be used only when the exposure to UV light is kept to a minimum (increase of the time of UV exposure decreased fluorescence intensity).

Gardyan and co-workers[12] conducted systematic investigations concerned with separation and detection of about 150 pesticides (mainly insecticides and fungicides). Most of pesticides mixtures were separated in NP systems with silica gel, the remaing ones in C_{18} RP systems. Silver nitrate–UV irradiation, chlorine–*o*-tolidine, and cholinesterase inhibition were used for visualization. It was indicated that the enzymatic reaction is realized only on a TLC plate with silica gel, and that computer-assisted densitometric evaluation allows direct quantitative determination of pesticides.

Investigations of stationary phases are usually called "behavior" studies of certain groups of pesticides on a given adsorbent. An example of such investigations is the work by Rathore and Begum.[13] They investigated the TLC behavior of carbaryl, carbendazim, carbofuran, propoxur, and some phenols on alumina, barium sulfate, calcium carbonate, calcium phosphate, cellulose, and silica gel. Acetone, benzene, carbon tetrachloride, chloroform, ethanol, and water were used as mobile phases. Values of R_f as a function of dielectric constant of the developers were determined. Knowledge of the results of such investigations allows forecast of chromatographic systems suitable for monitoring studies.

Among numerous studies devoted to TLC use in environmental investigations (monitoring) that have shown TLC to be comparable with GC and HPLC was Sherma's work on SPE combined with TLC.[14] Samples of water containing atrazine, 2,4-D, Silvex, and 2,4-T were treated by SPE and the 2,4,5-T eluates obtained were analyzed by TLC using preadsorbent silica gel plates impregnated with silver nitrate. After the chromatograms were developed, layers were exposed to UV light and densitometric scans were made. Recoveries were 70 to 88% for triazines and 93 to 100% for chlorophenoxy acid derivatives. It was shown that the SPE technique is much simpler and faster than LLE and had adequate recovery and precision for routine residue screening. In the case of TLC, a derivatization step was not required, as is necessary during analyses by means of gas chromatography.

Gries and Jork[15] quantitatively determined 15 urea herbicides in potable water after concentration (SPE on C_{18} columns, recovery 70 to 80%). Eluents were separated in three different chromatographic systems. Eleven of these compounds were separated on TLC plates coated with Al_2O_3; good separation was also obtained with HPTLC silica gel and C_{18} RP systems. The detection limits for each substance were in the range of 10 to 20 ng. Similar techniques of extracting pesticides from water were used by De la Vigne et al.[16] in analyses of atrazine and desethylatrazine derivatives (12 pesticides). The chromatograms were developed by the AMD technique, and the advantage of automated multiple development with gradient elution over classical TLC was indicated. SPE eluents were applied onto chromatographic plates by instrumental spray technique. It was proved that using 100 μm (instead of 200 μm) particle HPTLC silica gel layers and reducing the development distance increments from 3 mm to 1 mm increased the sensitivity, linearity, and speed of the method. The nonderivative plates were scanned by measuring the adsorbance at six or seven different wavelengths, and each substance was quantified by the absorbance very close to its maximum response wavelength. Experiments showed the complete usefulness of the method in environmental analyses, i.e., reliable identification of trace amounts of pesticides (e.g., 28 ppt of atrazine) in subsoil, surface, and drinking water. A similar technique (SPE, gradient AMD, application using a spray-on technique, and multiwavelength scans) was used to determine alachlor, chlorfenvinphos, and simazine pesticides (24 pesticides, four mixtures).

Kovác and co-workers[17] used the Hill reaction to analyze 18 herbicide residues in agricultural crops, foods, soil, and water. Pesticides were isolated from the matrix by liquid extraction. Concentrated extracts were cleaned up by column chromatography, and then the mixtures were separated in an NP system. The detection reagent was a mixture of a homogenate of bean leaves and of the redox indicator 2,6-dichloroindophenol. Dark blue inhibition zones were observed, which appeared on a pale yellow-green background during the exposure of the sprayed chromatographic plates to light. The dark blue zones disappeared again after some time; their lifetime was proportional to the amount of the herbicide in the zone. This chronometric method can be used for quantitation.

A simple HPTLC method was developed by Zehradnièková et al.[18] for the determination of atrazine and simazine herbicides in drinking and surface waters. It involved SPE on C_{18} cartridges, followed by development of the concentrated extracts on HPTLC silica gel, with the mobile phase nitromethane–tetrachloromethane (1:1). UV scanning densitometry was used for quantitation. The detection limits of the method were 30 and 60 ng/l for atrazine and simazine, respectively. The method was successfully applied to the analysis of top and surface waters with overall recoveries between 58 and 93% at the 80 to 400 ng/l fortification level.

Patil and Shingare[19] quantitatively determined carbaryl in biological materials, including stomach, intestine, liver, spleen, and kidney tissues. After liquid extraction and chromatogram development in an NP system, layers were sprayed with 1% ammonium cerium (IV) nitrate in 20% v/v hydrochloric acid. This reagent is specific; other insecticides do not give similarly colored spots. Moreover, organophosphorous, organochlorine, and pyrethroid insecticides and constituents of visceral extracts do not interfere. Sensitivity of the method is about 100 ng.

The publications mentioned in this section should be treated as examples of the use of TLC in pesticide investigations. They show the complexity of the problems and the variety of analytical tasks connected with environmental analyses by TLC. Thanks to investigations such as these, TLC is becoming a technique comparable with other methods of pesticides monitoring such as GC and HPLC.

2. Polycyclic Aromatic Hydrocarbons

Polycyclic aromatic hydrocarbons (PAHs) (see Figure 8.2 for structures) are compounds built of 2 to 13 aromatic rings arranged in linear, cluster, or angular shapes. They may contain some number of alkylic substitutes. PAHs are widespread environmental contaminants caused by emission from a variety of sources, including industrial combustion and discharge of fossil fuels, residential heating (both fossil fuel and wood burning), or motor vehicle exhaust. Because of their mutagenic and carcinogenic properties, PAHs have been measured in a variety of environmental matrices including air, water, soil (sediments), and tissue samples.

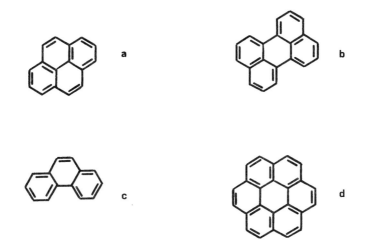

FIGURE 8.2 Polycyclic aromatic hydrocarbons: (a) pyrene, (b) perylene, (c) phenanthrene, (d) coronene.

The use of TLC for investigations of PAHs has a long history. As early as 1967, White and Howard[20] determined the behavior of many compounds of this group on cellulose layers modified by *N,N*-dimethylformamide in the presence of isooctane as the mobile phase and on cellulose acetate in the presence of a ternary mobile phase containing ethanol, toluene, and water. The method was first applied to investigations of PAHs in petrochemical plants and oil refineries. Hellmann[21] investigated hydrocarbons of biological as well as petroleum origin and showed differences in the composition of the aromatics fraction.

Fluorometric scanning has been used for quantitative analysis of benzo(a)pyrene, benzo(k)fluoranthene, and anthracene separated on plates with cellulose acetate. Polyamide stationary phase was used for separation of PAHs and their direct quantitation by laser mass spectrometry[22]; this adsorbent also proved suitable for the analysis of these compounds after their separation by urea-solubilized β-cyclodextrin.[23] Van de Nesse et al.[24] used on-line coupled HPLC-TLC for PAH quantitation in marine sediment using a fluorescence spectrometer for detection. Baranowska and co-workers[25] carried out chromatographic procedures involving various layers and developing systems covering both adsorption and reversed phase partitioning modes for PAH quantitations; this method was also used for monitoring soil extracts.

The studies (separation and detection methods and TLC coupling with other analytical techniques) presented here confirm the significance of TLC for the environmental monitoring of PAH content. Detailed information on this subject is available in numerous review papers as in Furton et al.[26]

3. Polycyclic Aromatic Heterocycles

In the natural environment, carbon atoms in PAH rings can be substituted with oxygen, sulfur, or nitrogen atoms; alkyl substituents can also be replaced (e.g., nitrogen derivatives of PAHs (see Figure 8.3 for structures)). In this way heteroaromatic compounds are formed, and they usually occur together with PAHs. The most dangerous of these, polychlorinated dibenzo-p-dioxins and polychlorinated dibenzofurans, are by-products formed during the manufacture of chlorophenols and related products; other sources include the pulp and paper industry and accidental fires that release polychlorinated biphenyls. Dibenzothiophene and some of its methyl-substituted compounds are persistent residues in sea environment after oil accidents. In the natural environment, polychlorinated thianthrenes and polychlorinated dibenzothiophenes also exist. As with their oxygen analogs, they are the hazardous chemicals. Azaarenes, mainly benz(c)acridine and many of its related compounds, have been shown to exhibit carcinogenic activity. These compounds are continually found in many natural and environmental samples. Nitro-polycyclic aromatic hydrocarbons are mutagenic and carcinogenic compounds.

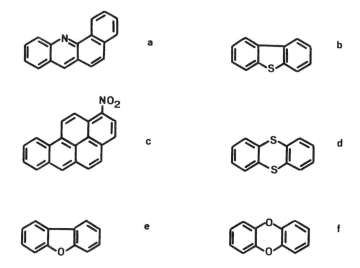

FIGURE 8.3 Polycyclic aromatic heterocycles: (a) benz(c)acridine, (c) 1-NO$_2$ - benzo(a)pyrene, b and d-f: the parent substances of polychlorinated — dibenzothiophenes (b), thianthrenes (d), dibenzofurans, (e) dibenzodioxins (f).

Benz(c)acridine was detected in air pollution sources of industrial effluents. Motohashi and co-workers[27] separated methyl-substituted benz(c)acridine mixtures by TLC in NP and RP systems. The fluorescent spots of benzacridine were treated with trifluoroacetic acid fumes and detected by spectrofluorometric technique. Handa et al.[28] reported methods for quantitation of benzacridines in both diesel and gasoline engine vehicle exhausts in air samples taken in a road tunnel. The collected samples were extracted in a Soxhlet apparatus, and compounds were separated on plates coated with alumina, kieselguhr, and acetylated cellulose by two-dimensional development. The TLC technique was also used for detection of these substances in coal-tar and river and marine sediments.

The quantitation of chlorinated aromatic thioethers by TLC was reported by Sinkkonen,[29] who quantitatively determined polychlorinated dibenzothiophenes in an RP system and proved that polychlorinated dibenzothiophenes elute more strongly than nonchlorinated parent compounds. Complete separation was achieved through prior oxidation of these compounds to corresponding sulfones (oxygen analogs of these compounds are mainly quantitatively determined by GC). An example of TLC analysis of nitrogen derivatives of PAHs is the work by Jäger.[30] Trypień[31] quantitatively determined carbazole, nitroanthracene, nitronaphthalene, and nitropyrene in airborne particulate matter (cellulose plates and N,N-dimethylformamide as mobile phase). The dried plates were observed under UV light both before and after reaction in a chamber containing trifluoroacetic acid fumes. The method was used for air monitoring.

4. Phenols

Phenols (see Figure 8.4 for structures) form a group of aromatic compounds with one or more hydroxy groups. Phenols and substituted phenols (products of manufacturing processes used in plastics, dyes, drugs, antioxidants, and pesticide industries) are a serious danger in the environment, especially when they enter the food chain as water pollutants. Even at very low concentrations (ppb), phenols affect the taste and odor of fish. Because of this, many phenol derivatives (mainly nitrophenols and chlorophenols, which are also poisons) are considered as priority pollutants, which should be monitored both in surface water and municipal water supplies.

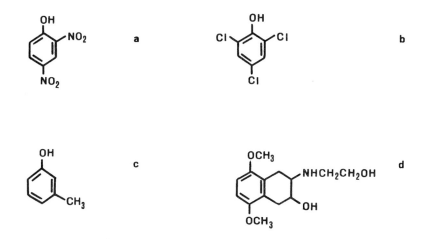

FIGURE 8.4 Representatives of phenols: (a) 2,4-Dinitrophenol, (b) 2,4,6-Trichlorophenol, (c) m-Cresol, (d) Tetraminol.

Considering the variation in polarity of phenolic compounds, they are difficult to monitor. These difficulties already appear in the stage of taking these compounds out of the water matrix, because they are retained on stationary phases commonly used in SPE only to a slight degree. The compounds are extracted usually by the LLE technique or liquid-solid extraction with anion-exchange materials. Di Corcia and co-workers[32] suggested a two-trap tandem extraction system for enhancing the selectivity of phenol separation (isolation). In the first cartridge, filled with graphitized black carbon (this adsorbent has on its surface chemical heterogeneities able to bind anions via electrostatic forces), most acidic phenols were retained. The least acidic phenols, washed through the first cartridge together with nonacidic compounds, were readsorbed in the second cartridge filled with strong anion-exchange material.

Among works devoted to phenol analysis, the Dietz et al.[33] publication attracts particular interest because it contains very comprehensive studies of over 100 phenolic compounds. The separations are conducted in NP systems (silica gel and aluminum oxide as adsorbents and three single-component eluents). Drandarov and Hais[34] used the same stationary phases in investigations of the positional isomers of some naphthol derivatives (detection with 4-nitrobenzenediazonium fluoroborate or measurement of fluorescence quenching). These authors separated 29 isomer pairs of tetrahydronaphthol and N-substituted 3-amino-tetrahydro-2-naphthols, including their lactone derivatives, by means of five solvent systems. The basic attention of this research was focused on assessment of separation mechanisms of more strongly retained compounds. Gumprecht's[35] experiments were significantly more practical; he used (among other compounds) phenols to appreciate commercially available TLC plates. He observed some very significant differences in R_f values as well as in the relative order in which the compounds were eluted. Also, some practical advice concerning environmental analyses of phenols can be found in a paper by Futer and Wall[36] on the quantitation of trace concentrations of phenols and cresols in contaminated land leaches. TLC has been used to separate and identify several commonly found phenols following derivatization as azo dyes and extraction into a dichloroethane. Detection was based on the oxidative coupling reaction between phenols and 3-methyl-2-benzothiazolinone hydrazone (detection limits were at the level of 10 ng).

Difficulties in obtaining adequate resolution of closely related structural isomers of phenols led to the application of modified sorbents (silica gel impregnated with inorganic or organic substances). In the literature, there are many publications devoted to this problem. An example is the work by Petrović and co-workers,[37] who separated 16 substituted phenols on silica gel layers modified with various concentrations of copper sulfate and aluminum chloride. A second direction of investigations leading to the improvement of the resolution of phenol mixtures is based on the choice of mobile phases. Typically, multicomponent solvent combinations containing eluents of widely varying polarity have been employed. Because of the long time required to develop chromatograms by means of such mixtures, studies were performed on the choice of mixtures that run rapidly and perform well in separating a very wide range of phenolic substances.[38] Investigations on OPLC applications were also conducted.[39]

5. Miscellaneous Pollutants

The examples given in this section prove common usage of TLC in investigation of various organic environmental contaminants. It is worth adding that the number of papers and analytes studied are growing systematically, confirming possibilities for use of TLC in this field.

a. Polychlorinated Biphenyls

In descriptions of environmental toxins, polychlorinated biphenyls (PCBs) (see Figure 8.5 for structures) are usually mentioned as being distinctly important from the analytical and toxicological points of view. In the case of TLC, such importance is not valid, because in scientific literature not many works on analyses of these compounds are available. This results probably from the fact that these substances are easy to detect and quantitate by means of electron capture detector used in GC.

There are 209 possible PCB configurations. These substances are different in their physicochemical characteristics as well as toxic potencies. Many representatives of this class of compounds (chlorobiphenyls) are persistent. Some of them are inducers of drug-metabolizing enzymes, also being able to affect various physiological processes such as reproduction, embryonic development, or carcinogenesis. The presence of PCBs in ecosystems is of anthropogenic origin only, and it is connected with applying mixtures (Arochlor, Clophen, and Pentachlor) used in transformers, condensers, heat exchangers, and thermoplastics. In the investigation of this group of compounds, TLC is mainly used to analyze changes of PCB products in the environment[40] or in techniques designated as screening methods.[41]

b. Warfare Agents

In Europe, increasing attention is focused on the need for monitoring warfare agents (see Figure 8.6 for structures), their by-products, and precursors. This fact stems not only from the need of observation of the chemical weapons convention[42] but also because certain toxic agents are already

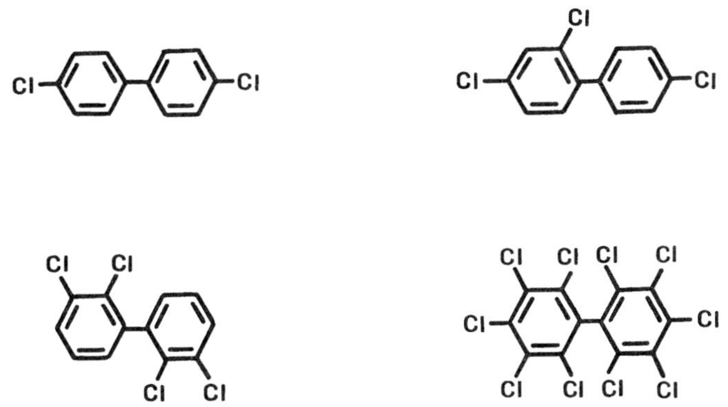

FIGURE 8.5 Examples of polychlorinated biphenyls.

spread in the environment (e.g., large amounts of mustard were sunk in the Baltic Sea). Out of this group of compounds, organophosphorous (O-ethyl S-2-diisopropylaminoethyl methyl phosphonothiolate–VX, O-pinacolyl methylphosphonofluoridate–soman) and the already mentioned bis-(2-chloroethyl) sulfide (mustard) have importance because of high toxicity. These substances can be successfully determined by the TLC technique. Munavalli and Panella[43] separated and identified mustard as well as its oxidation and hydrolysis products. Stachlewska[44] quantitatively determined phosphoroorganous (tabun, sarin, soman, and VX) in the presence of irritant warfare agents; visualization of chromatograms were conducted by the enzymatic method. Ludemann et al.[45] separated and identified irritant warfare agents (kamite, adamsite, and o-chlorobenzylidenemalonodinitryle). The separations were carried out in an NP system, using 4-nitrobenzopirydine as the visualizing reagent in the majority of cases. Of the more recent papers on this same topic, work by Mazurek and Witkiewicz[46] is worth mentioning. They used OPLC to analyze organophosphorous warfare agents in the presence of pesticides.

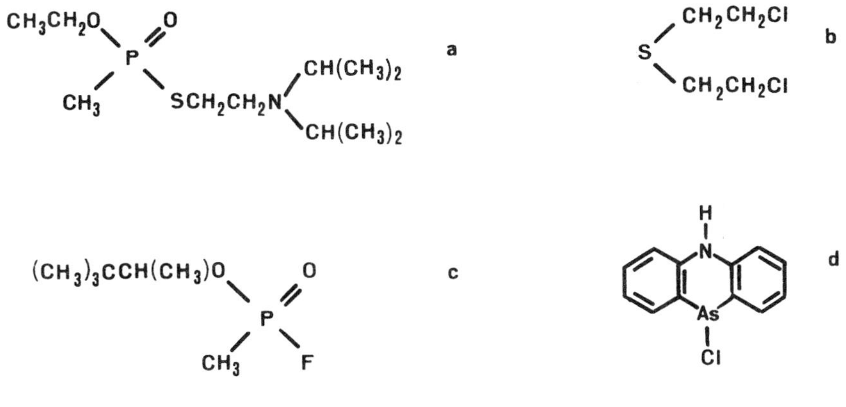

FIGURE 8.6 Warfare agents: (a) VX, (b) Mustard, (c) Soman, (d) Adamsite.

c. Trinitrotoluene

The explosive 2,4,6-trinitrotoluene (TNT) and its degradation products have been found as contaminants in water and soil and in the blood and urine of explosives manufacturing plant personnel. Because of the mutagenity of these compounds, environmental monitoring of TNT and its degradation products (2- and 4-monoaminodinitrotoluenes as well as 2,4- and 2,6-diaminonitrotoluenes) has been an important issue. Zou and co-workers[47] described an extraction method for these compounds from water (LLE), as well as separation (NP system) and determination (colors of the spots obtained in visible light and under UV illumination) by TLC.

d. Aldehydes and Ketones

Formaldehyde is a common environmental pollutant of anthropogenic and natural origin (it is formed in cells by oxidative demethylation of a variety of exogenous and endogenous N-, O-, and S-methyl compounds). Formaldehyde is an extremely reactive molecule, and it demonstrates genotoxic as well as carcinogenic properties. Tyihak et al.[48] described studies on the formaldehyde cycle in biological systems; the subject of investigation (by reversed phase OPLC) was potential natural formaldehyde generators, such as choline, betaine, and trigonelline, as well as dimedone adducts of formaldehyde and other aliphatic aldehydes. Quantitation was carried out by densitometry. In a subsequent work, he described separation of the 2,4-dinitrophenylhydrazine derivatives of eight saturated aldehydes and five saturated ketones in normal and reversed phase systems by means of OPLC.[49] These examples prove the possibilities of TLC usage in analyses of substances of relatively high volatility.

6. Inorganic Pollutants

Among inorganic environmental pollutants, TLC is used mostly to analyze heavy metals and radionuclides. Anions and other inorganic pollutants are also determined. Heavy metals (mainly Cd, Se, Hg, and Pb) pose serious threats to human health, even at very low concentrations in air and water. After polluting the soil, these pollutants can be incorporated into the food cycle via vegetables or, alternatively, be washed toward surface or underground water. Farming, industrial, and urban activities are most often mentioned as pollution sources of heavy metals. Rare earth elements and other fission products in freshly irradiated nuclear fuels are analyzed after their enrichment and separation by TLC. Rare earth elements and other fission products can be quantified by γ-spectroscopy. Jung and co-workers[50] isolated rare earth elements (two-dimensional elution in an NP system) from all other fission products.

Heavy metals are usually converted into appropriate colored derivatives. They can also be determined by enzymatic methods. Most effective separations are obtained on silica gel layers impregnated with inorganic salt solutions.[51] Funk and co-workers[52] gave a highly sensitive HPTLC method for the determination of selenium in water and biological matrices. Samples of drinking and surface water and biological materials were oxidized by a wet chemical digestion procedure, derivatized with 1,2,3,-naphthoselenodiazole, extracted on a column, and analyzed in an NP system involving a silica gel layer and chloroform mobile phase. Quantitations were carried out through fluorescence measurement at extinction and emission wavelengths of 365 nm and 560 nm, respectively. Sensitivity was 250 fg. Volynets[53] and co-workers described a rapid TLC method for the analysis of industrial and wastewaters for total heavy metal content. Ions of heavy metals (Fe^{2+}, Fe^{3+}, Co^{2+}, Ni^{2+}, and Cu^{2+}) were analyzed semiquantitatively as diethyldithiocarbamates after their concentration during the chromatographic process. Ajmal et al.[54] isolated Hg^{2+}, Hg^+, Pb^{2+}, Ni^{2+}, and Cu^{2+} on plates coated with silica gel impregnated with oxalic acid. Hg^{2+} was detected by dithizone and quantitation carried out by spectrophotometric method after spots were scraped and eluted from the adsorbent. Linearity was in the range of 5 to 10 mg/ml of Hg^{2+}. Myasoedowa and co-workers[55] used chelated cellulose sorbent with azopyrocatechol groups for separation and determination of heavy metal ions (Co^{2+}, Ni^{2+}, Cu^{2+}, Zn^{2+}, Cr^{3+}, Cr^{6+}, Fe^{2+}, and V^{5+}). Heavy metals reacted with the sorbent phase to produce colored products at pH 1 to 6. Detection limits were 2 ng to 5 mg. The method was used to analyze wastewaters from an electroplating process. Bruno et al.[56] detected mercury and some organomercury species at the nanogram level as dithizonates in tap and sea waters. HPTLC and densitometry were used to separate, detect, and determine these substances.

B. TLC AS A RESEARCH TECHNIQUE

It is possible to distinguish three types of TLC applications that comply with requirements of a research technique. The first is connected with the measurements of liquid-water partitioning of different pollutants. The second is related to the measurements of the mobilities of heavy metals, pesticides, and other pollutants in the environment. And third is the use of TLC as a pilot technique for column liquid chromatography.

1. Measurement of Lipid–Water Partitioning

Bioaccumulation is a function of lipid–water partitioning for a given substance and its refractivity toward degradation and biotransformation. Bioaccumulation potential rises due to growth of lipid solubility (in an environmental sense, bioaccumulation is usually higher in water than in terrestrial organisms). The lipid-water partitioning is characterized by value assessment of partitioning coefficient, P, of given pollution; this parameter is defined as the ratio of the equilibrium concentrations of dissolved substances in a two-phase system consisting of two largely immiscible solvents, e.g., *n*-octanol and water.

n-Octanol represents a substitute for a biotic lipid and hence gives an approximation to a biotic lipid–water partition coefficient. The ratio is reported as a logarithm, usually as *log P*. Compounds with high *log P* value have been classified as lipophilic.

The methods of determination of *log P* by RPTLC are described, among others, by Chessels and co-workers.[57] The idea of TLC application for such purposes is based on measuring R_f values of a given compound in chromatographic systems with a binary mobile phase (e.g., methanol–water and acetonitrile–water). It is a comparative method; in order to calculate *log P* values, the obtained results are compared with those obtained for standard substances of known *log P* values (the chemical structure of which does not differ significantly from the unknown) in the same chromatographic system.

2. Measurement of Pollutant Mobility in the Environment

Pollutants adsorbed onto soil particles may end as sediment at the bottom of lakes and rivers. Therefore, knowledge of mobility of pollutants in soil plays an important role in environmental investigations. The STLC (soil thin-layer chromatography) technique is used for such investigations. Soil samples taken from different regions are ground in mortar, then they are mixed with water to obtain a slurry and, from this slurry, chromatographic plates are prepared. A solution of substances whose mobility is to be studied is applied at the origin, and the chromatogram is developed using distilled water as the developer. Distribution of spots on chromatograms, which are visualized identically as in analytical investigations, is the measure of the mobility of the pollutant-soil system.

The need to perform investigations on the metabolism of pollutants in plants, animals, and soil in support of compound registration has led to an increase in the use of radiochromatographic techniques, including TLC. Radio-TLC is a technique used for investigations of changes of pollutants (mainly pesticides) in the environment. In such investigations, the content of radiolabeled pollutants and products of their reactions are measured. Examples of such TLC applications include work by Schneider and co-workers[58] (fate of rimsulfiron in the environment), Gallina and Stephenson[59] (dissipation of glufosinate measurements), Lee and co-workers[60] (bioaccumulation investigation), and many others. Such measurements require autoradiography detection and quantitation mostly by scraping and liquid scintillation counting.

In newer studies (since 1980), linear analyzers are also used, making *in situ* measurements possible.

The work of Klein and Clark[61] using a bio-imaging analyzer has particular importance. In this apparatus, an ultra-sensitive imaging plate is used, which, when exposed, accumulates and stores energy from radioactive zones. The plate is then inserted into an image-reading unit and scanned with a fine laser beam. The plate emits luminescence in proportion to the intensity of the recorded radiation, and this luminescence is measured by a photomultiplier tube and converted to electrical energy. This instrument was used to study plant metabolism. Radioactive solutions were investigated by one- and two-dimensional development in an NP system. Sensitivity, linearity, resolution, accuracy, and method reproducibility were investigated. It was shown that TLC with the bio-imaging analyzer is far more sensitive than any other radio-TLC technique currently available.

3. TLC as a Pilot Technique for HPLC

TLC is used for modeling systems in HPLC. Pre-investigation data from TLC experiments sometimes provides conditions suitable for separations in column liquid chromatography; in this way

TLC can be treated as a pilot technique. Rózyło and Janicka[62] gave theoretical aspects of such TLC usage in environmental investigations. Analyzing phenols, they proved that retention parameters obtained in sandwich chambers describe adsorption from solution in the same way as measurements from liquid chromatography. That means that if TLC is to be used as a pilot technique for HPLC, the results from sandwich chambers should be used, as only under these conditions do results mirror those obtained in HPLC. Gazdag et al.[63] carried out similar studies using OPLC. They obtained good agreement in selectivity between HPLC and OPLC ion-pair systems using the same eluent composition. However, it is worth emphasizing that the predictive power of TLC for HPLC depends strongly on the type of solutes and the experimental conditions. For instance, Cserháti and Forgács[64] showed that the correlation between the HPLC and TLC (on an alumina support) retention of a non-homologous series of pesticides is very poor.

REFERENCES

1. **Jork, H.,** TLC '92: Steady progress, *J. Planar Chromatogr.,* 5, 4, 1992.
2. **Hahn-Deinstrop, E.,** Stationary phases, sorbents, *J. Planar Chromatogr.–Mod.TLC,* 5, 57, 1992.
3. **Jork, H., Funk, W., Fischer, W., and Wimmer, H.,** *Thin-Layer Chromatography. Reagents and Detection Methods. Volume 1a: Physical and Chemical Detection Methods,* VCH, Weinheim, FRG, 1990; *Volume 1b: Activation Reactions Reagent Sequences,* VCH, Weinheim, FRG, 1994.
4. **Windhorst, G. and De Kleijn, J. P.,** Some technical recommendation for improving the performance of qualitative thin layer chromatography, *J. Planar Chromatogr.,* 5, 229, 1992.
5. **Zieloff, K.,** The actual state of quantitative TLC, *GIT Spez. Chromatogr.,* 12, 101, 1992.
6. **Rathore, H. S. and Begum, T.,** Thin-layer chromatographic methods for use in pesticide residue analysis, *J. Chromatogr.,* 643, 271, 1993.
7. **Sherma, J.,** Determination of pesticides by thin layer chromatography, *J. Planar Chromatogr.,* 7, 265, 1994.
8. **Sherma, J.,** Planar chromatography, *Anal. Chem.,* 66, 67R, 1994.
9. **Hurtubise, R. J.,** Application in pesticide analysis, in *Solid Surface Luminescence Analysis,* Hurtubise, R. J., ed., Guilbault, New York, 1981, Chap. 10.
10. **Judge, D. N., Mullins, D. E., and Young, R. W.,** HPTLC of pesticides and their by-products, *J. Planar Chromatogr.,* 6, 300, 1993.
11. **Sherma, J.,** Quantitation of benomyl and its metabolites by thin-layer densitometry, *J. Chromatogr.,* 104, 476, 1975.
12. **Gardyan, C. and Thier, P.,** Identifizierung und quantitative Bestimmung aktueller Pflanzenschutzmittel durch HPTLC, *Z. Lebensm. Unters. Forsch.,* 192, 40, 1991.
13. **Rathore, H. S. and Begum, T.,** Thin layer behavior of carbamate pesticides and related compounds, *J. Chromatogr.,* 643, 321, 1993.
14. **Sherma, J. and Miller, N. T.,** Quantitation of the triazine herbicides atriazine and simazine in water by thin layer chromatography with densitometry, *J. Liq. Chromatogr.,* 3, 901, 1980.
15. **Gries, W. and Jork, H.,** Urea herbicides in drinking water — HPTLC spectroscopic determination, *J. Planar Chromatogr.,* 2, 290, 1989.
16. **De la Vigne, U. and Jänchen, D.,** Determination of pesticides in water by HPTLC using AMD, *J. Planar Chromatogr.,* 3, 6, 1990.
17. **Kováč, J., Tekel', J., and Kurucová, M.,** Determination of herbicide residues in agricultural crops, foods, soil, and water by the chronometric method, *Z. Lebensm. Unters. Forsch.,* 184, 96, 1987.
18. **Zehradnièková, H., Marsalek, B., and Polisenska, M.,** Determination of the abscisic acid by TLC, *J. Planar Chromatogr.,* 3, 243, 1990.
19. **Patil, V. B. and Shingare, M. S.,** Thin-layer chromatographic spray reagent for the screening of biological materials for the presence of carbaryl, *Analyst,* 119, 415, 1994.
20. **White, R. H. and Howard, J. W.,** Thin-layer chromatography of polycyclic aromatic hydrocarbons, *J. Chromatogr.,* 29, 108, 1967.
21. **Hellmann, H.,** Zur Unterscheidung von biogenen und mineralölbürtigen Aromaten durch Fluorescenzspektroskopie, *Z. Anal. Chem.,* 272, 30, 1974.
22. **Kubis, A. J., Somyajula, K. V., Sharkey, A. G., and Hercules, D. M.,** Laser mass spectrometric analysis of compounds separated by TLC, *Anal. Chem.,* 61, 2516, 1989.
23. **Hinze, W. L., Pharr, D. Y., Fu, Z. S., and Burkert, W. G.,** Thin-layer chromatography with urea-solubilized β-cyclodextrin mobile phase, *Anal. Chem.,* 61, 422, 1989.

24. **Van de Nesse, R. J., Hoogland, G. J., De Moel, J. J. M., Gooijer, C., Brinkman, U. A. Th., and Velthorst, N. H.,** On-line coupling of liquid chromatography to TLC for the identification of PAHs in marine sediment by fluorescence excitation and emission spectroscopy, *J. Chromatogr.,* 552, 613, 1991.
25. **Baranowska, I., Szeja, W., and Wasilewski, P.,** The analysis of polycyclic aromatic hydrocarbons, *J. Planar Chromatogr.,* 7, 137, 1994.
26. **Furton, K. G., Jolly, E., and Pentzke, G.,** Recent advances in the analysis of PAHs and fullerenes, *J. Chromatogr.,* 642, 33, 1993.
27. **Motohashi, N. and Kamata, K.,** A convenient method for the purification of carcinogenic and noncarcinogenic methyl-substituted benz(c)acridines, *Yakugaku Zasshi,* 103, 795, 1983.
28. **Handa, T., Yamauchi, T., Sawai, K., Yamamura, T., Koseki, Y., and Ishii, T.,** *In situ* emission levels of carcinogenic and mutagenic compounds from diesel and gasoline engine vehicles on an expressway, *Environ. Sci. Technol.,* 18, 895, 1984.
29. **Sinkkonen, S.,** Isolation of chlorinated dibenzothiophenes by HPLC, *J. Chromatogr.,* 553, 453, 1991.
30. **Jäger, J.,** Detection and characterization of nitro derivatives of some hydrocarbons, *J.Chromatogr.,* 152, 575, 1978.
31. **Trypień, K.,** TLC Identification of some nitrogen derivatives of PAHs in airborne particulate matter, *J. Planar Chromatogr.,* 6, 413, 1993.
32. **Di Corcia, A., Marchese, S., and Samper, R.,** Selective determination of phenols in water by a two-trap tandem extraction system followed by liquid chromatography, *J.Chromatogr.,* 642, 175, 1993.
33. **Dietz, F., Traun, J., Koppe, P., and Rübelt, Ch.,** Systeme für die Identifizierung von Phenolen mittels Dünnschicht-Chromatographie, *Chromatographia,* 9, 380, 1976.
34. **Drandarov, K. and Hais, I. M.,** Thin-layer chromatography of the positional isomers of some naphthol derivatives, *J. Chromatogr.,* 628, 103, 1993.
35. **Grumprecht, D. L.,** Comparison of commercially available TLC plates with mixtures of dyes, analgesics, and phenols, *J. Chromatogr.,* 595, 368, 1992.
36. **Futter, J. E. and Wall, P.,** On-site identification of trace concentrations of phenols and cresols in contaminated land leachates using planar chromatography, *J. Planar Chromatogr.,* 6, 372, 1993.
37. **Petrović, M., Kastelan-Macan, M., and Horvat, A. J. M.,** Thin-layer chromatographic behavior of substituted phenolic compounds on silica gel layers impregnated with Al(III) and Cu(II), *J. Chromatogr.,* 607, 163, 1992.
38. **Ferry, J. and Larson, R. A.,** A mixed solvent for rapid TLC analysis of phenolic compounds, *J. Chromatogr. Sci.,* 29, 476, 1991.
39. **Senini, A., Fahimi, A., Moulounguit, Z., Delmas, M., and Gaset, A.,** Separation and preparative isolation of phenolic dialdehydes by on-line overpressured layer chromatography, *J.Chromatogr.,* 590, 369, 1992.
40. **Frei-Hausler, M., Frei, R. W., and Hutzinger, O.,** Detection of hydrobiphenyls as dansyl derivatives, *J. Chromatogr.,* 79, 209, 1973.
41. **Newborn, J. S. and Preston, J. S.,** Analytical field screening of soil by thin layer chromatography, *Hazard. Matter. Control,* 4, 56, 1991.
42. **United Nations,** *Convention on the Prohibition of the Development, Production, Stockpiling and Use of Chemical Weapons and Their Destruction,* 1993.
43. **Munavalli, S. and Pannella, M.,** Thin-layer chromatography of mustard and its metabolites, *J. Chromatogr.,* 437, 423, 1988.
44. **Stachlewska-Wróblowa, A.,** Rozdzielanie i identyfikacja grupy substancji toksycznych przy użyciu TLC, *Bull. WAT (Warsaw),* 7, 135, 1978.
45. **Ludemann, W. D., Stutz, M. H., and Sass, S.,** Thin-layer chromatography of some irritants, *Anal. Chem.,* 41, 679, 1969.
46. **Mazurek, M. and Witkiewicz, Z.,** The analysis of organophosphorous warfare agents in the presence of pesticides by OPLC, *J. Planar Chromatogr.,* 4, 379, 1991.
47. **Zou, H., Zhou, S., Hu, X., Zhang, Y., and Lu, P.,** Identification of 2,4,6-trinitrotoluene and its biodegeneration products by TLC, *J. Planar Chromatogr.,* 7, 461, 1994.
48. **Tyihák, E.,** Overpressured layer chromatographic methods in the study of the formaldehyde cycle in biological systems, *Trends Anal. Chem.,* 6, 90, 1987.
49. **Gersbeck, B., Schönbeck, F., and Tyihák, E.,** Measurement of formaldehyde and its main generators, *J. Planar Chromatogr.,* 2, 86, 1989.
50. **Jung, K. and Specker, H.,** Separation and determination of rare earth elements and other fission products, *Fresenius' Z. Anal. Chem.,* 300, 15, 1980.
51. **Mohammad, A. and Majid Khan, M. A.,** New surface-modified sorbent layer for the analysis of toxic metals in seawater and industrial wastewater, *J. Chromatogr.,* 642, 455, 1993.
52. **Funk, W., Dammann, V., Couturier, T., Schiller, J., and Volker, L.,** Quantitative HPTLC determination of selenium, *J. High Resolut. Chromatogr. Chromatogr. Commun.,* 9, 223, 1986.
53. **Volynets, M. P., Litaeva, L. P., and Timerbaer, A. P.,** An express test method for determination of total heavy metal contents of water by TLC, *Zh. Anal. Khim.,* 41, 1989, 1986.
54. **Ajmal, M., Mohammad, A., Fatima, N., and Khan, A. H.,** Reversed phase thin-layer chromatographic separation of heavy metal ions on silica gel layers loaded with tributylamine, *Microchem. J.,* 39, 361, 1989.

55. **Myasoedova, G. V., Volynets, M. P., Akimova, T. G., Bolshakova, L. I., Dubrova, T. V, Savrin, S. B., and Ermakov, A. N.**, Separation and determination of Cu, Co, Ni, Cr, Fe, and V by TLC on a cellulose sorbent containing azo catechol groups, *Zh. Anal. Khim.*, 41, 662, 1986.
56. **Bruno, P., Caselli, M., and Traini, A.**, Inorganic and organic mercury determination by HPTLC and *in situ* densitometric detection. Application to real samples, *J. High Resolut. Chromatogr. Chromatogr. Commun.*, 8, 135, 1985.
57. **Chessels, M., Hawker, D. W., and Connel, D. W.**, Critical evaluation of the measurement of the 1-octanol/water partitioning coefficient of hydrophobic compounds, *Chemosphere*, 22, 1175, 1991.
58. **Schneiders, G. E., Koeppe, M. K, Naidu, M. V., Horne, P., Brown, A. M., and Mucha, C. F.**, Fate of rimsulfuron in the environment, *J. Agric. Food Chem.*, 41, 2404, 1993.
59. **Gallina, M. A. and Stephenson, G. R.**, Dissipation of [^{14}C] glufosinate ammonium in two Ontario soils, *J. Agric. Food Chem.*, 40, 1992.
60. **Lee, P. W., Forbis, A. D., and Franklin, L.**, Bioaccumulation of cinmethylin in bluegill sunfish, *J. Agric. Food Chem.*, 38, 323, 1990.
61. **Klein, O. and Clark, T.**, The advantages of a new bio-imaging analyzer for investigation of the metabolism of ^{14}C – radiolabelled pesticides, *J. Planar Chromatogr.*, 6, 368, 1993.
62. **Rózyto, J. and Janicka, M.**, Some theoretical aspects of the use of TLC as a pilot technique for column liquid chromatography, *J. Planar Chromatogr.*, 3, 413, 1990.
63. **Gazdag, M., Szepesi, G., Hernyes, M., and Végh, Z.**, Optimization of reversed-phase ion-pair chromatography by OPLC. II. HPLC separation on silica by utilizing pre-investigation data from OPLC experiments, *J. Chromatogr.*, 290, 135, 1984.
64. **Cserháti, T. and Forgács, E.**, Relationship between the HPLC and TLC retention of non-homologous series of pesticides on an alumina support, *J. Chromatogr.*, 668, 495, 1994.

Chapter 9

THIN-LAYER CHROMATOGRAPHY IN FOOD ANALYSIS

Ali R. Shalaby

CONTENTS

I. Introduction 170

II. Sampling 171

III. Sample Preparation 172
    A. Extraction 172
    B. Cleanup 173

IV. TLC Techniques 174
    A. Layer Selection 175
    B. Layer Preparation 175
    C. Precoated Plates 176
    D. Sample Application 176
    E. Solvent Selection 177
    F. Development Techniques 177
    G. Qualitative Determination 177
    H. Quantitative Determination 178
        1. Assay on the Layer 178
        2. Assay after Extraction 179

V. Validation of the Method 179
    A. Sample Preparation 179
        1. Sampling 180
        2. Extraction 180
        3. Sample Cleanup and Concentration 180
        4. Stability of Sample in Solution 180
        5. Derivatization 180
    B. Chromatographic Process 180

VI. Selected Application Protocols 181
    A. Food Composition 181
        1. Proteins 181
        2. Liquids 182
        3. Carbohydrates 182
        4. Organic Acids 183
    B. Food Additives 183
        1. Antioxidants 183
        2. Benzoic Acid 183
        3. Nonnutritive Sweeteners 184
    C. Food Contaminants 184
        1. Pesticide Residues 184
        2. Antibiotic Residues 187
        3. Aflatoxins 188

D. Food Decomposition 188
     1. Biogenic Amines 188

References 189

I. INTRODUCTION

A primary objective of the food industry is to continue the supply of safe, wholesome foods for the public. To achieve this goal, food producers must be concerned about possible deterioration of foods, their contamination, and the safety of additives or other materials used in food processing. On the other hand, government is expected to protect the health and welfare of the consumers, as well as the interests of the fair and honest food producers, processors, and marketers against dishonest and unfair competition. This function is performed through governmental analysis of foods and food products to be sure that they are free from contamination, adulteration, or decomposition; moreover, there is the maintenance of general food quality and the control of the use of food additives. Therefore, a simple, rapid, accurate and inexpensive analytical method is urgently needed. Thin-layer chromatography (TLC) is by far one of the most common analytical methods used to achieve such purposes, and it has many valuable analytical applications in the field of food analysis and quality control.

As TLC has become such a widely used technique in food analysis laboratories, some examples of the major areas in which TLC can be used are seen. The problem of improving protein quality is one of major importance in the world today. A knowledge of the quality of individual proteins permits one to improve the nutritional adequacy of a given protein of low biological value by supplementing it with one or more amino acids, or with another protein that provides the limiting or missing essential amino acid.[1] Using TLC, the amino acid components of a given protein or peptide can be separated with high resolution and determined with many advantages over other methods.[2-5] The analysis of lipids and the identification of fatty acids derived from fats is frequently necessary to determine their quality and the presence of a foreign fat (adulteration).[1] Thin-layer chromatography is used extensively for lipid analysis and is a particularly valuable tool for the separation and tentative identification of neutral and complex lipid classes,[6-10] and it has the distinct advantage of allowing the separation of lipids of different polarities on the same plate with a single solvent system.[6] The determination of mono-, di-, and oligosaccharides occurs frequently in the food industry. Their quality in food and beverages is often a good indicator of product quality, contamination, and/or adulteration. It is also very important in monitoring different fermentation and other biotechnological processes.[11] Most analytical methods for specific carbohydrates require some type of separation scheme so that individual sugars may be isolated from each other, identified, and analyzed. As a result, TLC is ideally suited for carbohydrate analysis.[11-15]

TLC can be considered as a simple method for determining storage stability of foods and food products. Biogenic amines are natural antinutrition factors important from a hygienic point of view, since they have been implicated as the causative agents in a number of food poisoning episodes, and they are able to initiate various pharmacological reactions. Biogenic amines are present in foods as a result of the action of enzymes produced by a part of microflora on free amino acids. Therefore, biogenic amines may be used as freshness, decomposition, and microbial contamination indicators.[16-23] A simultaneous determination of eight biogenic amines in foods has been performed by TLC.[18,19] Storage stability of fat-containing food can also be studied using TLC.[24,25]

Thin-layer chromatography can be considered to be as powerful a determination method as gas liquid chromatography (GLC) and high performance liquid chromatography (HPLC) in determining minor components and residue analysis of food and food products.[26] Most of the minor components of plant origin food such as alkaloids, flavonoids, saponins, and steroids can be separated and determined by TLC techniques.[27-30] Some food additives have more than one function and/or nutritional significance; for example, carotenes and tocopherols have color and anti-oxidative properties as well as being vitamins. Such additives are susceptible to oxidation and degradation during preparation and storage of foods; therefore, the control of their level is recommended. TLC techniques offer a suitable method for such determination and for other vitamins,[31,32] natural pigments,[33,34] and other food additives.[35,36] The simultaneous detection of food preservatives such as benzoic, citric, and lactic acids in soft drinks and in foods has been affected by TLC.[35-37] Quantitative determination of antibiotics in fermentation broths and their residues in food samples, animal tissues, milk, and eggs has been performed using TLC.[38-42] Pesticide residues[26,43] and mycotoxins[44-46] in foods and food products have been determined by TLC. Some of the most striking recent applications of TLC in food analysis include the determination of selenium in foodstuffs and heavy metals in industrial and waste water.[47]

HPLC and GLC methods are quite valuable in food analysis, but many problems can occur during analysis. Moreover, the mobile phase, whether solvent or carrier gas, limits applications, since the mobile phase might reduce the detector response to an analyte by dilution or solvation.[48] TLC permits detection of analytes in the absence of the mobile phase; therefore, it can be more sensitive. On the other hand, although TLC is not as fast as GLC and HPLC, a number of samples can be applied on the same plate, and a number of plates can be run simultaneously; therefore, TLC will be faster per sample than the other chromatographic methods.[14,19] In addition, TLC is often the only method that can be used for particular compounds, such as carbamate insecticides,[49] where they are thermally labile or nonvolatile substances. It must be emphasized that TLC is most effective when it is employed in combination with other techniques utilizing complementary principles of chromatography.[50] Thus, TLC can be used as a preparative technique for many other methods such as GLC, IR spectroscopy, mass spectrometry, and thermal analysis methods. Moreover, TLC offers a good means of optimizing a separation as a pilot method for HPLC. The good correlation obtained between k' in HPLC and R_f data in TLC, as well as the minimal requirements for reagents and instrumentation, recommend the use of TLC as a preliminary step to a required HPLC separation.[51]

Thin layer densitometry is the choice for best quantitative results, while the visual comparison thin layer method is the choice for a rough and quick estimation.[14] Advantages of TLC over the other methods include simplicity, rapidity, versatility, applicability to a large number of samples in minimal time, and low cost in terms of reagents and equipment, as well as applicability to food in general[19] for a wide range of analysis.

II. SAMPLING

Sampling is an integral part of the analytical procedure. The objective of the sampling procedure is to obtain a laboratory sample (test portion) representative of the lot from which it is drawn.[45] Two types of sampling procedures are used: manual and continuous.[1,52] Manual sampling is accomplished with instruments such as triers, probes, or sampling tubes. Auger- or drill-type samplers can be used to remove a sample from solid materials such as cheese, butter, or frozen food products, while a syringe-type sampler may be used to obtain liquid samples.[1,52] In continuous sampling procedures, samplers or sample boxes mechanically divert a fraction of the material being sampled. For solid materials, a riffle cutter is often used, while a liquid sample may be obtained by "bleeding off" a fraction of the mainstream line through a smaller diversion line.[1]

The laboratory usually has no control over the field sampling of the food products and must assume that the portion received for analysis is representative of the lot of food sampled.[36] The laboratory sample as received may be bagged, packaged, tinned, or bottled and most often includes multiple units.[36] The sample received usually must be reduced to laboratory size, which should be representative of the whole. Therefore, the individual units from the same food lot might be combined and mixed to obtain a homogeneous material from which the laboratory sample will be drawn. Some liquid foods are reasonably homogeneous, but solid and semisolid foods are always heterogeneous. So, mechanical grinding, mixing, rolling, agitation, stirring, or any logical means of making the sample more homogeneous prior to subsampling is desired.[1,36,52] By one or more of these means the laboratory sample will be representative of the whole, and the constituent for which the food is being examined will be evenly distributed throughout the sample.

The type of sample preparation before subsampling will depend on the food product to be treated. A liquid food (e.g., milk) generally need only be well mixed or shaken before subsampling.[36] Semisolid foods are those containing a solid material plus a large portion of free liquid, such as canned foods. In the event that the solid or the liquid is to be analyzed individually, they are separated using a sieve or filter and individually mixed for subsampling. When both solid and liquid phases are to be analyzed as a unit, it is often advisable to blend or otherwise homogenize the two before subsampling.[36] Solid samples can be of three general types: finely divided (e.g., whole cereal grains or flour), an aggregate (e.g., solid mixtures such as sausage), or a whole unit (e.g., an entire fruit).[36] Finely divided dry products can be mixed for subsampling using commercial portioning equipment, or by spreading the sample over a large surface, quartering with a straight edge, and mixing opposite quarters.[1,36] The process is repeated one or more times to obtain a suitable representative portion.[1,36] An aggregate solid sample must be chopped or ground before mixing and subsampling.[36] The subsampling of the whole unit sample could be one quarter of a fruit, a piece of loin from a whole fish, or other similar sections.[36]

It is of interest to point out that proper sample preparation must not only produce a representative portion for analysis, but must also prevent changes in the sample that may result in biased analytical results.[36] The mechanical equipment usually generates heat during the processing, which can possibly change the sample composition, such as for fatty foods where the heat may be sufficient to partially melt the fat.[36] In such cases, hand chopping and mixing may be the best procedure.[36]

Freezing is often the only way to prevent changes in a food before analysis, especially for foods that were sampled while frozen, foods for decomposition analysis, or foods that are affected by the preparation process. Such foods may be thawed before analysis, and the food composition should remain unchanged during thawing. Therefore, slow thawing without heat and in a closed container should be done to prevent moisture loss by drying or gain by condensation, and any separated liquid must be mixed back into the thawed product before subsampling for analysis.[36]

III. SAMPLE PREPARATION

All chemical and residue analysis methods for the detection and determination of foods contain four basic steps: extraction of the components of interest from the sample; removal of interfering matter by purification and/or cleanup procedures; TLC separation of the extracted component; and detection and quantitation of the zones of interest. The best methods take into account the nature of the samples to be analyzed, e.g., fatty foods should be handled differently from nonfatty foodstuffs or plant matrices, and the chemical properties of the component to be determined.[43]

A. EXTRACTION

The first step in chemical analysis involves extraction of the test portion to separate the components of interest from the bulk of the matrix components and to obtain the materials of interest in a manageable form. Contact between solvent and solid substrate (liquid-solid extraction) is accomplished either for a short period (1–3 min) in a high-speed blender or for a longer period (30 min) by shaking in a flask. An overnight presoaking treatment of the samples, in special cases, prior to extraction gives the most efficient results. Liquid samples may be extracted in a separatory funnel (liquid-liquid extraction) or absorbed to a hydrophilic matrix that is prepacked in a column, after which extraction is accomplished by eluting the column with an extraction solvent. Food samples can be extracted by continuous extractors such as a Soxhlet unit. When using such a technique, the sample must be finely divided to prevent occlusion, and sufficient extractive solvents must be used to provide a good siphoning action plus enough in the reservoir flask to prevent complete dryness.[36] Extraction can also be done using a steam distillation unit (Figure 9.1), where solids such as benzoic and sorbic acids can be made sufficiently volatile to distill[36] and can be collected separately for TLC analysis.[35]

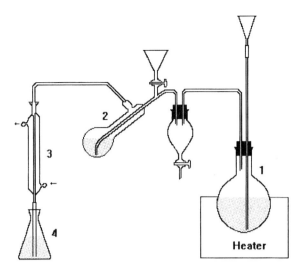

FIGURE 9.1 Steam distillation apparatus: 1, steam generator (3L flask); 2, sample flask (500–800 mL Kjeldahl flask); 3, condenser; and 4, receiver (1L flask).

All extractions, whether liquid-liquid or liquid-solid, involve partitioning of a material between two phases.[36] In a liquid-liquid extraction, the partitioning is governed basically by the relative solubility of the extractant in two liquids.[36] In a liquid-solid extraction, this is complicated further by possible physical occlusion of the extractant within some inert solid material.[36] Most liquid-solid extractions are exhaustive, where a material is to be completely extracted. In general, partitioning of a substance is rarely complete in one extraction. Therefore, multiple extractions are often necessary for quantitative results, where by using the same total extraction volume in a series of extractions, rather than all at once, more quantitative results can be obtained.[36] The chemical properties of both the matrix and the compound to be extracted affect the choice of the extracting solvent.[45] Often mixtures of solvents are found to be most effective. Lipids, for example, are a class of heterogeneous compounds; therefore, no single solvent is suitable for extracting lipids from naturally occurring materials. Solvents of low polarity will extract glycerides, sterols, and small amounts of complex lipids, whereas mixtures of low polarity with polar solvents will extract the majority of complex lipids.[1] If a substance is naturally acidic or alkaline, this feature can be used to affect the degree of partitioning between a water solution by adding small amounts of acid or alkali to the extraction solvent.[36]

Some components present no special problems in sample preparation, where samples have only to be diluted with an extracting solution to obtain a final assay.[53] On the contrary, in specific analysis procedures, hydrolysis is required by acid, alkaline, or enzymes. The amino acid component of a protein sample can be determined only after hydrolysis of the protein. Similarly, the amount of starch present in a food sample can be determined by hydrolyzing the starch with enzymes, and then the sugar is determined.[1] In other procedures, specific reactions may be done, e.g., separation of glycerides from the sterols by saponification.[54]

B. CLEANUP

The frequent presence of different substances that may interfere in the final detection makes it necessary to clean up the sample extract prior to determination. The cleanup techniques are, in fact, separation procedures in which groups of substances with certain physicochemical properties can be separated from one another.[45] In this way, the greater part of the co-extracted material can be removed. The choice of the cleanup procedure may depend on the method used for detection and determination, the required limit of detection, the speed of analysis, and the recovery.[45]

In some analytical procedures, precipitation can be done on the crude sample extract prior to cleanup. Precipitating reagents are normally used, e.g., lead acetate and fresh ferric hydroxide gel to precipitate gossypol pigments; cupric carbonate to remove chlorophyll; silver nitrate or ammonium hydroxide to remove alkaloids[45]; and borate ion for the complexing of glycols and sugars. Deproteinization may be done by acids, zinc ferrocyanide,[54] organic solvents,[54,55] and ultrafiltration.[56,57] Clarification of carbohydrate solutions may be done with basic lead carbonate, alumina cream, lead acetate, phosphotungstic acid, and/or activated charcoal.[1,54] Protein hydrolyzate is purified by desalting with barium hydroxide.

The crude sample extract may be subjected directly to cleanup without prior clarification. Different procedures, such as liquid-liquid partitioning or column cleanup, can be used to purify the sample extract. Liquid-liquid partitioning is often carried out in a separatory funnel. A partition system is manipulated by changing the nature of the two liquid phases, generally through the addition of solvents or buffers. In these cases, the more polar liquid is generally water, and its tendency to dissolve solutes can be changed by adding salts (to produce a salting-out effect) or buffers (to dissolve ionic materials such as amino acids). The less polar phase of a partition system is generally an organic solvent, and its ability to dissolve solutes can be varied by adding more or less polar solvents.

Column chromatographic cleanup is possible with various materials such as silica gel, aluminum oxide, polyamide, Florisil, and Sephadex. The sample extract is usually added to the column in an appropriate solvent, after which the column is washed with one or more solvents in which the components of interest are insoluble or less soluble than the impurities. Then the solvent composition is changed in such a way that the components of interest are selectively eluted from the column and the eluate is collected and concentrated. In studies of proteins, desalting can be accomplished by passing a protein solution through a Sephadex column or by dialysis[1] before hydrolysis.

Since many components of interest are normally present at low levels, a high concentration of the cleaned-up extract is necessary to make detection possible. The residue is redissolved in a small, specified volume of solvent prior to application for the TLC separation. Derivatization of the components of interest may be necessary to make them measurable or to optimize their chromatographic behavior.

The field of sample preparation has moved increasingly toward the use of disposable minicolumns (or cartridges) in order to speed and simplify extraction and cleanup. Sep-Pak cartridges enable the isolation of a narrower range of compounds than is possible by extraction alone and minimize the time used for liquid-liquid partitions and/or column chromatography. These sample preparation systems are of two basic types[58]: liquid-liquid extraction and solid-phase extraction. The latter is used widely in food analysis to concentrate solutes from dilute solution, purify concentrated solvent extracts, elute compounds with different polarities in separate fractions, or in microanalysis when the volume of sample is very low.

A diluted beer sample may be concentrated and cleaned up by use of a C_{18} Sep-Pak cartridge for the determination of inorganic anions and organic acids.[53] In carbohydrate analysis, when the samples contain large amounts of interfering compounds, it is best to remove them by use of disposable, ready-to-use columns.[11] Solid phase extraction was used to fractionate fat extracted from fish,[24] and for multi-residue determination of pesticides that differ sufficiently in hydrophobicity.[26] The simultaneous determination of 17 color additives in a food sample was performed using Sep-Pak NH_2 cartridges as a preparative technique.[59] For determination of antibiotics in foods and food products, a cleanup step with Sep-Pak C_{18} cartridges was necessary.[38] Sep-Pak C_{18} cartridges were also used to clean up the extract for determining aflatoxins in 0.2 g of liver sample.[60]

IV. TLC TECHNIQUES

Thin-layer chromatography was introduced in numerous laboratories for the separation of trace quantities of mixtures in research work and for routine investigations. Solutions of the mixtures are applied as spots or streaks on thin sorbent layers, and the plate is developed with various predominantly organic solvents. The separated substances are made visible by viewing under UV light or spraying with characteristic detection reagents, the identification being facilitated by the use of corresponding reference substances. Quantification can be made by comparison with standards on the plate, or after elution by different analytical methods.

There are three mechanisms of separation: partition (in which the compounds are moved between two liquid phases), adsorption (in which the components are reversibly bound on the surface of the adsorbent), and ion exchange (if true heteropolar chemical bonds are formed reversibly between the moving compound and the stationary phase). Therefore, when choosing a chromatographic method that will yield the desired result most quickly in a particular separation problem, the most important structural features of the compounds to be separated must be considered. The influence of molecular size is much less marked in adsorption chromatography than in the other methods, while differences in the nature and position of polar substituents are important. In partition chromatography, the effects of differences in the solubilities of compounds is very apparent. In practice, however, a combination of the three types of mechanisms is usually involved, with one type predominating. On the other hand, the choice of sorbent and solvent often enables the chromatographic behavior of a substance to be predicted. In general, if an active sorbent is used for a given substance, a correspondingly active elute is required and vice versa. Sorbents may be impregnated by various buffer ions during plate preparation to enhance the separation.

A. LAYER SELECTION

Sorbents for TLC are, in order of importance, silica gel, alumina, kieselguhr and cellulose, as well as many other substances.[50,58,61–65] They are finely divided and generally contain a binder. The letter G after the name of the substance indicates that calcium sulfate has been added to it as a binder, while the letter H indicates that no binder has been added, and the symbol F_{254} indicates that the thin layer fluoresces at 254 nm due to incorporation of a fluorescing agent.[36]

Silica gel is the most extensively used of all the TLC phases. If the compounds to be separated are neutral, containing one or two functional groups, or organic bases or acids, they may be resolved on layers of silica gel. Alumina (aluminum oxide basic) is frequently used for the separation of bases rather than silica gel. Because of its basic nature, additional base in the developing solvents is not required. Kieselguhr and cellulose both are supports for liquid films in partition chromatography. These types are always used for the separation of very polar molecules such as amino acids, carbohydrates, and various other naturally occurring hydrophilic compounds, and they are frequently required for the separation of closely related isomers.

A number of substances can be incorporated into thin layers or added to them to yield layers for specific separations.[36,61,64–66] The pH of layers can be varied widely by using aqueous solutions of acids, bases, and buffers (0.5 N in slurry) for the separation of acids, bases, and acidic or basic substances depending on the type of buffer used, respectively. Complexing agents are most important, e.g., boric acid (0.1 N in slurry) for the separation of sugars; silver nitrate (12.5% in slurry) for separation of saturated from unsaturated fatty acids; and, 2,4,7-trinitrofluorenone (0.3% in adsorbent) for the separation of aromatic hydrocarbons.[61] Chiral plates are composed of a reversed-phase layer impregnated with copper acetate.[67] If precoated plates are used, they can be dipped into a solution of the substance in a tray and then redried. Some care may be needed to ensure that impregnation occurs evenly.[36]

B. LAYER PREPARATION

Thin layers are usually prepared by casting a film of an adsorbent-water slurry (Table 9.1) onto a support plate (glass, aluminum sheets, plastic, or other inert material) by one of the methods described in several references.[36,62,63] Three standard sizes of plates can be used: 5 × 20, 10 × 20, and 20 × 20 cm. The coating of the glass plate is best carried out by means of a spreader, which is available in various designs from equipment manufacturers.[50,63–65] Coated plates must be dried in position for 15 min, then in an oven for 30 min at 80°C, removed, and cooled. The plates should be carefully examined in transmitted and reflected light for imperfections or irregularities in coating, and any plate showing extensive rippling or mottling of the layer should be discarded.[43]

TABLE 9.1
Recommended Amounts of Different Adsorbents and Distilled Water for Preparing Five TLC Plates, 20 × 20 cm

Absorbent	Amount of Adsorbent (g)	Amount of H_2O (mL)
Adsorbosil	35	50
Aluminium oxide	30	50
Cellulose microcrystalline	20	70
Cellulose powder	15	100
Florisil	25	45
Kieselguhr	30	60
Polyamide	15	60
Silica gel G	30	60

C. PRECOATED PLATES

Progress in TLC has been made largely due to the efforts of firms manufacturing chromatographic materials, which have, in a relatively short period, succeeded in producing thin layer plates (precoated) that are qualitatively superior to those prepared manually in the laboratory.[61] Precoated layers of silica gel, alumina, Florisil, polyamide, cellulose, ion exchangers, and chemically bonded layers are commercial available[67] in several standardized dimensions, such as 20 × 20, 10 × 20, and 5 × 20 cm. The precoated plates are also produced with 15 × 20, 20 × 40, 5 × 10, and 10 × 10 cm dimensions.[61] The most frequently used thicknesses for analytical purposes are 250 μm for silica gel, alumina, and kieselguhr, and 100 μm for cellulose and polyamide. Thin-layer support can be a glass plate, aluminum sheet, or plastic material.[61] A recent development is the commercial production of TLC phases (Empore) in the form of flexible sheets containing about 90% of silica gel or chemically bonded silica gel enmeshed in PTFE microfibrils with no supports.[67] Whatman, Merck, Analtech, and Machery Nagel are among the most important manufacturers of TLC plates, and literature from these companies should be consulted for details of availability, properties, and usage.[67]

Modern high performance TLC (HPTLC) began around 1975 with the introduction of commercially precoated high efficiency plates, which are smaller (10 × 10 or 10 × 20 cm); have a thinner (0.1–0.2 mm), more uniform layer composed of smaller-diameter particles (ca. 5 μm average), and are developed over shorter distances (ca. 3–6 cm) compared to classical 20 × 20 cm TLC plates.[66,67] Other characteristics of HPTLC compared to TLC include smaller sample volumes (0.1–0.2 μl), starting spot diameters (2–6 mm), development times (3–20 min), and detection limits (0.1–0.5 ng for absorption and 5–10 pg for fluorescence), and a greater number of samples lanes per plate (18–36).[67] Precoated TLC and HPTLC plates have been widely used in food analysis.

D. SAMPLE APPLICATION

Generally, 5 to 15 μl of extract is applied, in the form of spots or narrow bands, to the plate. However, in some cases, larger volumes can be applied. Depending on the desired accuracy and precision, different types of applicators are used. For screening purposes, disposable quantitative capillary pipets or precision syringes, which are more accurate and precise, are used. The latter allow the intermittent application of larger volumes by using them in combination with a repeating dispenser incorporated in a spotting device. This will keep the starting zones as small as possible for efficient separation. Evaporation of the sample solvent is aided and application made faster if a stream of cool or hot air or inert gas is blown gently across the plate being spotted.[58] The spotting of sample and standards can be carried out according to a spotting pattern (template), which serves as a guide for spotting. Apart from being an aid to accurate spotting, the template acts as a support for the hand and shields the surface of the plate from damage.[62]

E. SOLVENT SELECTION

In order to obtain reproducible results, only very pure solvents must be used in TLC.[65,68] The rate of migration of a compound on a given adsorbent depends upon the solvent used according to its polarity. The solvents have been arranged in the so-called elutropic series.[50,64,65,69] Mixtures of two or three solvents of different polarity often give better separation than chemically homogeneous solvents.[62] Solvents that undergo chemical changes should preferably be freshly prepared, such as the possibility of ester formation with solvents containing an alcohol and an acid.[65]

Solvent systems for TLC can be chosen by consulting the literature or by trial and error. The use of a micro-circular technique[62] or microscope-slide layer[64] is the best way to try to find satisfactory one- or two-component solvents for separation. There are also certain solvent additives that reduce streaking and give improved separation. When acidic materials are being chromatographed on silica gel, acetic acid (about 1%) may be added to the developing solvents. Conversely, ammonium hydroxide and diethylamine are used as additives for separation of basic materials. These acid and base additives help to buffer the materials being separated so that they remain completely in a nonionic form and give more compact spots. The addition of 2,5-hexane-dione to developing solvents in amounts of about 0.5% aids in the separation of inorganic ions.[64]

F. DEVELOPMENT TECHNIQUES

Regardless of whether the purpose of chromatography is qualitative analysis, quantitative analysis, or isolation of solute, the zones must be well separated without overlap.[70] The effectiveness of separation in chromatography depends on the compactness of zones as well as on the degree to which they migrate.

Chromatoplates are usually developed once by the ascending technique[59,61] in a rectangular glass tank (N-chamber) at room temperature, to a height of 15 to 18 cm. Tanks lined with filter paper are used to assure saturation of the atmosphere with solvent vapors in order to effect faster separation, a straight solvent front, and formation of rounder spots. However, an unequilibrated tank is required for specific separations. Descending development[62-65] does not offer advantages over the ascending technique, while horizontal development may be particularly useful for studying some theoretical aspects of thin-layer chromatography.[62-65] HPTLC plate development is carried out for shorter distances in smaller tanks.[67] The plate should not be left standing in the tank for a long time after the front has reached the stop-line, as diffusion and evaporation may cause spreading of the separated spots.[62]

Substances that are not resolved at ambient temperature may be separated by chromatography at higher or lower temperatures. Repeated chromatography in the same direction (multiple development) either with the same solvent or two different solvents yields better separations than single-pass development. Two-dimensional TLC can be used to effect a better separation of a mixture. Separations of the substances that have R_f values of less than 0.5 can be improved by continuous development.

G. QUALITATIVE DETERMINATION

Identification of the separated substances on TLC is best achieved by means of chemical locating reagents and color reactions, although the relative position of two or more substances on the same chromatogram can often be used with some confidence.[62] It is best to run reference substances or mixtures on the same plate as the unknown. Qualitative analysis is done by comparing R_f or relative R_f values and hue (or color) with those of standards that have been developed under identical experimental conditions. If the sample component absorbs UV radiation or fluoresces, the developed spot can be located by holding the TLC plate under a UV lamp. Some spots can be made visible, absorbing, or fluorescent with various chemical reagents (derivatization). Co-chromatography, where unknown spots and standard spots are superimposed, may be used as a confirmatory test rather than for determination. The presence on the plate of a single spot with R_f and color, absorption, or fluorescence pattern similar to that of the standard, but having more intensity than either sample or standard alone, is confirmation of the presence of analyte in the sample.

Derivatization refers to any process by which a compound is converted to a different chemical compound. It can be achieved by introducing chromophores or fluorophores into the molecules of the compounds to be determined. The purposes of the derivatization are to increase the stability of the sample and to improve the sensitivity and selectivity of detection. Depending on the moment of derivatization in the total procedure, pre- and postchromatographic derivatization can be distinguished. Derivatization reagents can be applied by means of spraying or dipping, but spraying is probably the preferred method.[48] Corrosive spray reagents can be used to char, in an oven, most nonvolatile organic substances on inorganic layers.[50] The consecutive application of two or more detection reagents on one chromatoplate increases the probability that no fraction remains undetected.[50] Derivatization prior to TLC can increase resolution as well as sensitivity of detection.[19,48]

H. QUANTITATIVE DETERMINATION

Two basic techniques are available for quantitative TLC. In the first technique, the substances to be determined are assayed directly on the layer. In the second one, the substances are removed from the layer and assayed in a convenient manner after extraction of the sample component into a suitable solvent.

1. Assay on the Layer

When substances are assayed directly on the layer, there are no extraction or transfer errors, and the procedures are quite simple. Quantitative analysis can be performed in any of several ways. The reflectance, transmittance (which are converted to absorbance units), fluorescence, or fluorescence quenching of a spot can be measured by a densitometer. The recorded detector signal can be plotted against the amounts of the reference sample, providing a calibration curve that is useful for quantitation over the desired range. Determination also may be done visually for quick semi-quantitative estimation.

Densitometric Measurement

The absorbance or fluorescence of the separated spots can be measured with a densitometer at appropriate excitation and emission wavelengths for the component of interest. The separated spots are scanned with a fixed sample light beam in the form of a rectangular slit and the diffusely reflected light is measured by the photosensor of the densitometer. The difference between the optical signal from the sample-free background and that from a sample zone is correlated with the amount of the respective fraction of calibration standards chromatographed on the plate.[71] A tungsten or halogen lamp is used as the source for scanning colored zones in the 400 to 800 nm range and a deuterium lamp for scanning UV-absorbing zones directly or as quenched zones on F-layers in the 190 to 400 nm range.[67] For fluorescence scanning, a high-intensity xenon or mercury lamp is used as a source,[67] where the substances are excited by UV light, and the emitted long-wavelength light is measured by the the photosensor of the densitometer. A cutoff filter that is interposed between sample and photosensor completely eliminates diffusely reflected light of the excitation wavelength. Accordingly, the measured light is directly proportional to the amount of fluorescing substance.[67,71]

A standard curve of the component of interest should be established on the same plate under the same conditions as for the sample. The quantity of the component can be determined by comparing the absorbance or the intensity of fluorescence of the sample spot (A) with that of the calibration curve. If the A is greater than the A of the highest standard in the standard curve, the extract should be diluted with the same solvent before repeating the TLC. When an internal standard is used, it is added in equal concentration to all samples and standards, and the calibration curve is constructed using the area ratios of the scans of the sample to the internal standard.

Visual Estimation

Semiquantitative visual estimation is frequently used for screening procedures. The intensity of the sample spot can be compared visually with those of standards to determine which of the standard spots most closely matches the sample spot. If the intensity of the extract spot absorbance or fluorescence is greater than that of the highest amount of the standard solution applied, the extract should be diluted with the same solvent and the TLC is repeated.

2. Assay after Extraction

In this procedure, the area of the spot is marked with pencil about 0.5 cm out from the edge of the spot. The encircled area is scraped off with a steel spatula onto a piece of glassine paper and carefully transferred to a centrifuge tube and extracted into a suitable solvent (5–10 ml) by shaking. Centrifugation is performed at high speed until the solution is clear (5 min). The clear solution is estimated using a spectrophotometer to determine the absorbance of the sample at the maximum absorbance or fluorescence wavelength of the component. Areas from the unused portion of the same plate, equal to that used for the sample, should be scraped and processed in the same manner, establishing the absorbance or fluorescence zero point. The standard curve can be obtained in the same manner by scraping, eluting, and measuring a series of standards.

V. VALIDATION OF THE METHOD

The primary goal of every analytical investigation is to obtain qualitative and quantitative information about the sample to be tested. Detailed analysis of the procedures can often lead to control of the technique by quite simple alterations.[56] Methods of analysis have scientific and practical characteristics that determine the reliability and applicability of the method, respectively.[45] Among the scientific characteristics of the analytical methods are precision, accuracy, detectability, and sensitivity; among the practical characteristics are the applicability, the cost of performance, the time and equipment required, and the level of training needed.[45]

Precision is a measure of variability of the results. A common description of precision is by standard deviation or relative standard deviation (coefficient of variation, c.v.) of a set of replicate results. The precision may relate to the within-laboratory error of a method (repeatability) or to the between-laboratory error of a method (reproducibility).[72] Accuracy is a measure of systematic deviation of the obtained results from the true value. The smaller the systematic part of the experimental error, the more accurate is the procedure. The accuracy of a method is usually expressed in terms of percentages of recovery.[45] Limit of detection of a method is the lowest concentration level that can be determined to be statistically different from an analytical blank.[45] IUPAC states that the limit of detection, expressed as a concentration, is derived from the smallest amount that can be detected with reasonable certainty for a given analytical procedure.[73] The sensitivity is the change in analytical signal per unit concentration change. It is represented by the slope of the analytical calibration curve.[45]

Desired accuracy and precision are quite important since the usability of the method can be excluded or confirmed depending on the accuracy and precision of the method.[70] Each step of a method, including both sample preparation stages and chromatographic process steps, must be carefully validated separately when developing methods or interpreting results.

A. SAMPLE PREPARATION

In general, the greater the complexity or number of preparatory stages applied to a sample, the greater are the opportunities for loss and other changes affecting recovery in the final analysis.[56] Samples usually require modification either to remove substances that interfere with chromatography, or to concentrate the compounds being analyzed.[56] Sample preparation errors can arise in at least five areas: taking a truly representative sample, extraction of sample components, cleanup of the extract and concentration, sample stability, and derivatization.[70]

1. Sampling
The result of a determination will be meaningless unless the material submitted for analysis is a statistically valid, representative sample of what is to be measured.[74]

2. Extraction
Extraction procedures should be both quantitative and specific. Efficiency of extraction is influenced by partition coefficients, and these may be modified by the presence of other constituents in the sample.[56] The errors that may arise from extraction can influence both accuracy and precision. To achieve high extraction efficiency requires great effort and expertise in many cases. When the organic solvents used for the extraction are not selected correctly, precision might be satisfactory but low accuracy could be obtained.[74] To validate the extraction procedure, a recovery study should be made. The most useful procedure is to "spike" samples with a series of increasing amounts of standard material to develop a standard recovery curve.[74] Recoveries should be determined at several concentrations over the range in which the results are expected to be found.

The recovery standard should be different from the internal standard, because the recovery standard serves to improve the accuracy of the analytical method, while the primary goal in the use of an internal standard is to enhance the precision of the analysis.[75]

3. Sample Cleanup and Concentration
During protein precipitation, constituents may be lost by adsorption onto the precipitate. Alternatively, exclusion of constituents from the space occupied by the protein precipitate can cause overestimation.[75] Desalting techniques may also affect the concentration of constituents in the treated samples; therefore, a careful standardization of techniques for reliable results is required.[56] On the other hand, similar principles, as were discussed for extraction, can be applied to carry out the liquid-liquid and/or solid-liquid cleanup step.[74]

Techniques involving solvent extraction or elution from a column usually result in dilution of constituents; therefore concentration, e.g., by evaporation, is often necessary prior to chromatography.[56] During the evaporation stage, some constituents may be lost by evaporation or decomposition, while others may appear as a result of the conversion or hydrolysis of complex molecules.[56] Allowance for volume changes may require the use of an internal marker technique.[56]

4. Stability of Sample in Solution
The other main source of error is likely to occur when the components are transformed in the sample solution. Errors can arise in two areas, during sample preparation and during storage of the sample solution used for chromatography.[74] Compound decomposition during sample preparation will lead to low accuracy, while decomposition during storage will decrease analysis precision.[74] To validate the stability of compounds in sample solution, the chromatographic process is repeated from time to time.[74]

5. Derivatization
The validation procedure for prechromatographic derivatization is carried out using a series of increasing amounts of standard materials, and subjecting them to chromatographic processing after derivatization to develop a standard calibration curve.[74] The precision of the derivatization can be calculated from the experimental results.

B. CHROMATOGRAPHIC PROCESS
Errors can arise from solvent selection for dissolving the sample, sample application to the chromatoplate, chromatographic separation, and quantitation. Proper preparation of the sample for the TLC separation may be one of the most important steps for success in quantitation. The solvent selected for dissolving the sample should be highly volatile and as nonpolar as possible[74] to obtain good spot shape and size.

Sample application is considered one of the most important sources of error.[56,74] To minimize its detrimental effects on the accuracy and precision of the method, special skill and expertise by the analyst are required.[74] The accuracy of microsyringes (micropipettes) must be taken into consideration.[56,74] The size and shape of the spot is important for increasing resolution and getting better conditions for densitometer evaluation, where precise quantitation requires spots with uniform shapes and sizes.

A further possible source of error is the decomposition that can take place on the active sites of a stationary phase surface prior to the run.[74] In such cases the drying conditions should be examined carefully to avoid or minimize transformation of the substance, and the standing time of the sample between spotting and development should be as short as possible.[74]

Good chromatographic resolution between the desired components is absolutely necessary for reliable results. The error related to chromatographic separation is influenced by many factors such as plate-to-plate variation, position of the initial spot on the plate, and the positions of the separated spots on the chromatogram.[56,74] Other factors, such as changes in temperature, mobile-phase composition, chamber saturation, and running distance, also have great importance with regard to reproducibility.

The position of the spots on the chromatogram can significantly influence the detector signal.[74] Concentrated spots at low R_f values and expanded spots at high R_f values will not be valid. The sample or component should ideally travel to an R_f position between 0.3 and 0.7 for reproducible quantitative results.[74] Spots on the edges of the plate may have different R_f values than spots in the center.[74] This error can be minimized by a data-pair technique, which significantly increases the precision of the determination.[69] The potential precision of an analytical method cannot be assessed without a densitometer of adequate precision.[56] Also, the densitometer directly influences the sensitivity of a method.

It is advisable to assess the performance of any preparatory procedure by means of recovery studies.[53] The concentration of desired components in the sample is of great importance; therefore, sample concentration or dilution, and/or prechromatographic derivatization may be applied.[70] Good chromatographic resolution between the desired components is absolutely necessary for reliable results.[70] Pure standards of the components desired are very important for identification and quantitative analysis.[70]

VI. SELECTED APPLICATION PROTOCOLS

A. FOOD COMPOSITION
1. Protein (Amino Acids)[1,2,76]
Hydrolysis of Proteins
1. Weigh 50 to 100 mg of the air-dried or lyophilized protein into a 10 ml ampule, add 8 ml of 6N HCl, degas the ampule with nitrogen, and seal under vacuum.
2. Place the sealed ampule in an oven at 110°C for 16 h. Open the ampule, remove the HCl by adding solid barium hydroxide, and remove the white precipitate by filtration.
3. Evaporate the filtrate under vacuum, dissolve the residue in 10% isopropanol solution, filter, and adjust the filtrate to 25 ml to use for subsequent chromatographic separation.

TLC Determination
1. Spot 10 µl of sample solution along with standard solutions of amino acids dissolved in 0.1 N HCl in 95% ethanol on a silica gel G plate 2 cm from bottom edge and at intervals of 2 cm, using a microsyringe.
2. Develop the plate in a well-saturated tank that has been pre-equilibrated for 30 min prior to inserting the plate, using *n*-butanol–acetic acid–water (3:1:1) as the developing system.
3. Spray the plate with ninhydrin reagent (5% ninhydrin, 5% acetic acid, and 1% cadmium acetate in 96% ethanol), and place it in an oven at 100°C for 5 min.

4. Identify spots with the help of co-developed standards (R_f values and colors), and quantify visually, by colorimetry after extraction of the separated spots, or by densitometry.

2. Lipids[6,24,77]

Extraction

1. Extract minced sample (1–10 g) with about 20 volumes of dichloromethane–methanol (2:1) by Vortex mixing for 2 min, and filter through Whatman No. 1 paper. Reextract the residue and collect the extract.
2. Evaporate the collected extract to dryness, reconstitute in 2 ml of chloroform, and filter through anhydrous sodium sulfate.
3. Evaporate the chloroform to dryness and reconstitute with 1 ml of chloroform.

TLC Determination

1. Spot about 0.5 μg of lipid as a chloroform solution on a silica gel G plate, 2 cm from lower edge and at intervals of 2 cm, along with standards.
2. Develop the plate for 15 cm with a solution of benzene–diethyl ether–ethyl acetate–acetic acid (80:10:10:0.2).
3. Dry the developed plate in air, dip in 10% cupric sulfate in 8% phosphoric acid solution for 30 s, and then air dry for 10 min.
4. Char the plate in an oven at 180°C for 5 min, and determine the charring density using a densitometer.

3. Carbohydrates (Sugars)[54]

Extraction

1. Extract an appropriate ground sample with hot 80% alcohol (taking into account the moisture of the sample) while stirring during boiling.
2. Decant the solution into a suitable flask, reextract the solids with 80% alcohol, transfer the mixture to the flask, and cool. Filter the combined extracts, and evaporate the filtrate until the alcohol odor disappears.
3. Filter the extract through a thin mat of Celite in a Buchner funnel, wash the Celite layer, and dilute the combined filtrate and washings to a suitable volume (ca. 200 ml), and use ca. 50 ml for purification.

Purification

1. Add just enough lead acetate to the solution to cause complete precipitation, shake thoroughly, and let the mixture stand a few minutes.
2. Filter into a beaker containing an excess of sodium oxalate crystals, filter off and wash the precipitate, collect the filtrate and washings in a volumetric flask, and dilute to the mark with water.

TLC Determination

1. Soak a precoated silica gel G plate in 0.3 N NaH_2PO_4 solution for 1 min, remove the plate, and dry at 100°C for 30 min.
2. Spot the samples and different volumes of standards (1% aqueous solutions of sucrose, glucose, lactose, maltose, and fructose) 3 cm from the bottom edge and at intervals of 1 cm.
3. Develop the plate in a well-saturated tank with n-butanol–acetone–0.3 N NaH_2PO_4 (4:5:1) to 15 cm above the spotting line.
4. Take the plate out of the tank and, after drying the chromatogram for 1 min in an oven, spray with ADP reagent (2 ml aniline + 2 g diphenylamine + 10 ml H_3PO_4 + 88 ml methanol), and heat at 110°C for 2 min.
5. Identify spots with the help of co-developed standards, and quantify visually or with a densitometer.

4. Organic Acids[54,78,79]

Sample Preparation
1. Extract 20 g of minced sample with 80 ml deionized water, filter through a cheesecloth, and wash the filter cake with 20 ml of deionized water.
2. Centrifuge the combined filtrate (or 100 ml of juice sample) at 27200 × g for 10 min, pass the supernatant through a 0.45 μm membrane filter, then through a C_{18} cartridge.

TLC Determination
1. Prepare TLC plates with cellulose powder or use precoated plates.
2. Spot 10 to 20 μl of sample extracts and standard solution (5 g of tartaric, malic, lactic, or succinic acids per liter of distilled water) 3 cm from the bottom edge, and at intervals of 2 cm.
3. Develop the plate in a well-saturated tank for 18 min, with one of the following developing solvent systems:
   - *n*-butanol (H_2O saturated)–formic acid (95:5)
   - phenol–formic acid–water (500 g:6.7 ml:167 ml)
   - *n*-butanol–*tert*-amyl alcohol–formic acid–water (45:15:18:18)
4. Remove the plate from the tank, dry, and spray the dried chromatogram with acridine solution (dissolve 250 mg of acridine in 200 ml of ethanol) and view under UV light after drying.
5. Estimate the concentration visually by comparing standard and sample spots size and intensity.

B. FOOD ADDITIVES

1. Antioxidants[36]

Sample Preparation
1. Dissolve 10 g of oil or fat in 80 ml of petroleum ether and transfer to a separatory funnel with 20 ml of petroleum ether. Extract the solution with three 25 ml portions of acetonitrile.
2. Evaporate the combined extracts to dryness at 40°C, and redissolve the residue in 2 ml of methanol.

TLC Determination
1. Prepare a silica gel G plate (0.25 mm) using 1% citric acid in methanol, dry, and heat for 1 hr at 130°C.
2. Spot 10 and 20 μl portions of the sample extract, and 2, 5, and 7 μl portions of appropriate antioxidant standard solution (1 mg/ml^{-1}).
3. Develop the plate for 15 cm in a well-saturated tank with petroleum ether–benzene–glacial acetic acid (2:2:1).
4. Take the plate out of the tank and after air drying, spray the plate with Gibb's reagent (2,6 dichloro-*p*-benzoquinine-4-chloroimine, 5% in ethanol), for detection of brown spots of propyl gallate, octyl gallate, or dodecyl gallate; a brown-red spot of BHA; and/or brown-violet spot of BHT.
5. Estimate the concentration visually by comparing standard and sample spots size and intensity.

2. Benzoic Acid[35,36]

Sample Preparation
1. Weigh a 50 to 60 g sample and homogenize semisolids and solid-liquid mixtures or blend solid samples with water before transferring to a distillation flask (Figure 9.1). Make the total volume 350 to 375 ml.
2. Distill the sample into 50 ml of NaOH solution (4 g/100 ml), until 725 to 750 ml of distillate are collected (about 90 min). Rinse the condenser with about 20 ml of H_2O.
3. Acidify the distillate using litmus paper with HCl and extract with 100 ml portions of $CHCl_3$–ether (2:1) in a separatory funnel.
4. Evaporate the collected extracts carefully on a steam bath to about 25 ml. Transfer to a 50 ml volumetric flask and dilute to the volume by washing the beaker with small portions of $CHCl_3$–ether (2:1) and adding the washings to the flask.

TLC Determination
1. Use a mixture of kieselguhr G–silica gel G F_{254} (1:1) plate (0.25 mm) that is activated at 100°C for 1 h for the separation.
2. Spot 100 μl each of sample and standard (1 mg ml^{-1} ethanol) solutions, 2.5 cm from the bottom and side edges, and 4 cm apart.
3. Develop the plate to 10 cm above the spotting line with *n*-hexane–acetic acid (96:4).
4. Take the plate out of the tank and after air drying for 5 min, view the plate under shortwave UV light (254 nm) for the detection of dark blue-purple benzoic acid spots on a light fluorescent background.
5. Extract the spots and determine A by spectrophotometry at the wavelength of maximum absorbance (about 272 nm).

$$\% \text{ Benzoic acid} = [(A/A') \times (\text{mg std}/50 \text{ ml}) (\text{g sample}/50 \text{ ml})] \times 100$$

$$\% \text{ Na benzoate} = \% \text{ benzoic acid} \times 1.180$$

where A = sample absorbance
       A' = standard absorbance

3. Nonnutritive Sweeteners[36,80]
Sample Preparation
1. Measure 50 ml of decarbonated beverage sample into a 125 ml separatory funnel. Continuously add 10 ml of aqueous H_2SO_4 (1:1). After cooling, extract with two 50 ml portions of petroleum ether and discard this extract.
2. Continuously add 5 ml of 50% NaOH solution (w/w) to the aqueous layer and, after cooling, extract with two 50 ml portions of ethyl acetate.
3. Combine the ethyl acetate extracts in a beaker with a pouring lip by filtration through ethyl acetate–washed cotton, and evaporate to 5 to 10 ml on a steam bath. Transfer to a graduated tube and evaporate to dryness.
4. Dilute the residue to 5 ml with NH_4OH–H_2O–ethanol (5:5:10) and mix thoroughly.

TLC Determination
1. Prepare a standard mixture of 50 mg calcium cyclamate, 10 mg sodium saccharin, 4 mg dulcin, and 4 mg 5-nitro-2-propoxyaniline (P-4000) in 10 ml aqueous ethanol (1:1).
2. Use a freshly prepared silica gel H plate (0.25 mm) without prior oven drying for the separation. Spot 5 μl of sample and different volumes of standard mixture 2.5 cm from bottom, 2 cm apart and 2 cm from edges, using a warm-air blower to dry the spots.
3. Develop the plate to 10 cm above the spotting line in a well-saturated tank with *n*-butanol–ethanol–NH_4OH–H_2O (40:4:1:9).
4. Take the plate out of the tank and after drying at room temperature (10 min), view the plate under shortwave UV light (254 mm) for detection of the fluorescent saccharin spot.
5. Spray the plate lightly with 5% bromine in CCl_4 (by vol.) followed by 0.25% fluorescein in dimethylformamide–ethanol (1:1) for detection of cyclamate (pink spot) and P-4000 (brown-pink spot).
6. Spray the plate with 2% *N*-1-naphthylethylenediamine · 2HCl in alcohol for detection of dulcin (brownish-pink or blue).
7. Compare standard and sample spots size and intensity, and estimate the concentration visually.

C. FOOD CONTAMINANTS
1. Pesticide Residues[36,43]
Extraction of Fat from Fat-Containing Food
- Milk and milk products

1. Shake 100 ml of milk, or blend 25 to 50 g of cheese, with 100 ml of ether–petroleum ether (1:1) in the presence of 100 ml of methanol and ca. 2 g of sodium oxalate.
2. Centrifuge for ca. 5 min at ca. 1500 rpm, blow off the solvent layer, and reextract the residue twice with 50 ml portions of ether–petroleum ether (1:1).
3. Wash the combined extracts in separatory funnel with a mixture of 600 ml of H_2O and 30 ml of saturated NaCl solution, followed by two washings with 100 ml portions of H_2O.
4. Pass the washed ether solution through an anhydrous Na_2SO_4 column, 50 × 25 mm o.d., and evaporate the eluate to obtain the fat.

- Fish and marine foods
5. Blend 25 to 50 g of thoroughly ground sample and mix with 150 ml of petroleum ether in the presence of 100 g of anhydrous Na_2SO_4.
6. Filter with the aid of suction, and reextract the residue twice with 100 ml portions of petroleum ether.
7. Pass the combined petroleum ether extracts through an anhydrous Na_2SO_4 column, 40 × 25 mm o.d., and evaporate the eluate to obtain the fat.

Extraction of Pesticides from Fat
1. Weigh 3 g of fat into a separatory funnel, make the volume 15 ml with petroleum ether, then extract with three 30 ml portions of petroleum ether–saturated acetonitrile (CH_3CN).
2. Combine the CH_3CN extracts in a separatory funnel containing 650 ml of H_2O and 40 ml of saturated NaCl solution, extract twice with 100 ml portions of petroleum ether, combine the extracts, and wash twice with 100 ml portions of H_2O.
3. Pass the washed petroleum ether layer through a 50 × 25 mm o.d. column of anhydrous Na_2SO_4, evaporate the petroleum ether eluate to ca. 10 ml, and transfer to a Florisil™ column (cleanup step).

Extraction of Pesticides from Nonfatty Foods
1. Blend the sample with the extraction mixture (Table 9.2) for 2 min at high speed, filter with the aid of suction, transfer the filtrate to a 250 ml graduated cylinder, and record the volume (F).
2. Transfer the filtrate to a separatory funnel, add 100 ml of petroleum ether, shake vigorously for 1 to 2 min, and add 10 ml of saturated NaCl solution and 600 ml of H_2O. Discard the aqueous layer and wash the solvent layer twice with 100 ml portions of H_2O. Transfer the solvent layer to a 100 ml graduated cylinder and measure the volume (P).
3. Add ca. 15 g of anhydrous Na_2SO_4 to the collected solvent, shake vigorously, and transfer the solution directly to a Florisil™ column (cleanup step).
4. Calculate weight of sample represented by the final volume of petroleum ether (G) from the following equation:

$$G = S \times (F/T) \times (P/100)$$

where S = g sample taken
T = ml total volume (ml H_2O in sample + ml CH_3CN added – correction in ml for volume contraction, Table 9.2)

Cleanup of the Extract
1. Prepare a column as in Figure 9.2a and transfer the petroleum ether extract to it. Operate the column with a flow rate of < 5 ml/min.
2. Elute the column with 200 ml of 6% ethyl ether in petroleum ether. This eluate will contain organochlorine pesticides (aldrin, BHC, DDE, DDD, *o,p'*- and *p,p'*-DDT, heptachlor, and mirex), industrial chemicals (polychlorinated biphenyls), and organophosphorus pesticides (ethion and ronnel).

TABLE 9.2
Nature of Sample, Sample Weight, and Extraction Solvent for Pesticides from Nonfatty Foods

Sample type	Sample weight (g)	Extraction mixture	Volume Contraction
High moisture > 70% H_2O			
Containing < 5% sugar	100	200 ml CH_3CN + 10 g Celite	5 ml
Containing 5–15% sugar	100	200 ml CH_3CN + 50 ml H_2O	10 ml
Containing 15–30% sugar	100	200 ml CH_3CN + 50 ml H_2O (75°C)	10 ml
Low moisture or dry	20–30	350 ml 35% H_2O – CH_3CN	0
Whole egg	25	200 ml CH_3CN + 10 g Celite	215*

Use 215 as T.

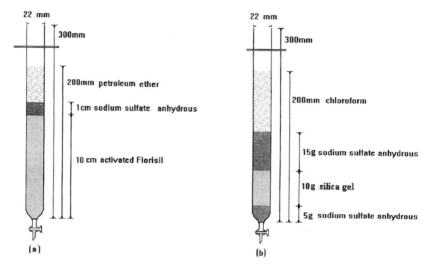

FIGURE 9.2 Column chromatography for cleanup of the extracts: (a) Column for cleanup of the pesticide extracts (b) Column for cleanup of the aflatoxin extracts.

3. Change the receiver and elute with 200 ml of 15% ethyl ether in petroleum ether. This eluate will contain additional organochlorine pesticides (dieldrin and endrin) and organophosphorus pesticides (diazinon, methyl parathion, and parathion).
4. Change the receiver and elute with 200 ml of 50% ethyl ether in petroleum ether. This eluate will contain the organophosphorus pesticide malathion.
5. Evaporate the eluates to 5 ml and use them for semiquantitation determination.

TLC Determination
- Organochlorine pesticides
1. Adjust aliquots of sample and standard mixtures to give residue spot within the range of 0.005 to 0.1 µg. Spot all 6% Florisil™ column eluates on one plate and 15% Florisil™ eluates on another plate.
2. Prepare TLC plates using Al_2O_3 "G," prewash the adsorbent layer by developing the prepared plates in 50% aqueous acetone, dry, cool, and store in desiccator.
3. Spot aliquots of samples (10 µl) and different volumes of standard mixture 4 cm from the bottom of the plate, 2 cm from side edges, and at 1 cm apart.
4. Develop the plate in a well-saturated tank, to 10 cm above spotting line, using *n*-heptane as the developing solvent for the 6% Florisil™ eluates and acetone–*n*-heptane(2:98) for the 15% Florisil™ eluates.

5. Dry the plates for 5 min in a hood and spray with chromogenic agent (dissolve 0.100 g of $AgNO_3$ in 1 ml of H_2O, add 20 ml of 2-phenoxyethanol, dilute to 200 ml with acetone, add a very small drop of 30% H_2O_2, mix, and store in the dark overnight before use) until the plate appears translucent.
6. Dry the chromatogram in a hood for 15 min, immediately place it under UV light until the spot of lowest concentration appears (15 to 20 min), and determine the quantity of organochlorine pesticides in the extract by visual estimation.

- Organophosphorus pesticides

1. Adjust aliquots of samples and standards to give residue spot within the range of 0.1 to 0.5 µg. Florisil™ eluates of 6, 15, and 50% may be spotted on the same plate. Apply aliquots of sample (10 µl) and different volumes of standard mixture on an Al_2O_3 TLC plate 4 cm from the bottom, 2 cm from side edge, and 1 cm apart.
2. Invert the plate and dip it just to the spotting line in the immobile solvent (dilute 100 ml of 20% *N,N*-dimethylformamide (DMF) in ether to 500 ml with ether).
3. Develop the plate, immediately after dipping, in a saturated tank to 10 cm above the spotting line with methylcyclohexane.
4. Dry the plate 5 min in a hood, then spray with dye reagent (dilute 10 ml of a 1 g tetrabromophenolphthalein ethyl ester solution in 100 ml acetone to 50 ml), followed by silver nitrate solution (dissolve 0.5 g of $AgNO_3$ in 25 ml of H_2O and dilute to 100 ml with acetone).
5. After 2 min, overspray the plate with citric acid solution (dissolve 5 g of granular citric acid in 50 ml of H_2O, and dilute to 100 ml with acetone). Respray the plate with the same solution after 10 min, and evaluate the chromatogram visually after 10 min more.

2. Antibiotic Residues(Tetracyclines)[41,42,82–85]

Extraction
1. Blend 5 g of food sample (animal tissues and milk) in a high-speed blender with 20, 20, and 10 ml of 0.1 M disodium ethylenediaminotetraacetate (Na_2EDTA)–McIlvaine buffer (33.62 g of Na_2EDTA + 614.5 ml of 0.1 M citric acid + 385.5 ml of 0.2 M disodium hydrogen phosphate, pH 4.0); centrifuge each time at 4000 rpm for 5 min.
2. Combine the supernatant, recentrifuge at 4000 rpm for 15 min, and filter.
3. For honey samples, dissolve 25 g in 50 ml of 0.1 M Na_2EDTA–McIlvaine buffer (pH 4.0), and filter.

Cleanup
1. Pass the filtrate through a C_{18}-cartridge pretreated with saturated aqueous Na_2EDTA.
2. Wash the cartridge with 20 ml of water, air-dry by aspiration for 5 min, and elute tetracyclines (TCs) with 10 ml of ethyl acetate followed by 20 ml of methanol–ethyl acetate (5:95).
3. Evaporate the eluate to dryness under reduced pressure at 30°C, and dissolve the residue in 0.1 ml of methanol.

TLC Determination
1. Use silica gel HPTLC plates or C_8-modified silica gel TLC plates for separation.
2. Predevelop the HPTLC plate with saturated aqueous Na_2EDTA solution, dry in air at room temperature for 1 h, and activate at 130°C for 2 h.
3. After applying the sample (5 µl) and different volumes of a solution standard (1 mg ml^{-1} methanol), develop the HPTLC plate with chloroform–methanol–5% Na_2EDTA solution (65:20:5, lower layer), or develop the C_8-TLC plate to 9 cm with methanol–acetonitrile–0.5 M oxalic acid solution, pH 2.0 (1:1:4).
4. Determine the quantity of TCs with a densitometer. The absorbance maxima for tetracycline, oxytetracycline, chlorotetracycline, doxycycline, and 4-epitetracycline are near 360 nm, for anhydrotetracycline and 4-epianhydrotetracycline, near 425 nm.

5. For screening, spray the developed HPTLC plate evenly with 0.2 M aqueous magnesium chloride, and air-dry at room temperature. Spray the developed C_8-TLC plate similarly, and after drying spray with 10% of triethanolamine in methanol. View both TLC plates under longwave UV light (360 nm) for detection of the fluorescent spots of TCs produced with Mg^{2+} and measure TCs by visual comparison of the intensity of the fluorescence with those of standards.

3. Aflatoxins[44,45]
Extraction
1. Weigh 50 g of ground sample into a conical flask, add 25 g of diatomaceous earth, 25 ml of water, and 250 ml of chloroform, then shake for 30 min or blend at high speed for 3 min.
2. Filter through fluted filter paper, discard the first 10 ml, and collect 50 ml.

Cleanup
1. Prepare a chromatographic column as in Figure (9.2b).
2. Mix 50 ml of the filtrate with 100 ml of hexane, and transfer the mixture to the column. Allow the liquid to flow until it is just above the upper surface of the sodium sulfate layer.
3. Wash the column with 100 ml of diethyl ether, and elute the toxins with 150 ml of chloroform–methanol (97:3).
4. Evaporate the eluate to dryness at 50°C, quantitatively transfer the residue with chloroform to a vial, and adjust the volume to 2 ml with chloroform.

TLC Determination
1. Spot on a TLC plate (silica gel G 60), 2 cm from the lower edge and at intervals of 1 cm, five different volumes (10, 15, 20, 30, and 40 µl) of standard solution (0.1 µg of aflatoxins B_1 and G_1, and 0.5 µg of B_2 and G_2), and two volumes (10 and 20 µl) of sample extracts using a microsyringe.
2. Develop the plate with toluene–ethyl acetate–formic acid (6:3:1) in an unsaturated tank to 15 cm above the spotting line.
3. Take the plate out of the tank, allow the solvent to evaporate in the dark, and examine the chromatogram under a UV lamp at 366 mm. The spots of aflatoxins B_1 and B_2 show blue fluorescence, while aflatoxins G_1 and G_2 show green fluorescence.
4. Determine the quantity of aflatoxins in the extract by visual estimation, or measure the intensity of fluorescence aflatoxin spots with the fluorodensitometer at an excitation wavelength of 365 nm and an emission wavelength of 443 nm.

D. FOOD DECOMPOSITION
1. Biogenic Amines[19]
Extraction
1. Extract 50 g of ground samples three times with 75 ml portions of 5% trichloroacetic acid (TCA) using a Waring blender at high speed for 2 min.
2. Centrifuge each blended mixture for 10 min at moderate speed (7000 rpm), combine the clear extracts in a 250 ml volumetric flask, and adjust the volume with 5% TCA solution.

Cleanup
1. To 10 ml of TCA extract add 4 g of NaCl and 1 ml of 50% NaOH. Extract three times with 5 ml portions of *n*-butanol–chloroform (1:1) by shaking, followed by centrifugation.
2. Combine the organic phases (upper), using a disposable Pasteur pipet, in a 60 ml separatory funnel.
3. Add to the combined organic phases (ca. 15 ml) an equal amount of *n*-heptane, and extract several times with 1 ml portions of 0.2 *N* HCl.

4. Collect the extracts in a glass-stoppered tube, and evaporate just until dryness using a current of air and a hot water bath.

Derivative Formation
1. Add to the residue 0.5 ml of saturated sodium bicarbonate solution and 1 ml of dansyl chloride solution (5 mg ml^{-1} acetone), and mix using a Vortex mixer.
2. After standing for ≥ 10 h, add 15 ml of purified water and extract the dansylated biogenic amines three times with 5 ml portions of ethyl ether.
3. Combine the ether extracts and carefully evaporate at 35°C in a water bath with the aid of a current of air.

TLC Determination
1. Spot on a TLC plate (silica gel G 60), 2 cm from lower edge and at intervals of 1 cm, five different volumes (10, 15, 20, 30, and 40 µl) of standard solution (10 mg/ml), and two volumes (10 and 20 µl) of sample extracts using a microsyringe.
2. Develop the plate in chloroform–benzene–triethylamine (6:4:1) to 15 cm. Take the plate out of the tank and allow it to dry at room temperature. Redevelop in the same direction with benzene–acetone–triethylamine (10:2:1) to 15 cm.
3. Allow the plate to dry at room temperature, then dry it with a hair dryer until the solvent has completely evaporated.
4. Examine the chromatogram under longwave (360 nm) ultraviolet light to establish whether or not the dansylamines of interest are present in the sample, comparing with co-developed standards.
5. Compare visually the intensity of each amine spot of the sample with that of the standard to determine which of the standard spots closely matches the sample spot.

REFERENCES

1. **Aurand, L. W., Woods, A. E., and Wells, M. R.**, *Food Composition and Analysis*, Van Nostrand Reinhold, New York, 1987.
2. **Bhushan, R.**, Amino acids and their derivatives, in *Handbook of Thin Layer Chromatography*, Sherma, J. and Fried, B., Eds., Marcel Dekker, New York, 1991, 353.
3. **Berry, H. K.**, Detection of amino acid abnormalities by quantitative thin layer chromatography, in *Quantitative Thin Layer Chromatography*, Touchstone, J. C., Ed., John Wiley and Sons, New York, 1973, 113.
4. **Ersser, R. S. and Smith, I.**, Amino acids and related compounds, in *Chromatographic and Electrophoretic Techniques*, Vol. I, 4th ed., Smith, I. and Seakins, J. W. T., Eds., William Heinemannedical Books Ltd., London, 1976, 75.
5. **Niederwieser, A.**, Chromatography of amino acids and oligopeptides, in *Chromatography: A Laboratory Handbook of Chromatographic and Electrophoretic Methods*, 3rd ed., Heftmann, E., Ed., Van Nostrand Reinhold, New York, 1975, 393.
6. **Fried, B.**, Lipids, in *Handbook of Thin Layer Chromatography*, Sherma, J. and Fried, B., Eds., Marcel Dekker, New York, 1991, 593.
7. **Christie, W. W., Ed.**, *Lipid Analysis*, Pergamon Press, Oxford, 1976.
8. **Dallas, M. S. J., Morris, L. J., and Nichols, B. W.**, Chromatography of lipids, in *Chromatography: A Laboratory Handbook of Chromatographic and Electrophoretic Methods*, Heftmann, E., Ed., Van Nostrand Reinhold, New York, 1975, 527.
9. **Privett, O. S., Dougherty, K. A., and Erdahl, W. L.**, Quantitative analysis of lipid classes by thin layer chromatography via charring and densitometry, in *Quantitative Thin Layer Chromatography*, Touchstone, J. C., Ed., John Wiley and Sons, New York, 1973, 57.
10. **Lake, B. D. and Goodwin, H. J.**, Lipids, in *Chromatographic and Elecrophoretic Techniques*, Vol. I, Smith, I. and Seakins, J. W. T., Eds., William Heinemann Medical Books Ltd., London, 1976, 345.
11. **Prosek, M., Bukl, M., and Jamnik, K.**, Carbohydrates, in *Handbook of Thin Layer Chromatography*, Sherma, J. and Fried, B., Eds., Marcel Dekker, New York, 1991, 439.

12. **Pruden, B. B. and Pineault, G.,** A thin layer chromatographic method for the quantitative determination of D-mannose, D-glucose and D-galactose in aqueous solution, *J. Chromatogr.*, 115, 477, 1975.
13. **Hansen, S. A.,** Thin layer chromatographic method for the identification of mono-, di- and trisaccharides, *J. Chromatogr.*, 107, 224, 1975.
14. **Mansfield, C. T.,** Use of thin layer chromatography in determination of carbohydrates, in *Quantitative Thin Layer Chromatography*, Touchstone, J. C., Ed., John Wiley and Sons, New York, 1973, 79.
15. **Menzies, I. S. and Seakins, J. W. T.,** Sugars, in *Chromatographic and Electrophoretic Techniques*, Vol. I, Smith, I. and Seakins, J. W. T., Eds., William Heinemann Medical Books Ltd., London, 1976, 183.
16. **Stratton, J. E., Hutkins, R. W., and Taylor, S. L.,** Biogenic amines in cheese and other fermented foods: a review, *J. Food Protection*, 54, 460, 1991.
17. **Shalaby, A. R.,** Survey on biogenic amines in Egyptian foods, *J. Sci. Food Agric.*, 62, 291, 1993.
18. **Shalaby, A. R.,** Separation, identification and estimation of biogenic amines in foods by thin layer chromatography, *Food Chem.*, 49, 305, 1994.
19. **Shalaby, A. R.,** Multidetection, semiquantitative method for determining biogenic amines in food, *Food Chem.*, 52, 367, 1995.
20. **Tawfik, N. F., Shalaby, A. R., and Effat, B. A.,** Biogenic amine contents of Ras cheese and incidence of their bacterial producers, *Egyptian J. Dairy Sci.*, 20, 219, 1992.
21. **Mietz, J. L. and Karmas, E.,** Polyamines and histamine content of rockfish, salmon, lobster and shrimp as an indicator of decomposition, *J. Assoc. Off. Anal. Chem.*, 61, 139, 1978.
22. **Staruszkiewicz, W. F. and Bond, J. E.,** Gas chromatographic determination of cadaverine, putrescine and histamine in foods, *J. Assoc. Off. Anal. Chem.*, 64, 584, 1981.
23. **Shalaby, A. R.,** Correlation between freshness indices and degree of fish decomposition, Ph.D. Thesis, Fac. Agric., Ain Shams Univ., Cairo, Egypt, 1990.
24. **Hwang, K. T. and Regenstein, J. M.,** Characteristics of mackerel mince lipid hydrolysis, *J. Food Sci.*, 58, 79, 1993.
25. **Anglo, A. J. and James, C., Jr.,** Analysis of lipids from cooked beef by thin layer chromatography with flame ionization detection, *JAOCS*, 70, 1245, 1993.
26. **Fodor-Csorba, K.,** Pesticides, in *Handbook of Thin Layer Chromatography*, Sherma, J. and Fried, B., Eds., Marcel Dekker, New York, 1991, 663.
27. **Tyihak, E. and Vagujfalvi, D.,** Thin layer chromatography of alkaloids, in *Progress in Thin Layer Chromatography and Related Methods*, Vol. III, Niederwieser, A. and Pataki, G., Eds., Ann Arbor Science Publishers, Ann Arbor, MI, 1972, 71.
28. **Abu-Raiia, S. H. and Shalaby, A. R.,** Influence of fenugreek seed extracts on aflatoxins production by *Aspergillus parasiticus*, *New Egyptian J. Medicine*, 6, 635, 1992.
29. **Ng, K. G., Price, K. R., and Fenwick, G. R.,** A TLC method for the analysis of quinoa (*Chenopodium quinoa*) saponins, *Food Chem.*, 49, 311, 1994.
30. **Szepesi, G. and Gazdag, M.,** Steroids, in *Handbook of Thin Layer Chromatography*, Sherma, J. and Fried, B., Eds., Marcel Dekker, New York, 1991, 907.
31. **De Leenheer, A. P., Lambert, W. E., and Nelis, N. J.,** Lipophilic vitamins, in *Handbook of Thin Layer Chromatography*, Sherma, J. and Fried, B., Eds., Marcel Dekker, New York, 1991, 993.
32. **Ruggeri, B. A., Watkins, T. R., Gray, R. J. H., and Tomlins, R. I.,** Comparative analysis of tocopherol by thin layer chromatography and high performance liquid chromatography, *J. Chromatogr.*, 291, 377, 1984.
33. **Isaksen, M.,** Natural pigments, in *Handbook of Thin Layer Chromatography*, Sherma, J. and Fried, B., Eds., Marcel Dekker, New York, 1991, 625.
34. **Shalberg, I. and Hynninen, P. H.,** Thin layer chromatography of chlorophylls and their derivatives on sucrose layer, *J. Chromatogr.*, 291, 331, 1984.
35. **Horwitz, W.,** *Official Methods of Analysis of the Association of Official Analytical Chemists*, 15th ed., AOAC, Arlington, VA, 1990, 1142.
36. **FAO,** FAO Food and Nutrition Paper, *Manual of Food Quality Control, 7. Food Analysis: General Techniques, Additives, Contaminants, and Composition*, Publication Division, United Nations Food and Agriculture Organization, Rome, 1986.
37. **Tyman, J. H. P.,** Phenols, aromatic carboxylic acids, and indoles, in *Handbook of Thin Layer Chromatography*, Sherma, J. and Fried, B., Eds., Marcel Dekker, New York, 1991, 757.
38. **Kreuzig, F.,** Antibiotics, in *Handbook of Thin Layer Chromatography*, Sherma, J. and Fried, B., Eds., Marcel Dekker, New York, 1991, 407.
39. **Hendrickx, S., Roets, E., Hoogmartens, J., and Vanderhaeghe, H.,** Identification of penicillins by thin layer chromatography, *J. Chromatogr.*, 291, 211, 1984.
40. **Okuyama, D., Okabe, M., Fukagawa, Y., and Ishikura, T.,** Thin layer chromatographic analysis of the OA-6129 group of carbapenem antibiotics in fermentation broths, *J. Chromatogr.*, 291, 464, 1984.
41. **Oka, H., Ikai, Y., Hayakawa, J., Masuda, K., Harada, K.-I., Suzuki, M., Martz, V., and MacNeil, J. D.,** Improvement of chemical analysis of antibiotics. 18: Identification of residual tetracyclines in bovine tissues by TLC/FABMS with a sample condensation technique, *J. Agric. Food Chem.*, 41, 410, 1993.

42. **Oka, H., Ikai, Y., Hayakawa, J., Masuda, K., Harada, K.-I., and Suzuki, M.,** Improvement of chemical analysis of antibiotics. 19: Determination of tetracycline in milk by liquid chromatography and thin-layer chromatography/fast atom bombardment mass spectrometry, *J. AOAC Int.*, 77, 891, 1994.
43. **Horwitz, W.,** *Official Methods of Analysis of the Association of Official Analytical Chemists*, 15th Ed., AOAC, Arlington, VA, 1990, 274.
44. **Horwitz, W.,** *Official Methods of Analysis of the Association of Official Analytical Chemists*, 15th Ed., AOAC, Arlington, VA, 1990, 1184.
45. **FAO,** FAO Food and Nutrition paper, *Manuals of Food Quality Control, 10. Training in Mycotoxins Analysis*, 1990.
46. **Takaskashi, T.,** Aflatoxin contamination of nutmeg: analysis of interfering TLC spots, *J. Food Sci.*, 58, 197, 1993.
47. **Mohammed, A. and Varshney, K. G.,** Inorganics and organometallics, in *Handbook of Thin Layer Chromatography*, Sherma, J. and Fried, B., Eds., Marcel Dekker, New York, 1991, 463.
48. **Touchstone, J. C.,** Perspectives in thin layer chromatography, in *Modern Thin Layer Chromatography*, Grinberg, N., Ed., Marcel Dekker, New York, 1990, 465.
49. **Dobbins, M. F. and Touchstone, J. C.,** Quantitation of pesticide using thin layer chromatography methods, in *Quantitative Thin Layer Chromatography*, Touchstone, J. C., Ed., John Wiley and Sons, New York, 1973, 293.
50. **Stahl, E. and Mangold, H. K.,** Techniques of thin layer chromatography, in *Chromatography: A Laboratory Handbook of Chromatographic and Electrophoretic Methods*, 3rd ed., Heftmann, E., Ed., Van Nostrand Reinhold, New York, 1975, 164.
51. **Grinberg, N.,** Relation between thin layer and column chromatography, in *Modern Thin Layer Chromatography*, Grinberg, N., Ed., Marcel Dekker, New York, 1990, 453.
52. **Pomeranz, Y. and Meloan, C. E.,** *Food Analysis: Theory and Practice*, AVI Publishing, Westport, CT, 1971.
53. **Boyles, S.,** Method for the analysis of inorganic and organic acid anions in all phases of beer production using gradient ion chromatography, *J. Am. Soc. Brew. Chem.*, 50, 61, 1992.
54. **FAO,** FAO Food and Nutrition paper, *Manuals of Food Quality Control, 8. Food Analysis: Quality, Adulteration, and Tests of Identity*, Publication Division, United Nations Food and Agriculture Organization, Rome, 1986.
55. **Moats, W. A. and Malisch, R.,** Determination of cloxacillin and penicillin V in milk using an automated liquid chromatography cleanup, *J. AOAC Int.*, 75, 257, 1992.
56. **Ersser, R. S. and Menzies, I. S.,** Source of error in paper and thin layer chromatographic, in *Chromatographic and Electrophoretic Techniques*, Vol. I, Smith, I. and Seakins, J. W. T., Eds., William Heinemann Medical Books Ltd., London, 1976, 57.
57. **Mathur, B. N., Whalen, P., Shahani, R., and Shahani, K. M.,** Quantification of lactulose in heat processed dairy products by HPLC, *Indian J. Dairy Sci.*, 45, 190, 1992.
58. **Sherma, J.,** Basic techniques, materials and apparatus, in *Handbook of Thin Layer Chromatography*, Sherma, J. and Fried, B., Eds., Marcel Dekker, New York, 1991, 3.
59. **Ishikawa, F., Saito, K., Nakazato, M., Fujinuma, K., Moriyasu, T., and Nishima, T.,** Determination of color additives by ion-pair high performance liquid chromatography, *Annual Report of Tokyo Metropoliton Research Laboratory of Public Health*, No. 41, 101, 1990.
60. **Naguib, Kh. M., Shalaby, A. R., and Badawy, A.,** Rapid micro-method for the extraction and determination of aflatoxins in animal or human liver tissue, *Bull. Nutr. Inst., Cairo, Egypt*, 13, 45, 1993.
61. **Gocan, S.,** Stationary phases in thin layer chromatography, in *Modern Thin Layer Chromatography*, Grinberg, N., Ed., Marcel, Dekker, New York, 1990, 5.
62. **Stock, R. and Rice, C. B. F., Eds.,** *Chromatographic Methods*, Chapman and Hall, London, UK, 1974, 279.
63. **Smith, I. and Ersser, R. S.,** Thin layer chromatography, in *Chromatographic and Electrophoretic Techniques*, Vol. I, Smith, I. and Seakins, J. V. T., Eds., William Heineman Medical Books Ltd., London, 1975, 40.
64. **Bobbitt, J. M., Schwarting, A. E., and Gritter, R. G.,** *Introduction to Chromatography*, Reinhold Book Corporation, New York, 1968, 39.
65. **Randerath, K.,** *Thin Layer Chromatography*, translated by Libman, 2nd ed., Academic Press, New York, 1966.
66. **Hauck, H. E., Mack, M., and Jost, W.,** Sorbents and precoated layers in thin layer chromatography, in *Handbook of Thin Layer Chromatography*, Sherma, J. and Fried, B., Eds., Marcel Dekker, New York, 1991, 87.
67. **Sherma, J.,** Modern high performance thin-layer chromatography, *J. AOAC Int.*, 77, 297, 1994.
68. **De Zeeuw, R. A.,** Reproducibility of R_f values in TLC, in *Progress in Thin Layer Chromatography and Related Methods*, Vol. III, Niederwieser, A. and Pataki, G., Eds., Ann Arbor Science, Ann Arbor, MI, 1972, 39.
69. **Kowalska, T.,** Theory and mechanism of thin layer chromatography, in *Handbook of Thin Layer Chromatography*, Sherma, J. and Fried, B., Eds., Marcel Dekker, New York, 1991, 43.
70. **Giddings, J. C.,** Theory of chromatography, in *Chromatography: A Laboratory Handbook of Chromatographic and Electrophoretic Methods*, 3rd ed., Heftmann, E., Ed., Van Nostrand Reinhold, New York, 1975, 27.
71. **Jaench, E. D.,** Instrumental thin-layer chromatography, in *Handbook of Thin Layer Chromatography*, Sherma, J. and Fried, B., Eds., Marcel Dekker, New York, 1991, 113.
72. **ISO,** International Standard Organization, Geneva, 5725, 1981.
73. **Long, G. L. and Winefordner, J. D.,** Limit of detection, a closer look at the IUPAC definition, *Anal. Chem.* 55, 712 A, 1983.

74. **Szepesi, G.,** Quantitation in thin layer chromatography, in *Modern Thin Layer Chromatography*, Grinberg, N., Ed., Marcel Dekker, New York, 1990, 249.
75. **Henry, R. J., Cannon, D. C., and Winkelman, J. W.,** *Clinical Chemistry, Principles and Techniques*, 2nd ed., Harper and Row, New York, 1974.
76. **Blackburn, S.,** Sample preparation and hydrolytic methods, in *Amino Acid Determination: Methods and Techniques*, Blackburn, S., Ed., Marcel Dekker, New York, 1978, 7.
77. **Erickson, C. M.,** Lipid extraction from channel muscle: composition of solvent systems, *J. Food Sci.*, 58, 84, 1993.
78. **Horwitz, W.,** *Official Methods of Analysis of the Association of Official Analytical Chemists*, 15th ed., AOAC, Arlington, VA, 1990, 921.
79. **Wang, T., Gonzalez, R. A., Gubur, E. E., and Aselage, M. J.,** Organic acid changes during ripening of processing peaches, *J. Food Sci.*, 58, 631, 1993.
80. **Horwitz, W.,** *Official Methods of Analysis of the Association of Official Analytical Chemists*, 15th ed., AOAC, Arlington, VA, 1990, 1167.
81. **Oka, H., Uno, K., Harada, K.-I., and Suzuki, M.,** Improvement of chemical analysis of antibiotics. III. Simple method for the analysis of tetracyclines on reversed-phase thin layer plates, *J. Chromatogr.*, 284, 227, 1984.
82. **Oka, H., Uno, K., Harada, K.-I., Hayashi, M., and Suzuki, M.,** Improvement of chemical analysis of antibiotics. VI. Detection reagents for tetracyclines in thin layer chromatography, *J. Chromatogr.*, 295, 129, 1984.
83. **Oka, H., Ikai, Y., Kawamura, N., Uno, K., Yamada, M., Harada, K.-I., Uchiyama, M., Asukabe, H. and Suzuki, M.,** Improvement of chemical analysis of antibiotics. X. Determination of eight tetracyclines using thin-layer and high-performance liquid chromatography, *J. Chromatogr.*, 393, 285, 1987.
84. **Oka, H., Ikai, Y., Kawamura, N., Uno, K., Yamada, M., Harada, K.-I., Uchiyama, M., Asukabe, H., and Suzuki, M.,** Improvement of chemical analysis of antibiotics. XII. Simultaneous analysis of seven tetracyclines in honey, *J. Chromatogr.*, 400, 253, 1987.
85. **Ikai, Y., Oka, H., Kawamura, N., Yamada, M., Harada, K.-I., and Suzuki, M.,** Improvement of chemical analysis of antibiotics. XIII. Systematic simultaneous analysis of tetracyclines in animal tissues using thin-layer and high-performance liquid chromatography, *J. Chromatogr.*, 411, 313, 1987.

Chapter 10

THIN-LAYER CHROMATOGRAPHY IN FORENSIC TOXICOLOGY

Ilkka Ojanperä

CONTENTS

I. Introduction 193
  A. General 193
  B. Toxicologically Relevant Substances 194
  C. Analytical Methods 194
  D. Utility of TLC 194
  E. Selected Literature 196

II. Screening Analysis 196
  A. General 196
  B. Specimens for Screening 196
  C. Extraction 197
  D. Chromatographic Systems 198
    1. Evaluation of Systems 198
    2. R_f Correction 209
    3. TIAFT Systems 210
    4. Toxi-Lab™ Systems 212
    5. UniTox Systems 212
  E. Identification 214
    1. Migration 214
    2. Visualization Reagents 214
    3. *In situ* UV Spectra 215

III. Target Analysis 218
  A. General 218
  B. Major Drugs of Abuse 218
    1. TLC vs. Immunoassay 218
    2. Amphetamines 220
    3. Cannabinoids 221
    4. Cocaine and Metabolites 221
    5. Lysergide 222
    6. Opiates 222
  C. Other Substances 223

References 224

I. INTRODUCTION

A. GENERAL
Forensic toxicology involves the analysis of potentially toxic xenobiotic compounds in human tissues or in related material and the toxicological interpretation of the results. Forensic toxicology

is influenced by the fact that the final reports produced may have to be introduced as evidence in a court of law. In forensic toxicological investigations, it is important to unequivocally determine, for example, which benzodiazepine or which opioid analgesic is in question instead of just testing for the presence or absence of the drug class.

Every finding, no matter how minor it may appear, deserves due attention because findings without direct toxicological significance may contribute important background information to the case under investigation. Such considerations make up the major differences between forensic and hospital toxicology, where a therapeutic point of view determines and often limits the extent of investigation. However, the analytical techniques used are closely related in the various fields of analytical toxicology, and no artificial categories are constructed in this text.

B. TOXICOLOGICALLY RELEVANT SUBSTANCES

Paracelsus (1493–1541) stated that all substances are poisons, and the right dose differentiates a poison from a remedy. Indeed, several people die annually due to the excess consumption of water. The enormous number of potential poisons makes it necessary to limit toxicological analysis to the most relevant substances. What is relevant depends on the duties of the laboratory as well as the geographical and socioeconomic status of the country.

In postmortem toxicology, a broad scale screening, comprising 200 to 500 substances, is usually necessary to clarify the circumstances of death. In human performance forensic toxicology, it is possible to limit tests to those compounds that may interfere with one's ability to drive or handle demanding machinery. Some workplace drug testing programs concentrate on alcohol and the major drugs of abuse, i.e., amphetamines, cannabis, cocaine, and opiates.

In most industrial countries, drugs of abuse and certain therapeutic drugs, such as antidepressants and benzodiazepines, are frequent findings in forensic investigations. The northern vodka zone produces more alcohol poisonings than the southern wine zone, where fatal alcohol poisonings may be rare. Developing countries suffer from poisonings caused by agricultural chemicals. Table 10.1 shows the number of fatal poisonings in Finland classified according to the most important causes.[1,2] Compared to other Nordic countries,[3] the number of fatal alcohol poisonings is high, and the number of fatal poisonings due to the drugs of abuse is low but increasing. For comparison, Flanagan et al.[4] gave a list of single agents associated with fatal poisoning in England and Wales, and Sunshine[5] compared substances found by different types of forensic laboratories in the United States. All the drugs in Table 10.1 except for digoxin can be detected at toxic concentrations in liver and/or urine samples by TLC.

C. ANALYTICAL METHODS

A modern forensic toxicology laboratory would find it difficult to operate without capillary gas chromatography with flame ionization (GC-FID), nitrogen specific (GC-NPD), and electron capture detection (GC-ECD), headspace GC, GC coupled with mass spectrometry (GC/MS), high-performance liquid chromatography with diode array UV (HPLC-DAD) and fluorescence detection (HPLC-FLD), atomic absorption spectrometry (AAS), UV spectrometry (UV/VIS), infrared spectrometry (IR), immunoassays, and thin-layer chromatography (TLC). How these methods should be utilized and combined for the optimum performance of the laboratory is a question that has no single answer, but Table 10.2 gives a rough outline of how they might be used. Unfortunately, there appears to be a tendency among analysts to apply expensive apparatuses requiring highly educated operators for determinations that can be solved more expediently by simpler and less expensive, but nevertheless powerful, techniques.

D. UTILITY OF TLC

Thin-layer chromatography has a creditable history in forensic toxicology, and numerous cases have been solved by this technique since Kirchner and Stahl demonstrated its potential in their books in the 1960s.[6] It is regrettable that many of today's laboratories have lost touch with modern

TABLE 10.1
Fatal Poisonings in Finland 1990-1992 According to the Most Significant Toxic Substance Found

	Cases/Year		
Substance	1990	1991	1992
Alcohols	450	404	370
Drugs	452	417	418
Dextropropoxyphene	53	57	39
Amitriptyline	48	38	39
Doxepin	39	38	30
Levomepromazine	27	24	27
Pentobarbital	21	18	21
Chlorprothixene	15	19	23
Promazine	16	20	16
Thioridazine	17	16	16
Melperone	12	12	13
Diltiazem	12	10	14
Chlorpromazine	13	13	10
Propranolol	17	5	7
Zopiclone	5	10	13
Temazepam	7	9	10
Codeine	6	8	9
Verapamil	7	8	8
Carbamazepine	6	8	7
Digoxin	8	8	4
Metoprolol	5	7	6
Oxazepam	6	5	3
...	...	...	...
Carbon monoxide	163	182	171
Other	16	13	14
Total	1,081	1,016	973

TABLE 10.2
Analytical Techniques in Forensic Toxicology

Technique	Application area
Head space GC	Alcohols and other volatiles
TLC	Screening for drugs and related substances
Immunoassays	Target screening for drugs of abuse
GC-NPD/ECD	Screening and quantification of drugs and related substances
GC-FID	Target screening for ethylene glycol and other non-nitrogen compounds
HPLC	Target screening for specific groups of compounds; quantification of polar compounds
GC-MS	Quantification of drugs of abuse; confirmation of specific findings
UV/VIS or CO-Oximetry	Carbon monoxide analysis
AAS	Metal analysis

TLC and have even given up this technique. A recent survey of the status of TLC pointed out that it has the following advantages over HPLC[7]: TLC plates are disposable, sample preparation is easier; samples are developed in parallel, throughput is more rapid; solvent consumption and operation costs are lower; detection can be performed in the absence of solvent independent of time and place; and detection possibilities are greater.

Other advantages[8] include the fact that the whole sample remains visible to the analyst, instead of some of it being lost in the HPLC pre-column or GC injection port, or eluted with the solvent front. Modern TLC can certainly also be carried out by instruments, with automated application, development, quantification, and measurement of spectra.[9,10] The disadvantages of TLC include poorer separation efficiency, limitations in quantification and automation, and a smaller variety of available separation media.[7]

As Table 10.1 indicates, certain drugs take a prominent position among the substances encountered in poisonings. TLC is exceptionally well-suited for the analysis of these types of organic compounds, which have functional groups amenable to visualization, and consequently the present chapter will focus on drugs of abuse and therapeutic drugs. While the utility of modern TLC is emphasized, it is evident that cases in forensic toxicology can only be solved by combining several analytical techniques.

E. SELECTED LITERATURE

Some invaluable reference books for the TLC-oriented forensic toxicologist are Clarke's *Isolation and Identification of Drugs*, edited by Moffat,[11] *Toxicological Analysis*, edited by Müller,[12] *Thin-Layer Chromatographic R_f Values of Toxicologically Relevant Substances on Standardized Systems*, edited by De Zeeuw et al.,[13] and *Thin-Layer Chromatography: Reagents and Detection Methods*, by Jork et al.[14]

Selected reading on TLC in forensic analysis includes the chapters in the books by Minty,[15] McDonald and Gough,[16] Franke and De Zeeuw,[17] Roberson,[18] and Ng.[19] Useful reviews include those by Kwong et al.,[20] Singh et al.,[21] Ojanperä,[22] and de Zeeuw et al.[23] Regularly appearing literature reviews include Sherma on planar chromatography,[24] Gilpin and Pachla on pharmaceuticals and related drugs[25]; Brettell and Saferstein on forensic science[26] also contains concise information on TLC methods in toxicology.

II. SCREENING ANALYSIS

A. GENERAL

Screening analysis in forensic toxicology is, instead of being preliminary testing with modest equipment, as with homemade TLC plates or spot reactions, probably the most important stage of the whole investigation. Confirmation and quantification steps are usually only possible when a potentially toxic xenobiotic has first been detected among the number of endogenous substances by an efficient screening method.

It is a recommended practice to use two independent analytical techniques and two different biological fluids or tissues for screening. If urine screening by TLC is the first approach, the other can well be blood screening by advanced GC methods.[27-29] Both of these two techniques are capable of detecting nonpolar and medium polar substances, while polar substances are only detected in urine by TLC. The GC screening may also involve simultaneous quantification of many substances.

HPLC is dependent on the UV absorption of the analytes, and this makes the detection of many substances at very low levels impossible. However, automated HPLC may be practical in hospital toxicology, providing a single compromise screening and quantification method.[30,31] GC/MS has been succesfully utilized in the toxicological screening of urine samples,[32] but many forensic toxicologists reserve the valuable analysis time of this instrument for the confirmation of results obtained by other methods and difficult cases. Immunoassays cannot compete with TLC in general screening, as assay kits are not available for all relevant classes of compounds, and even the coverage of the most important classes by immunoassays is expensive. However, immunoassays can supplement general chromatographic screening in the area of drugs of abuse.

Despite the obvious advantages of TLC, the technique has been criticized as being labor intensive and very much dependent on the skill and experience of the analyst. Furthermore, the raw data in TLC has not been usually retained for subsequent scrutiny by another investigator.[20] Fortunately, the recent advances of instrumental TLC have largely eliminated these problems.

B. SPECIMENS FOR SCREENING

In postmortem toxicology, a wide selection of biofluids and tissues are available for screening.[33] The most useful materials are liver tissue and urine, where xenobiotic concentrations tend to be higher than in the blood. Findings in the author's laboratory show that hydrophobic substances are better detected in the liver and hydrophilic substances in urine (Figure 10.1). By reversed phase TLC, the hydrophobic character of a substance can be easily estimated. Thus, sotalol with a high R_f value on the reversed phase system is found exclusively in urine, whereas chlorprothixene with a low R_f value can be seen, as a parent drug, better in the liver. In an ideal situation, both liver and urine should be screened in order to cover a broad range of drugs. In clinical forensic toxicology, urine is clearly the first specimen for screening analysis. Hair has recently gained popularity as sample material,[34] particularly in connection with workplace drug testing programs.

Rf	Liver ++	Liver +	Liver+ Urine +	Urine +	Urine ++
1.0					
0.9					
0.8					Sotalol Atenolol
0.7				Sulpiride	Codeine
0.6				Amphetamine Ethylmorphine	
0.5					
0.4			Acebutolol	Metoprolol Moclobemide Zopiclone	
0.3			Melperone		
0.2		Propranolol Doxepin Verapamil Propoxyphene			
0.1					
0.0	Fluoxetine Chlorprothixene				

FIGURE 10.1 Correlation of drug findings in the liver and in unhydrolyzed urine with the lipophilicity of drug as estimated by the R_f value on the RP system (Tables 10.4 and 10.5). ++ Found nearly exclusively in this material, + found predominantly in this material.

Although urine is often extracted directly, an acidic, basic, or enzyme hydrolysis can precede the extraction in order to liberate drugs from their conjugates.[35-37] This is because conjugated drugs are too hydrophilic to be extracted with most conventional techniques. The recoveries of narcotic analgesics and benzodiazepines can especially be improved by using β-glucuronidase enzymes. Benzodiazepines are often analyzed after acid hydrolysis as their corresponding aminobenzophenones.[13]

Liver samples require homogenization and protein removal before extraction. Early methods, such as the traditional Stas-Otto method, involved mechanical homogenization and protein precipitation by acid, ammonium sulfate, or water-miscible solvents. After the pioneering studies of Osselton et al. with subtilisin Carlsberg,[38] various enzyme digestion methods involving subtilisin, trypsin, and papain have proved to be superior to the earlier methods in terms of efficiency and hygienic aspects. Recently, McCurdy[36] gave a comprehensive review on enzyme digestion of biological specimens for drug analysis.

C. EXTRACTION

Classical liquid-liquid extraction is still widely utilized in TLC screening procedures. From the intermolecular interactions in liquid-liquid systems, such as hydrogen bonding, it follows that hydrogen-accepting solvents such as ethers, esters, or ketones should be used for hydrogen-donating solutes like organic acids. Weakly hydrogen-donating solvents, such as chloroform and dichloromethane, are suitable for amines and other hydrogen-accepting solutes. Alcohols have a high extracting ability for both acids and bases, and they can be added to widen the extraction ability of the main solvent.[39] However, it is evident that classical liquid-liquid extraction techniques are insufficient for the extraction of hydrophilic drugs.

Ionic compounds can be extracted from an aqueous solution if they are accompanied by an equivalent amount of a compound of the opposite charge, forming an ion-pair in the organic phase. Ion-pair reagents, such as inorganic ions, benzoates, sulfonates, sulfates, picrates, and water-soluble dyes, have been used to improve the extraction of quaternary ammonium compounds and other hydrophilic substances.[39,40]

Ojanperä and Vuori have developed a TLC method that involves the extraction of basic and quaternary drugs as bis(2-ethylhexyl)phosphate (HDEHP) ion-pairs.[22,41,42] Based on these investigations, a general extraction scheme for acidic, neutral, basic, amphoteric, and quaternary drugs in liver and urine samples is presented in Figure 10.2. Table 10.3 compares the recoveries of common drugs extracted as HDEHP ion-pairs and conventionally. It is evident that this hydrophobic counterion allows the isolation of even the most hydrophilic drugs, particularly from urine samples, while the recoveries of more hydrophobic drugs are nearly equal to those obtained with the conventional extraction. In contrast to less hydrophobic counterions, an excess of HDEHP remains in the organic phase and is concentrated in the evaporation residue. Consequently, the TLC systems used should be able to separate the ion-pair. Such systems, a normal phase system and a reversed phase system, are described under the subhead "UniTox Systems" below (Tables 10.4 and 10.5).

In liquid-solid extraction (solid phase extraction), the extracts are usually clean, but the technique seldom covers a polarity range wide enough for screening purposes. The classical sorbents used, such as celite, charcoal, and XAD-2 resin, have been augmented by a wide variety of bonded phase silica columns.[43,44] In recent studies, promising results were achieved with columns exhibiting mixed properties, such as hydrophobic and ion-pair, in the general extraction of acidic, neutral, and basic drugs.[45] A problem with solid phase extraction is column clogging associated with turbid samples, such as hemolyzed whole blood, but this phenomenon can be reduced by dilution, sonication, and centrifugation.[46]

D. CHROMATOGRAPHIC SYSTEMS
1. Evaluation of Systems

The optimization of screening systems differs from the optimization of the separation of a few-component mixture. In the case of screening systems, the most important features are the distribution of R_f values across the plate, the reproducibility of the measurement of those values, and the correlation of chromatographic properties between systems.[13] There are several computational methods capable of taking into account these features[13]: discriminating power,[47,48] mean list length,[49] information content,[50,51] quotient of distribution equality,[52] and principal component analysis.[53,54] The mean list length method (MLL) of de Zeeuw et al. has been chosen as the standard in this treatise because the method is perspicuous, it can be used in computerized substance identification, and it has found widespread use.

In the MLL method it is assumed that the R_f of a substance varies with a normal distribution and that its standard deviation (SD) is dependent on the chromatographic system used. For each substance in the data set, the probability is calculated such that the substance is a candidate for identification for a given R_f value. The probabilities are normalized so that the sum of the probabilities becomes 1, and the substances are ranked in decreasing order of probability. The list length for a given drug is the number of drugs, including the drug itself, that would qualify for identification at a certain cumulative probability level (0.95 is used in this text). The MLL is obtained by averaging the individual list length values. The shorter the MLL, the better the chromatographic system.[49]

It should be noted that the MLL values depend on the quality and size of the substance population used, as well as on the manner in which the standard deviation (SD) was obtained. The MLL does

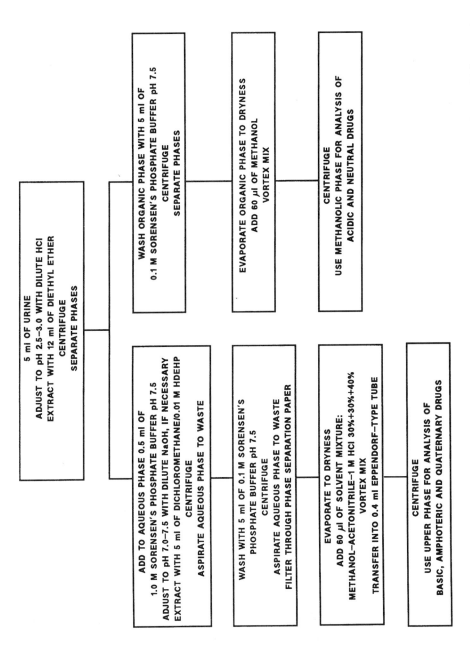

FIGURE 10.2(a) General extraction scheme for acidic, neutral, basic, amphoteric, and quaternary substances in (a) urine and (b) liver samples.

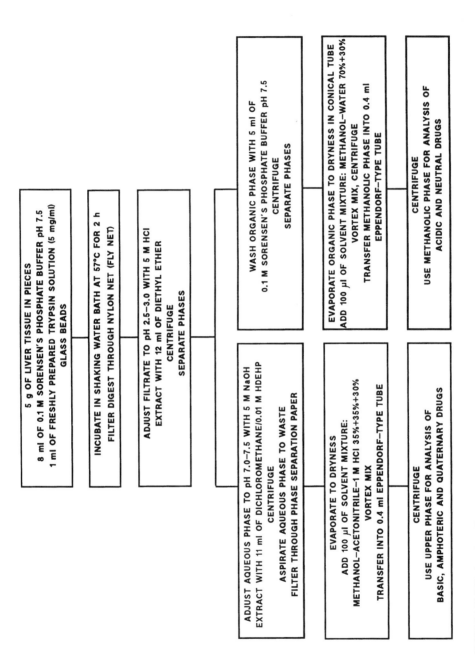

FIGURE 10.2(b) (continued)

Table 10.3
Comparison of the Extraction Efficiency of the HDEHP Ion-Pair Method with a Conventional Basic Extraction Method[a]

Drug	Character	Visualization[b]	Ion-pair advantage[c] Urine	Liver
Acebutolol	basic	FBK	+ −	+ −
Alcuronium	quaternary	Dr/IPt	+ +	+
Amantadine	basic	FBK	+ −	+ −
Amiloride	basic	Fl	+ +	+ +
Amitriptyline	basic	Dr/Ipt	+ −	+ −
Amphetamine	basic	FBK	+ −	−
Atenolol	basic	FBK	+ −	−
Buprenorphine	amphoteric	Dr/IPt	+ −	+ −
Butylscopolammonium	quaternary	Dr/IPt	+ +	+ +
Chloroquine	basic	Dr/IPt	+ −	+ −
Chlorprothixene	basic	Sal/Fl	−	+ −
Cimetidine	basic	FBK	+ +	+
Cocaine	basic	Dr/IPt	+ −	+ −
Codeine	basic	Dr/IPt	+ −	+ −
Debrisoquine	basic	Dr/IPt	+ +	+
Dextropropoxyphene	basic	Dr/IPt	+ −	+ −
Diltiazem	basic	Dr/IPt	+	+
Diphenhydramine	basic	Dr/IPt	+ −	+ −
Disopyramide	basic	Dr/IPt	+	+ −
Doxepin	basic	Dr/IPt	+ −	+ −
Emepronium	quaternary	Dr/IPt	+	+
Ephedrine	basic	FBK	+ −	+ −
Fencamfamin	basic	FBK	+ −	+ −
Fenoterol	amphoteric	FBK	+ +	+ +
Flecainide	basic	FBK	+ −	+ −
Hydroxyzine	basic	Dr/IPt	+ −	+ −
Labetalol	amphoteric	FBK	+ +	+ +
Levomepromazine	basic	Sal/Fl	−	−
Lidocaine	basic	Dr/IPt	+ −	+ −
Melperone	basic	Dr/IPt	+ −	−
Methadone	basic	Dr/IPt	+ −	+ −
Metoclopramide	basic	Dr/IPt	+ −	+ −
Metoprolol	basic	FBK	+ −	+ −
Mexiletine	basic	FBK	+ −	+ −
Maprotiline	basic	FBK	+ −	+ −
Mianserin	basic	Dr/IPt	+ −	+ −
Minoxidil	basic	FBK	+	+
Morphine	amphoteric	Dr/IPt	+ +	−
Nalorphine	amphoteric	Dr/IPt	+ +	+
Nortriptyline	basic	FBK	−	+ −
Orciprenaline	amphoteric	FBK	+ +	+ +
Orphenadrine	basic	Dr/IPt	+ −	+ −
Oxycodone	basic	Dr/IPt	+ −	+ −
Perphenazine	basic	Sal	+ −	+ −
Phentermine	basic	FBK	+ −	+ −
Phenylpropanolamine	basic	FBK	+	+
Practolol	basic	FBK	+	+ −
Prilocaine	basic	FBK	+ −	+ −
Procainamide	basic	FBK	+	+ −
Procaine	basic	FBK	+	+
Propranolol	basic	FBK	+ −	+ −
Quinidine	basic	Sal/Fl	+ −	+ −
Ranitidine	basic	FBK	+ −	+ −
Salbutamol	amphoteric	FBK	+ +	+ +

Table 10.3 (continued)
Comparison of the Extraction Efficiency of the HDEHP Ion-Pair Method with a Conventional Basic Extraction Method[a]

Drug	Character	Visualization[b]	Ion-pair advantage[c]	
			Urine	Liver
Sotalol	amphoteric	FBK	+ +	+ +
Sulpiride	amphoteric	Dr/IPt	+	+ –
Terbutaline	amphoteric	FBK	+ +	+ +
Tetracycline	amphoteric	FBK	+ +	+
Tocainide	basic	FBK	+ –	+ –
Triamterene	basic	Fl	+ –	–
Tubocurarine	quaternary	Dr/IPt	+ +	+ +
Verapamil	basic	Dr/IPt	+ –	+ –

[a] Conventional extraction with dichloromethane/5% isopropyl alcohol at pH 11.5.
[b] Visualization by Fast Black K salt reagent (FBK), Dragendorff's reagent followed by acidified iodoplatinate reagent (Dr/IPt), fluorescence at 366 nm (Fl), Salkowski reagent (Sal).
[c] Detection limit by the ion–pair method ≥ 5–fold lower (+ +), ≥ 2–fold lower (+), no difference (+ –), ≤ 2–fold higher (–), compared to the conventional method.

TABLE 10.4
The UniTox Systems for General TLC Screening

Name	Mobile phase[a]	Sorbent	Reference compounds	hR_f^c	SD[b]
Acineut	Methanol–water 65:35	Octadecyl silica	Diazepam	16	1.3
			Secobarbital	35	
			Phenobarbital	54	
			Paracetamol	74	
Krim	Toluene–acetone–ethanol–conc.ammonia 45:45:7:3	Silica	Codeine	16	1.7
			Promazine	36	
			Clomipramine	49	
			Cocaine	66	
RP	Methanol–water–conc. hydrochloric acid 50:50:1	Octadecyl silica	Hydroxyzine	20	1.3
			Lidocaine	46	
			Codeine	66	
			Morphine	81	

[a] Mobile phase composition: volume+volume. Krim system is used with a saturated chamber.
[b] The intralaboratory standard deviation of measurement of hR_f values.

Table 10.5
Corrected hR_f Values (hR_f^c) of Drugs on the UniTox Systems[a]

Drug	hR_f^c		Acineut
	Krim	RP	
Acebutolol	11	43	
Acetamidonitrazepam, 7-	43	63	
Acetazolamide	8	66	
Acetophenazine	18	17	
Acetylsalicylic acid			55
Aconitine	59	13	
Ajmaline	29	39	

Table 10.5 (continued)
Corrected hR$_f$ Values (hR$_f^c$) of Drugs on the UniTox Systems[a]

Drug	Krim	hR$_f^c$ RP	Acineut
Alcuronium	0	29	
Allobarbital			54
Alprazolam	29	33	16
Alprenolol	27	21	
Amfepramone	75	48	
Amidephrine	27	83	
Amiloride	4	55	
Aminoflunitrazepam, 7-			46
Aminonitrazepam, 7-			50
Aminophenazine	46	67	31
Amiodarone	80	0	
Amiphenazole	55	64	
Amitriptyline	46	9	
Amlodipine	41	8	
Amobarbital			37
Amphetamine	37	57	
Antazoline	17	21	
Apomorphine	45	48	
Aprobarbital	50		
Astemizole	40	19	
Atenolol	4	77	
Azapropazone			33
Barbital			63
Bemegride			55
Benactyzine	62	25	
Benorilate	48	10	30
Benzhexol	81	10	
Benzoylecgonine	0	42	
Benztropine	9	10	
Betahistine	2	91	
Betaxolol	22	24	
Biperiden	79	11	
Bisacodyl	65	29	
Bisoprolol	21	30	
Brallobarbital			51
Bromazepam	55	46	29
Bromhexine	83	11	
Bromisoval			44
Brotizolam	43	19	7
Brucine	8	41	
Buclizine	89	2	
Bumetanide			39
Bupivacaine	73	23	
Buprenorphine	72	17	
Butalbital			45
Butaperazine	30	6	
Butobarbital			46
Butylscopolammonium	0	34	
Caffeine	36	31	45
Captopril			68
Carbamazepine			32
Carbinoxamine	24	56	
Carbromal			35
Carbutamide	11	44	
Chloramphenicol	31	34	56

Table 10.5 (continued)
Corrected hR$_f$ Values (hR$_f^c$) of Drugs on the UniTox Systems[a]

Drug	Krim	hR$_f^c$ RP	Acineut
Chlorcyclizine	48	20	
Chlordiazepoxide	42	34	16
Chlormezanone			46
Chlorobenzoxamine	82	13	
Chloroquine	14	48	
Chlorphentermine	24	33	
Chlorpromazine	50	5	
Chlorpropamide	9	19	41
Chlorprothixene	56	4	
Chlorpyramine	45	54	
Chlortalidone			65
Chlorzoxazone			36
Cimetidine	15	76	
Cinchocaine	49	14	
Cinnarizine	78	15	
Cisapride	44	12	
Citalopram	32	22	
Clemastine	30	4	
Clidinium	0	26	
Clobazam	56	10	29
Clobutinol	59	21	
Clomethiazole	62	55	31
Clomipramine	49	5	
Clonazepam	57	32	16
Clonidine	53	56	
Clothiapine	50	36	
Clozapine	36	49	
Cocaine	66	33	
Codeine	16	66	
Cyclizine	46	43	
Cyclobarbital			46
Cyproheptadine	46	10	
Debrisoquine	0	49	
Demoxepam	49	14	
Desipramine	16	10	
Dexbrompheniramine	22	53	
Dexchlorpheniramine	20	59	
Dextromethorphan	21	18	
Dextromoramide	67	13	
Dextropropoxyphene	67	11	
Diacetylmorphine	31	43	
Diazepam	67	29	16
Dibenzepin	33	27	
Diclofenac			19
Diflunisal			22
Diltiazem	42	13	
Dimethindene	22	54	
Diphenhydramine	39	20	
Diphenoxylate	84	5	
Diphenylpyraline	28	15	
Diprophylline	14	56	
Dipyridamole	36	9	
Dipyrocetyl	4	37	
Disopyramide	30	61	
Disulfiram	83	4	

Table 10.5 (continued)
Corrected hR$_f$ Values (hR$_f^c$) of Drugs on the UniTox Systems[a]

Drug	Krim	hR$_f^c$ RP	Acineut
Dixyrazine	33	18	
DMA, 2,5-[b]	34	46	
DOB[c]	35	27	
DOET[d]	34	19	
DOM[e]	34	30	
Doxepin	44	16	
Drofenine	70	4	
Dropropizine	25	65	
Emepronium	1	16	
Ephedrine	7	63	
Ergotamine	32	11	
Ethenzamide	46	24	44
Ethosuximide			63
Ethyl-MDA, N-[f]	26	47	
Ethylamphetamine, N-	28	49	
Ethylmorphine	17	55	
Etodroxizine	35	18	
Famotidine	14	81	
Felodipine	63	1	
Fencamfamin	57	25	
Fenfluramine	33	27	
Fenoprofen			21
Fenoterol	11	73	
Fentanyl	67	20	
Flecainide	27	21	
Fluanisone	70	19	
Fluconazole	24	51	
Flumazenil	47	24	
Flunitrazepam	61	13	27
Fluoxetine	20	7	
Flupenthixol	32	6	
Fluphenazine	29	8	
Flurazepam	51	24	11
Fluvoxamine	44	8	
Furosemide			58
Glibenclamide	7	2	13
Glipizide	5	7	27
Glutethimide			36
Glycopyrronium	0	18	
Guaifenesin	29	37	
Haloperidol	53	13	
Heptabarbital			38
Hexobarbital			38
Histapyrrodine	59	18	
Hydralazine	68	75	
Hydrochlorothiazide			81
Hydrocone	15	59	
Hydroxychloroquine	13	53	
Hydroxyzine	41	20	
Ibuprofen			15
Imipramine	40	11	
Indapamide	54	22	
Indometacine			17
Iprindole	44	5	
Isoetarine	19	81	

Table 10.5 (continued)
Corrected hR$_f$ Values (hR$_f^c$) of Drugs on the UniTox Systems[a]

Drug	Krim	hR$_f^c$ RP	Acineut
Isoniazid	18	91	
Isoprenaline	4	88	
Isoxsuprine	45	37	
Isradipine	68	3	
Ketamine	61	45	
Ketoprofen			32
Labetalol	21	28	
Levomepromazine	60	7	
Lidocaine	66	46	
Loperamide	49	6	
Lorazepam	39	11	31
Lormetazepam			24
Lysergide	44	27	
Maprotiline	11	9	
MDA[g]	34	54	
MDMA[h]	13	52	
Meclozine	81	6	
Medazepam	69	5	20
Mefenamic acid			11
Mefruside			47
Melperone	57	28	
Mephenesin	43		
Mepivacaine	55	42	
Mepyramine	37	68	
Mescaline	26	63	
Metformin	0	92	
Methadone	54	9	
Methamphetamine	12	53	
Methaqualone			23
Methaqualone	66	26	
Methocarbamol	30	35	57
Methohexital			29
Methoxyphenamine	12	44	
Methyl phenidate	43	41	
Methylaminophenazone, 4-	32		
Methylephedrine	16	59	
Methylphenobarbital			40
Metixene	41	6	
Metoclopramide	25	39	
Metolazone	48	26	50
Metoprolol	21	41	
Metronidazole	30	82	
Mexiletine	47	34	
Mianserin	50	17	
Midazolam			9
Minoxidil	3	47	
MMDA[i]	32	51	
Moclobemide	36	41	
Molindone	53	36	
Moperone	50	16	
Morphine	8	81	
Nalorphine	30	73	
Naloxone	48	75	
Naproxen			26
Neostigmine	0	60	

Table 10.5 (continued)
Corrected hR$_f$ Values (hR$_f^c$) of Drugs on the UniTox Systems[a]

Drug	Krim	hR$_f^c$ RP	Acineut
Nialamide	15	60	
Nicotinamide			69
Nicotine	36	87	
Nicotinyl alcohol	29	88	
Nifedipine	61	6	
Nikethamide	43	73	
Nitrazepam	54	43	32
Nitrofurantoin	7	50	
Nizatidine	14	85	
Nomifensine	40	61	
Nordiazepam	59	42	20
Normethadone	42	14	
Norpseudoephedrine	54	69	
Nortriptyline	20	9	
Noscapine	69	34	
Omeprazole	32	31	
Opipramol	20	23	
Orciprenaline	6	89	
Orphenadrine	45	13	
Oxazepam	47	16	30
Oxprenolol	21	32	
Oxycarbazepin			40
Oxycodone	54	64	
Oxypertine	61	27	
Oxyphencyclimine	1	11	
Papaverine	55	30	
Paracetamol			74
Paroxetine	14	11	
Pemoline	30	50	
Penfluridol	72	1	
Pentazocine	55	31	
Pentifylline	54	2	9
Pentobarbital			38
Pentoxifylline	42	37	20
Pentoxyverine	40	10	
Periciazine	36	12	
Perphenazine	27	11	
Pethidine	34	34	
Phenacetin			42
Phenazocine	59	23	
Phenazone	31	43	45
Phenethylamine	36	65	
Phenindione			38
Pheniramine	18	76	
Phenmetrazine	23	56	
Phenobarbital			53
Phentermine	21	47	
Phenylbutazone			27
Phenylephrine	25	86	
Phenylpropanolamine	47	67	
Phenytoin			43
Pilocarpine	25	76	
Pimozide	56	5	
Pindolol	21	58	
Pipotiazine	74	1	

Table 10.5 (continued)
Corrected hR$_f$ Values (hR$_f^c$) of Drugs on the UniTox Systems[a]

Drug	Krim	hR$_f^c$ RP	Acineut
Pipradrol	67	33	
Pirmenol	33	64	
Piroxicam	18	23	33
Pitofenone	56	26	
Pizotifen	43	12	
PMA[j]	33	54	
Practolol	8	71	
Prazosin	43	23	
Prenylamine	70	6	
Prilocaine	64	47	
Primidone			54
Procainamide	13	92	
Procaine	50	88	
Prochlorperazine	31	10	
Procyclidine	53	13	
Prolintane	60	28	
Promazine	36	10	
Promethazine	41	11	
Propafenone	58	11	
Propranolol	24	21	
Propyphenazone	62	19	31
Proquazone	60	10	13
Protriptyline	10	11	
Proxyphylline	30	39	
Pseudoephedrine	65	64	
Pyrazinamide	41	62	
Quinidine	24	66	
Quinine	24	66	
Ranitidine	10	84	
Remoxipride	32	32	
Reserpine	67	5	
Rimiterol	2	84	
Salbutamol	6	83	
Salicylamide			56
Salicylic acid			60
Secbutabarbital			47
Secobarbital			35
Selegiline	69	51	
Sotalol	13	82	
Spironolactone	67	3	
Strychnine	15	42	
Sulindac			23
Sulpiride	17	71	
Sulthiame			71
Sumatriptan	8	67	
Temazepam	53	11	25
Terbutaline	8	84	
Terfenadine	54	3	
Terodiline	57	14	
Tetracaine	38	43	
Theophylline	16	45	58
Thiethylperazine	33	6	
Thiopental			32
Thioproperazine	21	19	
Thioridazine	45	3	

Table 10.5 (continued)
Corrected hR$_f$ Values (hR$_f^c$) of Drugs on the UniTox Systems[a]

Drug	Krim	hR$_f^c$ RP	Acineut
Thiothixene	24	17	
Tiaprofenic acid			40
Tilidine	85	36	
Timolol	27	40	
Tizanidine	46	59	
TMA, 3,4,5-[k]	26	59	
Tocainide	42	55	
Tolbutamide	8	14	37
Tolfenamic acid			10
Toremifene	49	2	
Trazodone	53	23	
Triazolam	29	7	20
Trifluoperazine	31	8	
Trimeprazine	59	9	
Trimethoprim	20	61	
Trimipramine	65	9	
Tripelennamine	41	75	
Triprolidine	25	64	
Tubocurarine	0	55	
Verapamil	57	12	
Vinbarbital			50
Vinylbital			39
Warfarin			23
Xanthinol nicotinate	7	75	
Xylometazoline	5	14	
Yohimbine	56	35	
Zopiclone	30	35	
Zuclopenthixol	30	9	

[a] Compiled in part from References 69–71.
[b] 2,5-Dimethoxyamphetamine
[c] 4-Bromo-2,5-dimethoxyamphetamine
[d] 2,5-Dimethoxy-4-ethylamphetamine
[e] 2,5-Dimethoxy-4-methylamphetamine
[f] N-Ethyl-3,4-methylenedioxyamphetamine
[g] 3,4-Methylenedioxyamphetamine
[h] N-Methyl-3,4-methylenedioxyamphetamine
[i] 3-Methoxy-4,5-methylenedioxyamphetamine
[j] 4-Methoxyamphetamine
[k] 3,4,5-Trimethoxyamphetamine

not tell anything about the separation number, i.e., the number of spots that can be separated by a system with a certain resolution.

2. R$_f$ Correction

The use of R$_f$ correction to compensate for the effects of environmental and operational variations is an old concept in TLC.[55–57] In contrast to column chromatography, the use of a single R$_f$ standard, corresponding to the relative retention time, may produce erroneous results in an open system such as TLC. The relative R$_f$ values are often less reproducible than the R$_f$ values themselves and can be misleading as they may simulate safe R$_f$ values.[58] The method, which uses four standards that are structurally similar to the analytes and linear interpolation between the standards, has attained an established position in drug and pesticide screening (Figure 10.3).[13,59,60] The resulting corrected R$_f$ values (hR$_f^c$) are highly reproducible, and TLC can be practiced even in extreme conditions,

$$hR_f^c(X) = hR_f^c(A) + \frac{\Delta^c}{\Delta}[hR_f(X) - hR_f(A)], \text{ where}$$

$$\Delta^c = hR_f^c(B) - hR_f^c(A) \text{ and}$$

$$\Delta = hR_f(B) - hR_f(A)$$

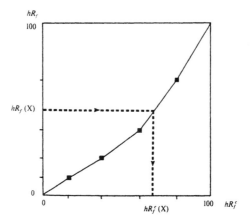

FIGURE 10.3 Correction of R_f values (a) by calculation or (b) by means of the six-point correction graph. X is the unknown, A and B are the bracketing standards (A may also be the starting point and B may also be the solvent front).[13] (Reproduced by permission of VCH Verlagsgesellschaft mbH.)

such as in the tropics at high temperatures and humidities.[61,62] It is noteworthy that the R_f correction described above is based on essentially the same formula as the calculation of linear retention indices in GC.

Bogusz et al.[63,64] showed that both the precision and, to a lesser extent, accuracy of the R_f values were lowered with biological extracts. These effects can be partly eliminated by using drugs extracted from biological material as correction standards, but whether the extraction step is needed depends on the TLC system in question. However, a slightly higher SD of the hR_f^c values is to be expected in the analysis of toxicological samples compared with pure drug substances.

In the following section, several established screening systems are described that have proved to be applicable on an interlaboratory basis.

3. TIAFT Systems

The R_f values were collected for 1600 toxicologically relevant substances in 10 standardized TLC systems by The Committee for Systematic Toxicological Analysis of The International Association of Forensic Toxicologists (TIAFT) (Table 10.6). Four of these systems are for acidic and neutral drugs (Systems 1–4a) and seven for basic drugs (Systems 4b–10). The systems were chosen based on their separating power measurements, widespread use, and their large existing databases.[13]

A single TLC system does not generally allow unequivocal identification of a substance, but the identification power can be significantly improved by combining the information from a second system that has a low correlation with the first one. It is the author's opinion that instead of trying to obtain further reliability with a third system, one should utilize *in situ* spectrometric methods and visualization reactions. Using a data set of 100 acidic and neutral drugs and 100 basic drugs, the lowest MLL values were obtained for systems 2 (MLL = 12) and 5 (MLL = 15), respectively.[17] The best combination for acidic and neutral drugs was the pair 2–4 (MLL = 5.2). For basic drugs, there were two equally good alternatives, the pair 5–8 and the pair 7–8 (MLL = 5.0).[17] The above MLL values were calculated using interlaboratory SD values (Table 10.6).

TABLE 10.6
The TIAFT Systems for General TLC Screening
(Reproduced by permission of VCH Verlagsgesellschaft mbH)

	Mobile phase[a]	Sorbent	Reference compounds[b]	hR_f^c	SD[c]
1	Chloroform–acetone 80:20	Silica	Paracetamol	15	2.3
			Clonazepam	35	
			Secobarbital	55	
			Methylphenobarbital	70	
2	Ethyl acetate	Silica	Sulfathiazole	20	2.7
			Phenacetin	38	
			Salicylamide	55	
			Secobarbital	68	
3	Chloroform–methanol 90:10	Silica	Hydrochlorothiazide	11	2.7
			Sulfafurazole	33	
			Phenacetin	52	
			Prazepam	72	
4a	Ethyl acetate–methanol–conc.ammonia 85:10:5	Silica	Sulfadimidine	13	3.7
			Hydrochlorothiazide	34	
			Temazepam	63	
			Prazepam	81	
4b	Ethyl acetate–methanol–conc.ammonia 85:10:5	Silica	Morphine	20	3.3
			Codeine	35	
			Hydroxyzine	53	
			Trimipramine	80	
5	Methanol	Silica	Codeine	20	2.7
			Trimipramine	36	
			Hydroxyzine	56	
			Diazepam	82	
6	Methanol–n-butanol 60:40, 0.1 mol/L NaBr	Silica	Codeine	22	3.0
			Diphenhydramine	48	
			Quinine	65	
			Diazepam	85	
7	Methanol–conc.ammonia 100:1.5	Silica[d]	Atropine	18	3.0
			Codeine	33	
			Chlorprothixene	56	
			Diazepam	75	
8	Cyclohexane–toluene–diethylamine 75:15:10	Silica[d]	Codeine	6	2.7
			Desipramine	20	
			Prazepam	36	
			Trimipramine	62	
9	Chloroform–methanol 90:10	Silica[d]	Desipramine	11	3.7
			Physostigmine	36	
			Trimipramine	54	
			Lidocaine	71	
10	Acetone	Silica[d]	Amitriptyline	15	3.0
			Procaine	30	
			Papaverine	47	
			Cinnarizine	65	

[a] Mobile phase composition: volume+volume; Saturated systems are used except for systems 5 and 6 which are used with unsaturated chambers. System 4 is split: 4a for acidic and neutral substances and 4b for basic and neutral substances.
[b] Solutions of the four reference compounds at a concentration of approximately 2 mg/ml of each substance.
[c] The interlaboratory standard deviation of measurement of hR_f values.
[d] Silica impregnated with 0.1 mol/L KOH and dried.

4. Toxi-Lab™ Systems

A TLC kit for toxicological drug screening, containing equipment for extraction, development, detection, and identification, was developed by Toxi-Lab Inc. (Irvine, CA).[65–67] Special glass fiber plates impregnated with the sorbent are used. The plates have holes at the origin for the inoculation of factory-made standard substance disks and disks containing the evaporated samples. The Toxi-Lab™ A system is designed for basic and neutral substances and the Toxi-Lab™ B for acidic and neutral substances. Because of its widespread use in North America, the Toxi-Lab™ A was recommended by TIAFT, and the R_f values of several hundred drugs and metabolites run on this system are available.[13] The Toxi-Lab™ A mobile phase consists of 3 ml of ethyl acetate–methanol–water (87:3:1.5) and 15 µl of concentrated ammonia; and the reference compounds are strychnine (hR_f^c 12), amitriptyline (hR_f^c 58), and methaqualone (hR_f^c 81).[13] Toxi-Lab™ also offers a reversed phase system, Toxi-Lab™ C8, to be used in combination with Toxi-Lab™ A. The combination significantly enhances the identification power as shown by the MLL method. For a set of 81 basic and neutral drugs, the MLL values were 8.9, 8.2, and 1.9 for the Toxi-Lab™ A, for the Toxi-Lab™ C8, and for their combination, respectively.[68] Intralaboratory SD values were used in these calculations.

A remarkable feature of the Toxi-Lab™ procedure is its four-stage standardized visualization sequence developed to aid identification (see "Visualization" below). On the other hand, the resolution and separation number are obviously lower than with ordinary TLC systems.

5. UniTox Systems

A system for acidic and neutral drugs (Acineut) and two systems for basic, amphoteric, and quaternary drugs (Krim and RP) were developed to meet the requirements of comprehensive drug screening in forensic toxicology (Table 10.4).[41,69–71] These systems, named here as UniTox systems, are compatible with the broad scale extraction scheme shown in Figure 10.2. Table 10.5 lists the hR_f^c values of 365 drugs, of which 100 were measured on the Acineut, 300 on both the Krim and RP, and 35 on all three systems. For a set of 100 acidic and neutral drugs, the MLL value was 9.7.[71] For a set of 140 basic, amphoteric, and quaternary drugs, the MLL values were 10.9, 9.7, and 1.8 for the Krim, for the RP, and for their combination, respectively.[69] Intralaboratory SDs of uncorrected R_f values were used in these calculations, but an interlaboratory study involving 15 pure drug substances and five laboratories revealed comparable SDs for hR_f^c values.[60] The chromatographic behavior of selected basic drugs run on the Krim and RP systems is illustrated in Figures 10.4 and 10.5.

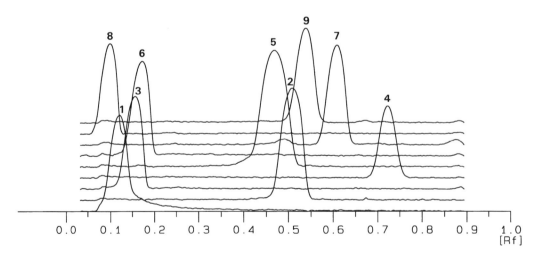

FIGURE 10.4 Selected basic and amphoteric drugs run on the Krim system (Table 10.4): (1) acebutolol, (2) amitriptyline, (3) codeine, (4) dextropropoxyphene, (5) doxepin, (6) ethylmorphine, (7) methadone, (8) morphine, and (9) procaine.

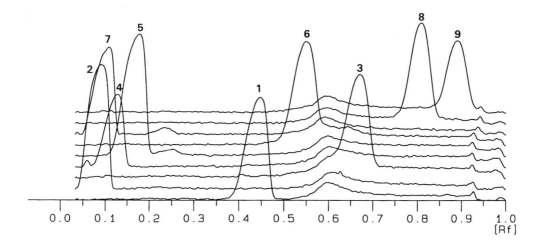

FIGURE 10.5 Selected basic and amphoteric drugs run on the RP system (Table 10.4). Compounds as in Figure 10.4.

Most TLC separations in forensic toxicology are still carried out using the ascending free-flow technique, and applications of promising newer techniques such as the automatic multiple development by Camag (Muttenz, Switzerland) have been scarce.[72] Large existing databases created using conventional methods may hinder the introduction of more efficient techniques that usually produce R_f values incompatible with the old ones. The Camag Automatic Developing Chamber (ADC), which is an automated version of the ascending free-flow chamber, produced values compatible with the twin-trough chamber after the R_f correction (Table 10.7). The ADC was superior to the manual chamber in terms of convenience and documentation facilities, and it also offered better precision than the manual chamber on the Krim system. It is interesting that the reversed phase systems, Acineut and RP, produced poor results with the ADC. However, by modifying the structure of the ADC and adding a piece of filter paper for better chamber saturation, it was possible to obtain correct results also with these systems

TABLE 10.7
Comparison of the Precision Obtained by the Automatic Developing Chamber (ADC) with the Twin-Trough Chamber for the Krim System[a]

	Literature[b] hR_f^c	Twin-trough Chamber				Automatic Chamber (ADC)			
		hR_f		hR_f^c		hR_f		hR_f^c	
		Mean	CV%	Mean	CV%	Mean	CV%	Mean	CV%
Acebutolol	11	13	24.7	13	9.5	18	6.8	13	2.4
Amitriptyline	46	53	9.5	47	1.2	54	4.8	47	1.3
Amphetamine	37	43	13.5	36	2.4	43	3.4	37	4.5
Buprenorphine	72	74	4.4	69	5.4	85	2.1	82	1.8
Chlorpromazine	50	55	8.6	48	0.9	56	4.6	49	1.2
Dextropropoxyphene	67	72	4.5	67	1.7	73	1.2	67	1.5
Fencamfamin	57	63	7.1	57	2.1	60	2.6	54	2.0
Orciprenaline	6	7	10.1	7	12.2	10	9.2	7	3.9
Sulpiride	17	21	20.4	20	5.1	25	5.4	19	3.5
Zopiclone	30	36	10.7	32	5.6	42	6.2	36	2.3

[a] Based on six measurements with both methods during a six-week period
[b] Obtained with the twin-trough chamber (Reference 69)

E. IDENTIFICATION
1. Migration

In the past, many forensic toxicologists considered the migration distance or the R_f value just a rough means of identification using a procedure in which they divided each chromatographic plate into four "R_f zones." However, by carefully adjusting the chromatographic conditions, such as the temperature, layer activity, and chamber saturation, very reproducible R_f values can be obtained with today's precoated silica gel plates.[73] Aqueous reversed phase systems are less influenced by environmental factors, but the lot-to-lot reproducibility of the sorbent is important. In routine toxicological analysis it is not possible to spend much time in keeping the analytical conditions exactly reproducible, and, consequently, R_f correction is a way to achieve excellent results with minimum effort. Without correction, it would be useless to create large R_f libraries for interlaboratory use.

Substance identification by searching R_f libraries can be performed manually or with the help of computers.[54,74,75] Commercial programs are available to facilitate the search of libraries containing hundreds of substances. The Merck Tox Screening System (MTSS) (Merck, Darmstadt, Germany) is statistical search software that utilizes the MLL method, giving each candidate in the identification process a probability value. The MTSS allows the user to create in-house libraries based on their own TLC systems or to utilize the built-in libraries of the TIAFT systems (Table 10.6). The Toxi-Search software by Toxi-Lab™ also utilizes the MLL approach, and it is designed to be used together with the company's other products. Both the MTSS and the Toxi-Search can incorporate numerical color information from visualization reactions.

The CATS software by Camag combines instrumental evaluation of chromatographic plates with substance identification by hR_f^c values and *in situ* UV spectra. The hR_f^c presearch uses an adjustable search window that limits the number of substance candidates for the spectrum correlation search (see "*In situ* UV Spectra" below).[60]

2. Visualization Reagents

Visualization of the immobilized analytes, independent of time and place, is a unique advantage of TLC. Even poor UV-absorbing substances, such as amphetamines, can be detected at low concentrations with suitable reagents. There are good collections of visualization reagents for the general location of chromatographic spots as well as for the functional group-specific or substance-specific visualization.[11-14,76-81] A selection of useful visualization reagents is listed in Table 10.8. In screening analysis, visualization sequences involving several reagents that can be oversprayed on the previous ones have proved to be particularly powerful. An example of such a sequence is ninhydrin, FPN reagent, Dragendorff's reagent, and acidified iodoplatinate, which can be used to detect a wide variety of basic, amphoteric, and quaternary drugs.[11] The area of visualization sequences still provides a challenge for forensic toxicologists.

The interpretation of colors in TLC has been criticized as being an art that requires a lot of experience in order to produce reliable results. Hegge et al.[106] introduced a numerical color-coding system to make the color reactions amenable to computer handling. The system was based on the Toxi-Lab™ sequence of four color reactions, and encoding of the observed color was done by means of a wheel of reference colors. The Toxi-Lab™ sequence consists of formaldehyde vapor + Mandelin's reagent, water, fluorescence under 366 nm UV light, and modified Dragendorff's reagent. The combined information of R_f values and color reactions resulted in an eightfold reduction of the MLL values compared with the information from R_f values alone. The option of utilizing numerical color codes for identification is available in the Toxi-Search and MTSS software. Basically, the color wheel developed can be used to digitize colors from any detection reaction or sequence of detection reactions.

A tri-stimulus reflection colorimeter was evaluated for the standardization of color measurement in TLC.[107] The identification power, expressed as the MLL for 100 drugs, was 8.3 and 6.3 for visual and colorimetric methods, respectively, when the Toxi-Lab™ color reaction sequence was used. However, when the fluorescence inspection stage was included in the encoding system,

TABLE 10.8
Visualization Reagents for Screening Analysis

Reagent	Application	Reference[a]
Bratton-Marshall reagent	Aromatic *prim*-amines, sulfonamides	13
7-Chloro-4-nitrobenzo-2-oxa-1,3-diazole (NBD-Cl)	Aliph. *prim*- and *sec*-amines, amino acids, sulfonamides, alkaloids, phenols	82,83
2,6-Dichlorophenol-indophenol (Tilmann's reagent)	Organic acids	14
2,6-Dichloroquinone-4-chloroimide	Aliph. *prim*- and *sec*-amines, arylamines, phenols, heterocycles, various drugs	84
2,4-Dinitrophenylhydrazine	Aldehydes and ketones	14
Dragendorff's reagent	Alkaloids and drugs	11,85,86
Fast Black K salt	Aliph. *prim*- and *sec*-amines, arylamines, phenols, heterocycles	70,87,88
Fast Blue B salt, Fast Blue BB salt	Aliph. *prim*- and *sec*-amines, arylamines, phenols, cannabinoids, heterocycles	89-91
Fluorescamine	*Prim*-amines, amino acids, sulfonamides	92-94
Forrest reagent	Phenothiazines, antidepressants	95
FPN reagent	Phenothiazines, dibenzazepines	96
Iodoplatinate, acidic	Alkaloids, quaternary ammonium compounds	96-98
Mandelin's reagent	Various drugs	99
Marquis reagent	Various drugs	99
Mercuric chloride-diphenylcarbazone	Barbiturates	11
Mercurous nitrate	Barbiturates	11
Ninhydrin	*Prim*-amines, amino acids, (*sec*-amines)	100,101
Salkowski reagent (FeCl$_3$ + H$_2$SO$_4$)	Phenothiazines, thioxanthenes	77,102
Sodium nitroprusside	Aliph. *sec*-amines	94,103
3,3′,5,5′-Tetramethylbenzidine, *o*-tolidine (after Cl$_2$)	Pesticides, various acidic and neutral drugs, indoles	13,71,77, 104,105
Van Urk's reagent	Indoles, amines, sulfonamides, pesticides	14

[a] The reference is not necessarily made to the original work; see also the general references mentioned above.

the MLL obtained from visual coding improved to 2.9. The colorimeter was unable to measure UV reflection.

3. *In situ* UV Spectra

Spectrometric techniques have been traditionally coupled with TLC off-line: the spots are scraped off the plate, the analyte is extracted with a solvent, and the sample is analyzed by UV/VIS, IR, mass, or NMR spectrometry. Automatic measurement of *in situ* UV spectra on TLC plates has been possible for over 10 years, but the coupling with other spectrometric techniques is still too cumbersome for routine screening purposes.

Ojanperä and Jänchen[60] described an instrumental method for drug screening, based on the hR_f^c values obtained by a built-in correction procedure and *in situ* UV spectra. Libraries of spectra, created using the UniTox systems, were found to be reproducible on an international basis. According to a study involving 15 standard drugs analyzed in five laboratories against a library of 100 drugs, the correct result was among the two first positions on the hit list in 98.7% and 97.3% of cases, using the Krim and RP systems, respectively.[60]

Figures 10.6 and 10.7 show an example of the analysis of autopsy liver extracts by instrumental qualitative TLC, using the CATS software (Camag). The hR_f^c presearch window used, (+/−) 7 units, has been found to be wide enough to cope with the deviations caused by biological background and varying drug concentrations. According to the hit lists and spectra shown in Figure 10.7, the antidepressant trimipramine was identified without problems by combining the information from the two complementary chromatographic systems. The finding was confirmed by dropping nitric acid on the spot and observing the blue-green color produced. By GC, the trimipramine blood concentration was

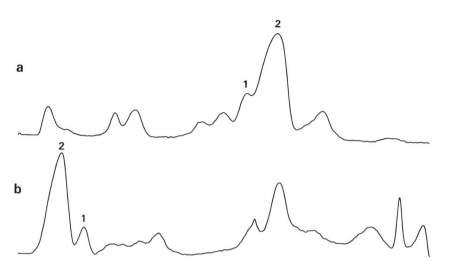

FIGURE 10.6 An autopsy liver extract analyzed on (a) the Krim system and (b) the RP system. Peaks: (1) haloperidol, (2) trimipramine. See Figure 10.7.

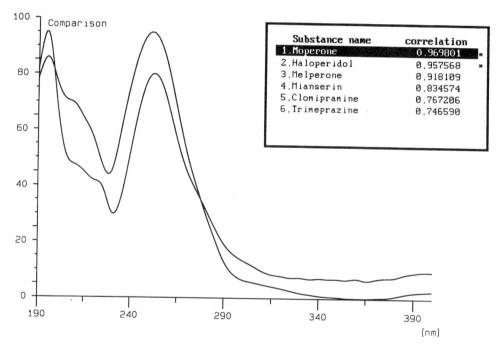

FIGURE 10.7 (a) Results of the hR_f^c/spectrum library search for the findings in Figure 10.6, with the best match superimposed with the sample spectrum: Peak 1 (a) Krim system, (b) RP system, peak 2 (c) Krim system, and (d) RP system.

found to be 0.7 mg/l. The other finding was most probably haloperidol, a low-dose antipsychotic, but the other candidate, moperone, could not be eliminated by the hR_f^c/spectrum search or visualization. Using GC, the substance proved to be haloperidol with a blood concentration of 0.1 mg/l.

In situ UV spectra are known to be dependent on the amount of substance on the plate,[108] and this phenomenon is more pronounced with small amounts of substance. However, with toxicological samples containing drugs at moderate or high concentrations, sufficiently high correlation values are usually obtained. Figure 10.8 shows spectra for various amounts (0.5–15 μg) of melperone, propranolol, and sulpiride run on the Krim system, and the correlation values with the library

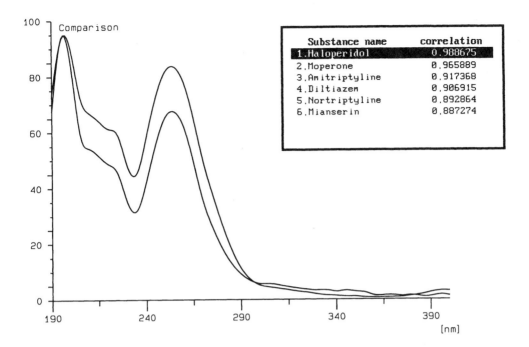

FIGURE 10.7(b)

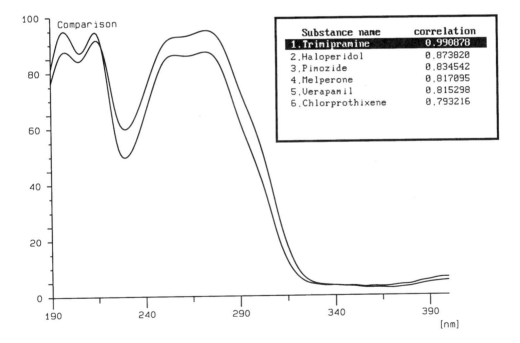

FIGURE 10.7(c)

spectra are shown in Table 10.9. Spectrum libraries created with one chromatographic system can sometimes be used with another closely related system, certainly without the R_f information. This is the case with, e.g., the Krim system and the TIAFT system 4b, which show highly correlated

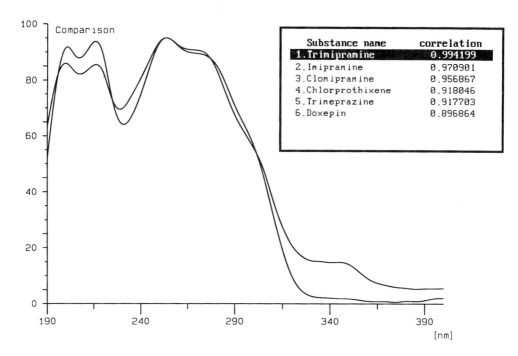

FIGURE 10.7(d)

UV spectra. Computerized hR_f^c/spectrum libraries containing hundreds of drugs based on the UniTox systems were recently made available by Camag.

III. TARGET ANALYSIS

A. GENERAL

An enormous number of TLC methods are available for various analyses in forensic toxicology: *Toxicological Analysis*[12] alone lists 1,453 published TLC systems for potentially toxic compounds, some of which are dedicated to single substances and others to groups of compounds. Although the utility of TLC can be most clearly seen in the general screening for hundreds of substances, TLC can also be the method of choice in target screening or in quantitative analysis. This is especially true in circumstances where high sample throughput is required and a selective and sensitive derivatization reaction is available to improve detection.

In the following section, the more recent methods for the analysis of the major drugs of abuse and for some other important groups of substances are reviewed. Finally, a list of recent papers utilizing densitometry in the quantification of toxicologically interesting substances is presented.

B. MAJOR DRUGS OF ABUSE

1. TLC vs. Immunoassay

Screening and confirmation of the major drugs of abuse in urine samples are the basic duties of forensic toxicology laboratories, and particularly in the United States, many laboratories specialize in this area.[20,109] Previously TLC was very much involved in the screening step, but immunoassays, especially the enzyme-multiplied immunoassay technique (Emit™ by Syva, San Jose, CA) and the fluorescence polarization immunoassay (TDx™ by Abbott, Irving, TX), largely replaced TLC in the 1980s because they are easier to use and standardize than manual TLC.

However, immunoassays are not usually specific for a certain compound but, rather, for a class of compounds, e.g., amphetamines or opiates. Immunoassays show a different degree of reactivity with each compound in the particular class and may also show cross-reactivity with

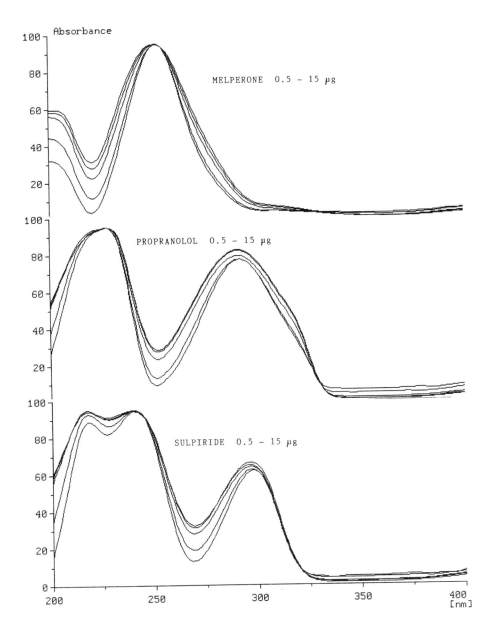

FIGURE 10.8 Dependence of *in situ* UV spectra on the substance amount on the plate (see Table 10.9).

other compounds outside the class, and this may lead to apparent discrepancies with chromatographic methods that measure well-defined compounds. Today, a major application area of TLC is in confirming immunoassay results in situations where GC-MS is not considered legally imperative.

Ferrara et al.[110] published a comprehensive and detailed statistical evaluation of several immunochemical and chromatographic techniques applied to drugs of abuse testing in urine. The Toxi-Lab™ systems that represented TLC in the study showed generally higher specificity and produced less false-positive findings than the immunoassays. On the other hand, the sensitivity of Toxi-Lab™ was generally lower; consequently, the rate of false-negatives was higher than with the immunoassays. The study stressed the high reliability that can be achieved by combining an immunoassay and a chromatographic technique.

TABLE 10.9
Concentration Dependence of in situ UV Spectra on the Krim System: Correlation with Library Spectra (4 µg)

Drug	Amount (µg)	Correlation value	Hit list Position[a]
Melperone			
	0.5	0.960	1st
	1.0	0.987	1st
	5.0	0.999	1st
	10.0	0.998	1st
	15.0	0.996	1st
Propranolol			
	0.5	0.975	1st
	1.0	0.990	1st
	5.0	0.999	1st
	10.0	0.997	1st
	15.0	0.996	1st
Sulpiride			
	0.5	0.973	1st
	1.0	0.992	1st
	5.0	0.999	1st
	10.0	0.999	1st
	15.0	0.999	1st

Hit list generated against a 100-compound drug library by hR_f pre-search (+/– 7 units) followed by spectrum correlation search.

A feature that favors TLC is that urine adulteration with numerous chemicals, such as salt, bleach, soap, and eyedrops containing tetryzoline, may cause false-negative findings by immunoassays. There is also an efficient commercial adulterant product available containing glutaraldehyde (UrinAid).[111]

2. Amphetamines

Gübitz and Wintersteiger[112] described sensitive methods for amphetamine-like compounds by combining high performance TLC with group specific fluorescent reagents, i.e., fluorescamine, NBD-Cl or DANS-Cl. By carrying out the reactions *in situ* at the origin, detection limits ranging from 20 to 50 ng per spot were obtained for pure compounds. The methods were stated to be applicable also to urine samples.

Various ring-substituted amphetamine derivatives were differentiated from each other and from over-the-counter drugs using the Toxi-Lab™ procedure with two solvent systems.[113] The four-stage Toxi-Lab™ sequence was used to visualize the 19 test substances. The detection limit of amphetamine in urine was 0.5 µg/ml, and the detection limits of the other derivatives were 0.5 to 3.0 µg/ml. Methylenedioxy derivatives could be detected at a level of 0.25 µg/ml with the use of two additional reagents, gallic acid and chromotropic acid.

Soine et al.[94] reported that the 11 side chain positional amphetamine isomers studied could be distinguished using a combination of spot tests, TLC, and MS. The visualization reagents used in the TLC procedure were iodine vapor, fluorescamine, iodoplatinate, ceric sulfate in sulfuric acid, iodoplatinate oversprayed with ceric sulfate, and potassium permanganate. Neuninger[114] described a method for the identification of 29 stimulants and hallucinogens by spot tests and TLC. The TLC procedure was comprised of six TLC systems and six visualization reagents: ninhydrin, ninhydrin/cadmium acetate, iodoplatinate, potassium permanganate, sulfuric acid, and Marquis reagent. The detection limits were 0.3 to 0.5 µg per spot for pure substances.

Ojanperä et al.[70] described a screening procedure for 25 amphetamine-type drugs using the UniTox systems and the visualization reagent Fast Black K salt (FBK). All the derivatives were equally reactive with FBK, the detection limit being 0.1 µg per spot with pure substances, and could be readily classified into primary and secondary amines by color. With a set of 78 drugs visualizable with FBK, the number of potential candidates for identification could be reduced to an average of two, as evaluated by the MLL method. Especially, the method enabled the differentiation of abused amphetamines from therapeutic drugs. The general characterization of amines by FBK has been thoroughly studied elsewhere.[87,88]

In conclusion, there are many powerful TLC-based methods available for the detection and identification of amphetamines. However, except for the Toxi-Lab™ system, there are no recent papers that apply these methods to amphetamine testing in biological material. The TLC methods could obviously compete succesfully with immunoassays as the cut-off limit for amphetamine in urine by both Emit™ and TDx™ is 0.3 µg/ml.

3. Cannabinoids

Willson and Kogan et al.[115,116] reported a method for the confirmation of Emit™ cannabinoid assay results using solid phase extraction with a special bonded phase column followed by TLC. The visualization of the main urinary cannabinoid, 11-nor-delta^9-tetrahydrocannabinol-9-carboxylic acid (THCA), was carried out with Fast Blue RR salt. A 100% agreement between the Emit™ test, with a cut-off concentration of 20 ng/ml, and TLC results was observed when 20 ml of alkali-hydrolyzed urine was used with the latter method. Later evaluations of this TLC method, applied to 10 ml urine samples, verified its applicability to the confirmation of immunoassay-positive samples in agreement with the GC-MS technique.[117,118] By utilizing scanning densitometry, the detection limit could be as low as 1 ng/ml.[119] An analogous octadecyl bonded phase extraction TLC method was described by von Meyer,[120] reporting a THCA detection limit of 10 ng/ml in 10 ml of urine.

Sutheimer et al.[121] compared the Toxi-Lab™ cannabinoid procedure with two Emit™ immunoassays. The Toxi-Lab™ method utilized liquid extraction and visualization with Fast Blue BB salt and diethylamine vapors, followed by hydrochloric acid vapors. The evaluation of several hundred urine specimens indicated that the combination of Emit™ and Toxi-Lab™ assure a reliable urinary cannabinoid analysis above a concentration of 25 ng/ml, using 10 ml of urine for TLC. Another comparison of the TLC method with various immunoassays and with GC-MS confirmed a good correlation.[122]

In an improved sample preparation method by Toxi-Lab™,[123] the hydrolyzed urine samples were aspirated directly through a porous bonded silica layer at the origin of the TLC plate. The concentrated THCA was then chromatographed and visualized with Fast Blue BB salt. The TLC method had a 92% agreement with GC-MS in confirming Emit™-positive findings obtained with a 20 ng/ml cut-off concentration. It had a throughput of about 60 samples per hour when performed by one technician.

Vinson and Lopatofsky[124] described a procedure for the detection and quantitative analysis of THCA in urine, utilizing automated solid phase extraction with XAD-2 resin, chromatography on high performance plates with a preadsorbent area, and scanning densitometry at 485 nm. The method had an average recovery of 96% and an average precision of 6.3%. The visual detection limit, using Fast Blue BB salt, was as low as 2 ng/ml using 2 ml of urine. A good correlation between the TLC method and GC-MS existed.

Another method for THCA also used high performance plates and Fast Blue BB salt but relied on liquid extraction.[125] The detection limit was 5 ng/ml when 2 ml of urine was used, and the average extraction recovery was 71%, as measured by densitometry at 520 nm. No drugs, metabolites, or endogenous urinary substances were found to interfere with the procedure. A limited evaluation showed good agreement with GC-MS.

4. Cocaine and Metabolites

Screening for cocaine abuse is usually based on the detection of its metabolite benzoylecgonine (BE) in urine, the cut-off concentration of immunoassays being generally 0.3 µg/ml. Sherma et al.[126] reported a method for the screening of BE in 5 ml urine specimens at 0.2 µg/ml, using cyclodextrin solid phase extraction and high performance TLC with a preadsorbent area. Spraying with Dragendorff's reagent followed by sulfuric acid visualized BE as a narrow orange zone. The recovery was approximately 50%, as measured by densitometry.

Kogan et al.[127] developed a high performance TLC method with solid phase extraction for confirming the Emit™ urine cocaine metabolite assays. The authors stated that Ludy Tenger reagent followed by sulfuric acid is the chromogenic reagent of choice for visualizing cocaine metabolites. Like Dragendorff's reagent, it contains bismuth, which produces a more intense color reaction to BE than iodoplatinate. Unlike Dragendorff's reagent, which produces an orange background with orange to brown metabolite bands, Ludy Tenger reagent produces highly colored red to orange bands on a bright yellow background. The method allowed the confirmation of 97% of the Emit™-positive findings obtained with a cut-off concentration of 0.3 µg/ml.

Funk and Dröschel[128] described a quantitative high performance TLC method for cocaine and its principal metabolites. The mobile phase was optimized according to the Prisma model. Qualitative detection was performed with a modified Dragendorff's reagent. Quantitative determination at 50 ng per spot was performed by fluorescence at 313 nm/> 390 nm after heating the developed chromatogram without previous drying. No biological samples were studied.

Cocaethylene (ethylcocaine) is a toxicologically significant derivative of cocaine, formed by transesterification of cocaine with ethanol when they are consumed together. Bailey[129] described a TLC method for cocaethylene and cocaine in 35 ml urine samples using XAD-2 solid phase extraction and visualization with iodoplatinate. The average recovery was 70%, and the detection limit of both cocaethylene and cocaine was 0.2 µg/ml. The method was succesfully applied to urine samples of 13 patients who had ingested cocaine with ethanol.

5. Lysergide

Blum et al.[130] developed a method for lysergic acid diethylamide or lysergide (LSD) in urine, using high performance TLC plates with a preadsorptive region. Before the chromatographic separation, a single-step liquid extraction with alkaline wash was performed. Instrumental detection was by fluorescence at 313 nm/320–400 nm, followed by chromogenic visualization using the p-dimethylaminobenzaldehyde reagent. The detection limit was less than 1 ng/ml. Of the 48 urine specimens examined, 28 were positive by both TLC and radioimmunoassay (RIA), seven were negative by both methods, and 13 specimens gave a positive RIA response with negative TLC findings. The RIA cut-off concentration was 0.5 ng/ml, but as with immunoassays in general there is cross reactivity with metabolites and possibly with other substances.

6. Opiates

Wang and Tham[131] reported a semiquantitative screening method for morphine in 1 ml of enzyme hydrolyzed urine. The sample was allowed to react with 4-dimethylaminoazobenzene-4'-sulfonyl chloride (DAB-Cl), and the orange-colored morphine derivative was analyzed by TLC after liquid extraction. The visual detection limit was 75 ng/ml, and the limit of quantification by densitometry was 375 ng/ml. The analysis of 20 urine samples of heroin-abuse suspects revealed a good correlation with another TLC method using iodoplatinate for visualization.

Blakesley et al.[132] described a TLC method for the detection of commonly abused basic drugs, including morphine, the specific heroin metabolite 6-acetylmorphine, codeine, and dihydrocodeine, in 20 ml urine samples. The extraction was based on the ToxElut tubes, and visualization was performed with dilute sulfuric acid followed by iodoplatinate, utilizing the various color shades developed. Low concentrations of morphine were visualized only after 10 to 20 min. The detection limit was 1 µg/ml.

Nadkarni et al.[133] evaluated the Toxi-Lab™ system for morphine and heroin abuse screening in urine. The authors noticed that the detection limit of 1 µg/ml for morphine, stated by the manufacturer, was obtained at stage four only in the Toxi-Lab™ A visualization procedure. A definite morphine reaction at all four detection stages was observed only at concentrations of 4 µg/ml and above. The Emit™ cut-off concentration for morphine is 0.3 µg/ml. When urine samples from heroin addicts were treated with β-glucuronidase, more positives were found by the TLC method, but some other drugs initially found were not detected on re-analysis.

Wolff et al.[134] presented a method for detecting opioids, cocaine, and amphetamine in 2 ml of urine, using octadecyl silica solid phase extraction. Two horizontal TLC systems were used, and visualization was carried out with iodoplatinate. The detection limit for morphine and codeine was 50 ng per spot with pure substances. The authors also evaluated their method with case urine samples and compared it with the Toxi-Lab™ A procedure, Emit™, and HPLC.[135] The Wolff method was in good agreement with Emit™ and HPLC but disagreed with the Toxi-Lab™ method in opiate analysis. This was partially due to the lower detection limits of the Wolff method, 0.1 to 0.5 µg/ml for morphine, codeine, and dihydrocodeine, while the Toxi-Lab™ detection limit was 0.5 to 1.0 µg/ml. Furthermore, in some cases the Toxi-Lab™ method was unable to differentiate between various opiates and between opiates and matrix components.

C. OTHER SUBSTANCES

In addition to drugs of abuse, there are specific groups of substances that may not be satisfactorily analyzed with the general screening systems. Dedicated TLC methods are available for such groups, including alkaloids,[86] anabolic steroids,[136] benzodiazepines,[137–140] β-adrenergic blocking drugs,[141–144] laxatives,[145] pesticides,[146,147] phenols,[148] phenothiazines,[149] solvent metabolites,[150] sulfa drugs,[151] and sympathomimetics.[152] TIAFT recommends a special system for benzodiazepines and two special systems for pesticides that can be used to supplement the general screening systems.[13] These systems utilize the R_f correction procedure described earlier. In the benzodiazepine system, urine is hydrolyzed to yield the corresponding aminobenzophenones, which are then extracted, separated by TLC, and photolytically dealkylated. The resulting products are diazotized and coupled with an azo-dye using the Bratton-Marshall reagent. The pesticide systems include R_f values and visualization data, obtained using six reagents, for 170 pesticides.

Table 10.10 lists recent instrumental quantitative TLC methods for selected substances in biological materials. Although it is not comprehensive, the table is intended to provide examples of various quantitative determinations realized by TLC and promote further investigations in this field. Hyphenated techniques other than TLC-UV are also available for target analysis. TLC has been successfully coupled with Fourier transform IR,[179] Raman spectrometry,[180] and MS.[181] These techniques allow an interesting spot, which is detected but not identified at the screening stage, to be directly submitted for further structural elucidation.

TABLE 10.10
Selected Instrumental Quantitative TLC Analyses in Biological Materials[a]

Substance	Category	Matrix	Visualization reagent	Quantification	Reference
Astemizole	Antihistamine	Plasma	—	UV	153
Baclofen and its fluoro-analog	Muscle relaxant	Plasma, urine	Benoxaprofen chloride	pre-chromatographic derivatization, fl.	154
Dexamethasone	Glucocorticoid	Plasma	—	UV	155
Diazepam and metabolites	Anxiolytic	Urine	—	UV	156
Erythromycin and metabolites	Antibacterial	Urine, plasma	Xanthydrol	post-chromatographic visualization, VIS(525 nm)	157
Gentiopicroside	Chinese herb ingredient	Biological fluids and tissues	—	UV	158
Halofantrine	Antimalarial	Erythrocytes, plasma, serum	—	UV	159

TABLE 10.10 (continued)
Selected Instrumental Quantitative TLC Analyses in Biological Materials[a]

Substance	Category	Matrix	Visualization reagent	Quantification	Reference
Hydrochlorothiazide	Diuretic	Urine	Diazotization+naphthylethylenediamine	UV	160
Ifosfamide and metabolites	Antineoplastic	Urine, cerebrospinal fluid, plasma	Nitrobenzylpyridine + KOH	post-chromatographic visualization, photography-VIS(500 nm)	161
Levoprotiline	Antidepressant	Plasma	NBD–chloride	pre-chromatographic derivatization, fl.	162
Nicotine and metabolites	Tobacco ingredient	Serum	—	pre-chromatographic derivatization, UV	163
Norfloxacin	Antibacterial	Plasma, urine	—	fl., fl. enhancement	164
Omeprazole	Antiulcerative	Plasma	—	UV	165
Oxprenolol	Antihypertensive	Blood	Fast Blue B salt, naphtoquinonesulfonic acid, formaldehyde-sulfuric acid	UV	166
Paraquat	Herbicide	Plasma, urine	—	TLC-FID	167
Pentoxifylline	Vasodilator	Dog serum	—	UV	168
Phosacetim	Rodenticide	Rabbit blood and tissues	Bromophenol blue and others	UV	169
Pyrazinamide	Antibacterial	Serum	—	UV	170
Quinine, quinidine, hydroquinidine[b]	Antimalarial	Plasma	Natural fl.	fl., fl. enhancement	171
Salbutamol	Bronchodilator	Plasma, urine	Dimethyl-*p*-phenylenediamine	pre-chromatographic visualization, VIS(650 nm)	172
Salicylates and metabolites[c]	Analgesic	Urine	Ferric chloride	post-chromatographic visualization, VIS(550 nm)	173
Tetrodotoxin	Marine toxin	Serum, urine	—	TLC-FID	174
Thiothixene	Antipsychotic	Plasma	Ozone-oxygen	post-chromatographic derivatization, fl.	175
Trazodone	Antidepressant	Serum	Natural fl.	fl.	176
Trinitrotoluene and metabolites	Explosive	Urine	ethylenediamine	UV	177
Valproic acid	Antiepileptic	Plasma	2-naphtacyl bromide	pre-chromatographic derivatization, UV	178

[a] UV absorbance detection (UV), fluorescence detection (fl.), absorbance detection at the visible range (VIS).
[b] Over-pressure layer chromatography was used.
[c] The method is amenable to quantification.

REFERENCES

1. **Vuori, E., Ojanperä, I., Ruohonen, A., and Penttilä, A.,** Myrkytyskuolemat vuosina 1988–1990, *Suomen Lääkäril.*, 47, 1217, 1992.
2. **Vuori, E. and Ojanperä, I.,** Myrkytyskuolemat ja niiden oikeuskemiallinen tutkiminen, *Kem.-Kemi*, 20, 302, 1994.
3. **Ceder, G., Holmgren, P., Steentoft, A., Vuori, E., Teige, B., Kaa, E., Kristinsson, J., Pikkarainen, J., and Wethe, G.,** Dödliga förgiftningar i Norden – narkomaner en utsatt grupp, *Nord. Med.*, 104, 224, 1989.
4. **Flanagan, R. J., Widdop, B., Ramsey, J. D., and Loveland, M.,** Analytical toxicology, *Human Toxicol.*, 7, 489, 1988.

5. **Sunshine, I.,** Analytical toxicology, in *Casarett and Doull's Toxicology,* 3rd ed., Klaassen, C. D., Amdur, M. O., and Doull, J., Eds., Macmillan, New York, 1986, Chap. 27.
6. **Touchstone, J. C.,** History of chromatography, *J. Liq. Chromatogr.,* 16, 1647, 1993.
7. **Majors, R. E.,** Thin-layer chromatography — a survey of the experts, *LC-GC Int.,* 3, 8, 1990.
8. **Treiber, L. R.,** Utility of thin layer chromatography as an analytical tool, *J. Chromatogr. Sci.,* 24, 220, 1986.
9. **Poole, C. F. and Poole, S. K.,** Instrumental thin-layer chromatography, *Anal. Chem.,* 66, 27A, 1994.
10. **Sherma, J.,** Modern high performance thin-layer chromatography, *J. AOAC Int.,* 77, 297, 1994.
11. **Moffat, A. C., Ed.,** *Clarke's Isolation and Identification of Drugs,* 2nd ed., Pharmaceutical Press, London, 1986.
12. **Müller, R. K., Ed.,** *Toxicological Analysis,* Edition Molinapress, Leipzig, 1991.
13. **De Zeeuw, R. A., Franke, J. P., Degel, F., Machbert, G., Schütz, H., and Wijsbeek, J., Eds.,** *Thin Layer Chromatographic R_f Values of Toxicologically Relevant Substances on Standardized Systems,* 2nd ed., DFG/TIAFT, VCH, Weinheim, 1992.
14. **Jork, H., Funk, W., Fischer, W., and Wimmer, H.,** *Thin Layer Chromatography: Reagents and Detection Methods,* VCH, Weinheim, Vol. 1a, 1990, and Vol. 1b, 1994.
15. **Minty, P. S. B.,** Drug analyses in a department of forensic medicine and toxicology, in *Analytical Methods in Human Toxicology,* Part 2, Curry, A. S., Ed., Verlag Chemie, Weinheim, 1986, Chap. 6.
16. **McDonald, P. A. and Gough, T. A.,** The use of chromatography in the analysis of illicit drugs, in *Analytical Methods in Forensic Chemistry,* Ho, M. H., Ed., Ellis Horwood, Chichester, 1990, Chap. 11.
17. **Franke, J. P. and De Zeeuw, R. A.,** The multi-technique approach, in *The Analysis of Drugs of Abuse,* Gough, T. A., Ed., Wiley, Chichester, 1991, Chap. 3.
18. **Roberson, J. C.,** The use of thin-layer chromatography in the analysis of drugs of abuse, in *The Analysis of Drugs of Abuse,* Gough, T. A., Ed., Wiley, Chichester, 1991, Chap. 1.
19. **Ng, L. L.,** Pharmaceuticals and drugs, in *Handbook of Thin-Layer Chromatography,* Sherma, J. and Fried, B., Eds., Marcel Dekker, New York, 1991, Chap. 23.
20. **Kwong, T. C., Chamberlain, R. T., Frederick, D. L., Kapur, B., and Sunshine, I.,** Critical issues in urinalysis of abused substances: report of the substance-abuse testing committee, *Clin. Chem.,* 34, 605, 1988.
21. **Singh, A. K., Granley, K., Ashraf, M., and Mishra, U.,** Drug screening and confirmation by thin layer chromatography, *J. Planar Chromatogr.–Mod. TLC,* 2, 410, 1989.
22. **Ojanperä, I.,** Toxicological drug screening by thin-layer chromatography, *Trends Anal. Chem.,* 11, 222, 1992.
23. **De Zeeuw, R. A., Hartstra, J., and Franke, J. P.,** Potential and pitfalls of chromatographic techniques and detection modes in substance identification for systematic toxicological analysis, *J. Chromatogr.,* 674, 3, 1994.
24. **Sherma, J.,** Planar chromatography, *Anal. Chem.,* 66, 67R, 1994.
25. **Gilpin, R. K. and Pachla, L. A.,** Pharmaceuticals and related drugs, *Anal. Chem.,* 65, 117R, 1993.
26. **Brettell, T. A. and Saferstein, R.,** Forensic science, *Anal. Chem.,* 65, 293R, 1993.
27. **Ojanperä, I., Rasanen, I., and Vuori, E.,** Automated quantitative screening for acidic and neutral drugs in whole blood by dual-column capillary gas chromatography, *J. Anal. Toxicol.,* 15, 204, 1991.
28. **Rasanen, I., Ojanperä, I., Vuori, E., and Hase, T. A.,** An homologous series of benzodiazepine retention index standards for gas chromatography, *J. High Resol. Chromatogr.,* 16, 495, 1993.
29. **Rasanen, I., Ojanperä, I., and Vuori, E.,** Comparison of four homologous retention index standard series for gas chromatography of basic drugs, *J. Chromatogr.,* 693, 69, 1995.
30. **Binder, S. R., Regalia, M., Biaggi-McEachern, M., and Mazhar, M.,** Automated liquid chromatographic analysis of drugs in urine by on-line sample cleanup and isocratic multicolumn separation, *J. Chromatogr.,* 473, 325, 1989.
31. **Bogusz, M. and Erkens, M.,** Reversed-phase high-performance liquid chromatographic database of retention indices and UV spectra of toxicologically relevant substances and its interlaboratory use, *J. Chromatogr.,* 674, 97, 1994.
32. **Maurer, H. H.,** Systematic toxicological analysis of drugs and their metabolites by gas chromatography–mass spectrometry, *J. Chromatogr.,* 580, 3, 1992.
33. **Evans, M. A. and Baselt, R. C.,** Principles of toxicant disposition, in *Introduction to Forensic Toxicology,* Cravey, R. H. and Baselt, R. C., Biomedical Publications, Davis, 1991, Chap. 4.
34. **Tagliaro, F., Chiarotti, M., and Deyl, Z., Eds.,** *Forensic Sci. Int.,* 63 (1–3), 1993 (Special issue: Hair analysis as a diagnostic tool for drugs of abuse investigation).
35. **McDowall, R. D.,** Sample preparation for biomedical analysis, *J. Chromatogr.,* 492, 3, 1989.
36. **McCurdy, H. H.,** Enzymic digestion of biological specimens for drug analysis, *Forensic Sci. Rev.,* 5, 68, 1993.
37. **Meatherall, R.,** Optimal enzymatic hydrolysis of urinary benzodiazepine conjugates, *J. Anal. Toxicol.,* 18, 382, 1994.
38. **Osselton, M. D., Hammond, M. D., and Twitchett, P. J.,** The extraction and analysis of benzodiazepines in tissues by enzymic digestion and high-performance liquid chromatography, *J. Pharm. Pharmacol.,* 29, 460, 1977.
39. **Schill, G., Ehrsson, H., Vessman, J., and Westerlund, D.,** *Separation Methods for Drugs and Related Organic Compounds,* 2nd ed., Swedish Pharmaceutical Press, Stockholm, 1984, p. 6.
40. **Tomlinson, E.,** Ion-pair extraction and high-performance liquid chromatography in pharmaceutical and biomedical analysis, *J. Pharm. Biomed. Anal.,* 1, 11, 1983.
41. **Ojanperä, I. and Vuori, E.,** Thin layer chromatographic analysis of basic and quaternary drugs extracted as bis(2-ethylhexyl)phosphate ion-pairs, *J. Liq. Chromatogr.,* 10, 3595, 1987.

42. Ojanperä, I. and Vuori, E., Identification of drugs in autopsy liver samples by instrumental qualitative thin-layer chromatography, *J. Chromatogr.*, 674, 147, 1994.
43. McDowall, R. D., Pearce, J. C., and Murkitt, G. S., Liquid-solid sample preparation in drug analysis, *J. Pharm. Biomed. Anal.*, 4, 3, 1986.
44. Scheurer, J. and Moore, C. M., Solid-phase extraction of drugs from biological tissues – a review, *J. Anal. Toxicol.*, 16, 264, 1992.
45. Chen, X.-H., Wijsbeek, J., Franke, J.-P., and De Zeeuw, R. A., A single-column procedure on bond elut certify for systematic toxicological analysis of drugs in plasma and urine, *J. Forensic Sci.*, 37, 61, 1992.
46. Chen, X.-H., Franke, J.-P., Wijsbeek, J., and De Zeeuw, R. A., Isolation of acidic, neutral, and basic drugs from whole blood using a single mixed-mode solid-phase extraction column, *J. Anal. Toxicol.*, 16, 351, 1992.
47. Moffat, A. C. and Smalldon, K. W., Optimum use of paper, thin-layer and gas chromatography for the identification of basic drugs. 2. Paper and thin-layer chromatography., *J. Chromatogr.*, 90, 9, 1974.
48. Owen, P., Pendlebury, A., and Moffat, A. C., Choice of thin-layer chromatographic systems for the routine screening for neutral drugs during toxicological analyses, *J. Chromatogr.*, 161, 187, 1978.
49. Schepers, P. G. A. M., Franke, J. P., and de Zeeuw, R. A., System evaluation and substance identification in systematic toxicological analysis by the mean list length approach, *J. Anal. Toxicol.*, 7, 272, 1983.
50. Massart, D. L., The use of information theory for evaluating the quality of thin-layer chromatographic separations, *J. Chromatogr.*, 79, 157, 1973.
51. De Clerq, H. and Massart, D. L., Evaluation and selection of optimal solvents and solvent combinations in thin-layer chromatography. Application of the method to basic drugs, *J. Chromatogr.*, 115, 1, 1975.
52. Müller, R. K., Mückel, W., Wallenborn, H., Weihermüller, A., Weihermüller, C., and Lauermann, I., Objektive Kriterien zur Auswahl optimaler chromatographischer Systeme. 1. Mitteilung: Optimierung von DC-Fliessmitteln auf Grund der R_f Verteilung grösserer Stoffgruppen, *Beitr. Gerichtl. Med.*, 34, 265, 1976.
53. Musumarra, G., Scarlata, G., Romano, G., and Clementi, S., Identification of drugs by principal components analysis of R_f data obtained by TLC in different eluent systems, *J. Anal. Toxicol.*, 7, 286, 1983.
54. Romano, G., Caruso, G., Musumarra, G., Pavone, D., and Cruciani, G., Qualitative organic analysis. Part 3. Identification of drugs and their metabolites by PCA of standardized TLC data, *J. Planar Chromatogr.*, 7, 233, 1994.
55. Galanos, D. S. and Kapoulas, V. M., The paper chromatographic identification of compounds using two reference compounds, *J. Chromatogr.*, 13, 128, 1964.
56. Moffat, A. C., The standardization of thin-layer chromatographic systems for the identification of basic drugs, *J. Chromatogr.*, 110, 341, 1975.
57. Van Wendel de Joode, M. D., Hindriks, H., and Lakeman, J., R_f correction in thin-layer chromatography, *J. Chromatogr.*, 170, 412, 1979.
58. Geiss, F., *Fundamentals of Thin Layer Chromatography*, Hüthig, Heidelberg, 1987, 120.
59. Erdmann, F., Brose, C., and Schütz, H., TLC screening program for 170 commonly used pesticides using the corrected R_f value, *Int. J. Legal Med.*, 104, 25, 1990.
60. Ojanperä, I. and Jänchen, P., The application of instrumental qualitative thin-layer chromatography to drug screening, *LC-GC Int.*, 7, 164, 1994.
61. De Zeeuw, R. A., Franke, J. P., Dik, E., ten Dolle, W., and Kam, B. L., Impact of tropical conditions on thin layer chromatography in analytical toxicology: high temperatures and moderate humidities, *J. Forensic Sci.*, 37, 984, 1992.
62. De Zeeuw, R. A., Franke, J. P., van Halem, M., Schaapman, S., Logawa, E., and Siregar, C. J. P., Impact of hot and humid tropical conditions on Toxi-Lab™ thin-layer chromatographic systems for the screening of basic and neutral drugs, *J. Anal. Toxicol.*, 18, 402, 1994.
63. Bogusz, M., Klys, M., Wijsbeek, J., Franke, J. P., and De Zeeuw, R. A., Impact of biological matrix and isolation methods on detectability and interlaboratory variations of TLC R_f-values in systematic toxicological analysis, *J. Anal. Toxicol.*, 8, 149, 1984.
64. Bogusz, M., Gierz, J., De Zeeuw, R. A., and Franke, J. P., Influence of the biological matrix on retention behavior in thin-layer chromatography: evidence of systematic differences between pure and extracted drugs, *J. Chromatogr.*, 342, 241, 1985.
65. Jones, D. W., Adams, D., Martel, P. A., and Rousseau, R. J., Drug population in one thousand geographically distributed urine specimens, *J. Anal. Toxicol.*, 9, 125, 1985.
66. Jarvie, D. R. and Simpson, D., Drug screening: evaluation of the Toxi-Lab™ TLC system, *Ann. Clin. Biochem.*, 23, 76, 1986.
67. Whitter, P. D. and Cary, P. L., A rapid method for the identification of acidic, neutral, and basic drugs in post-mortem liver specimens by Toxi-Lab™, *J. Anal. Toxicol.*, 10, 68, 1986.
68. Harper, J. D., Martel, P. A., and O'Donnel, C. M., Evaluation of a multiple-variable thin-layer and reversed-phase thin-layer chromatographic scheme for the identification of basic and neutral drugs in an emergency toxicology setting, *J. Anal. Toxicol.*, 13, 31, 1989.
69. Ojanperä, I., Vartiovaara, J., Ruohonen, A., and Vuori, E., Combined use of normal and reversed phase thin layer chromatography in the screening for basic and quaternary drugs, *J. Liq. Chromatogr.*, 14, 1435, 1991.

70. **Ojanperä, I., Lillsunde, P., Vartiovaara, J., and Vuori, E.,** Screening for amphetamines with a combination of normal and reversed phase thin layer chromatography and visualization with Fast Black K salt, *J. Planar Chromatogr.*, 4, 373, 1991.
71. **Ojanperä, I., Hyppölä, R., Vartiovaara, J., and Vuori, E.,** Screening for acidic and neutral drugs in liver and urine samples by instrumental qualitative thin-layer chromatography, *J. Anal. Toxicol.* (submitted).
72. **Bigalke, H.-J., Ebel, S., Ullrich, W., and Voelkl, S.,** Identification of drugs by AMD/HPTLC and spectral properties, in *Proceedings of the Fourth International Symposium on Instrumental High Performance Thin-Layer Chromatography* (Selvino/Bergamo, Italy, September 22-25, 1987), Traitler, H., Studer, A., and Kaiser, R. E., Eds., Institute for Chromatography, Bad Durkheim, 1987, Chap. 8.
73. **Hauck, H. E., Junker-Buchheit, A., and Wenig, R.,** Reproducibility of Silica Gel 60 precoated plates in TLC and HPTLC, *LC-GC Int.*, 8, 34, 1995.
74. **Franke, J.-P., De Zeeuw, R. A., and Schepers, P. G. A. M.,** Retrieval of analytical data and substance identification in systematic toxicological analysis by the mean list length approach, *J. Forensic Sci.*, 30, 1074, 1985.
75. **Gill, R., Law, B., Brown, C., and Moffat, A. C.,** A computer search system for the identification of drugs using a combination of thin-layer chromatographic, gas-liquid chromatographic and ultraviolet spectroscopic data, *Analyst*, 110, 1059, 1985.
76. **Clarke, E. G. C.,** Isolation and identification of alkaloids, in *Methods of Forensic Science*, Vol 1., Lundquist, F., Ed., Interscience (Wiley), 1962, p. 1.
77. **Stahl, E.,** *Dünnschicht-Chromatographie. Ein Laboratoriumshandbuch*, 2nd ed., Springer-Verlag, Berlin, 1967, Chap. Z.
78. *Dyeing Reagents for Thin Layer and Paper Chromatography,* E. Merck, Darmstadt, 1980.
79. **Daldrup, T. and Rickert, A.,** Arzneimittel–und Drogenscreening aus Urin mittels DC unter besonderer Berücksichtigung von Reagentien mit geringer toxischer Belastung für Laborpersonal und Umwelt, *Fresenius Z. Anal. Chem.*, 334, 349, 1989.
80. **Lillsunde, P. and Korte, T.,** Comprehensive drug screening in urine using solid-phase extraction and combined TLC and GC/MS identification, *J. Anal. Toxicol.*, 15, 71, 1991.
81. **Touchstone, J. C.,** *Practice of Thin Layer Chromatography*, 3rd ed., Wiley, New York, 1992, Chap. 9.
82. **Klimisch, H.-J. and Stadler, L.,** Microquantitative determination of aliphatic amines with 7-chloro-4-nitrobenzo-2-oxa-1,3-diazole, *J. Chromatogr.*, 90, 141, 1974.
83. **Reisch, J., Kommert, H.-J., and Clasing, D.,** Tüpfelnachweis für einige Alkaloide und N-haltige Arzneistoffe mit 7-Chlor-4-nitrobenzo-oxa-1,3-diazol (NBD-Chlorid), *Pharm. Ztg.*, 115, 752, 1970.
84. **Ross, J. H.,** 2,6-Dichloroquinone 4-chloroimide as a reagent for amines and aromatic hydrocarbons on thin-layer chromatograms, *Anal. Chem.*, 40, 2138, 1968.
85. **Habib, A-A. M.,** False-positive alkaloid reactions, *J. Pharm. Sci.*, 69, 37, 1980.
86. **Svendsen, A. B.,** Thin layer chromatography of alkaloids, *J. Planar Chromatogr.*, 2, 8, 1989.
87. **Ojanperä, I., Wähälä, K., and Hase, T. A.,** Fast Black K salt: a versatile thin-layer chromatographic visualisation reagent for the differentiation of aliphatic amines, *Analyst*, 115, 263, 1990.
88. **Ojanperä, I., Wähälä, K., and Hase, T. A.,** Characterization of amines by Fast Black K salt in thin-layer chromatography, *Analyst*, 117, 1559, 1992.
89. **Gough, T. A. and Baker, P. B.,** Identification of major drugs of abuse using chromatography, *J. Chromatogr. Sci.*, 20, 289, 1982.
90. **De Faubert Maunder, M. J.,** Preservation of cannabis thin-layer chromatograms, *J. Chromatogr.*, 100, 196, 1974.
91. **Eberhardt, H. and Debackere, M.,** Der Nachweis von zentral stimulierenden Substanzen in reiner Form und nach Körperpassage mit der Dünnschichtchromatographie, *Arzneim.-Forsch.*, 15, 929, 1965.
92. **Udenfried, S., Stein, S., Böhlen, B., Dairman, W., Leimgruber, W., and Weigele, M.,** Fluorescamine: a reagent for assay of amino acids, peptides, proteins, and primary amines in the picomole range, *Science*, 178, 871, 1972.
93. **Shaw, M. A. and Peel, H. W.,** Thin-layer chromatography of 3,4-methylenedioxyamphetamine, 3,4-methylenedioxymethamphetamine and other phenethylamine derivatives, *J. Chromatogr.*, 104, 201, 1975.
94. **Soine, W. H., Thomas, M. N., Shark, R. E., Scott, J., and Agee, D. T.,** Differentiation of side chain positional isomers of amphetamine, *J. Forensic Sci.*, 29, 177, 1984.
95. **Riemer, F. and Daldrup, T.,** Das Forrest-reagens als Detektionsmittel für die Dünnschichtchromatographie, *Pharmazie*, 47, 559, 1992.
96. **Stead, A. H., Gill, R., Wright, T., Gibbs, J. P., and Moffat, A. C.,** Standardized thin-layer chromatographic systems for the identification of drugs and poisons, *Analyst*, 107, 1106, 1982.
97. **Stevens, H. M. and Moffat, A. C.,** A rapid screening procedure for quaternary ammonium compounds in fluids and tissues with special reference to suxamethonium (succinylcholine), *J. Forensic Sci. Soc.*, 14, 141, 1974.
98. **Masoud, A. N.,** Systematic identification of drugs of abuse II: TLC, *J. Pharm. Sci.*, 65, 1585, 1976.
99. **Stevens, H. M.,** Color tests, in *Clarke's Isolation and Identification of Drugs*, 2nd ed., Moffat, A. C., Ed., Pharmaceutical Press, London, 1986, p. 128.
100. **Bastos, M. L., Kananen, G. E., Young, R. M., Monforte, J. R., and Sunshine, I.,** Detection of basic organic drugs and their metabolites in urine, *Clin. Chem.*, 16, 931, 1970.

101. **Dutt, M. C. and Poh, T. T.**, Use of ninhydrin as a spray reagent for the detection of some basic drugs on thin-layer chromatograms, *J. Chromatogr.*, 195, 133, 1980.
102. **Ruohonen, A.**, Salkowski reagent for the visualization of phenothiazines and thioxanthenes in TLC (Personal communication).
103. **Macek, K., Hacaperkova, J., and Kakac, B.**, Systematische Analyse von Alkaloiden mittels Papierchromatographie, *Pharmazie*, 11, 533, 1956.
104. **Ruohonen, A. and Ojansivu, R.-L.**, Substitution of o-tolidine by 3,3′,5,5′-tetramethylbenzidine in the visualization of acidic and neutral drugs in TLC (Personal communication).
105. **Pasha, A. and Vijayashankar, Y. N.**, Thin-layer chromatographic detection of pyrethroid insecticides using o-tolidine, *Analyst*, 118, 777, 1993.
106. **Hegge, H. F. J., Franke, J. P., and De Zeeuw, R. A.**, Combined information from retardation factor (R_f) values and color reactions on the plate greatly enhances the identification power of thin-layer chromatography in systematic toxicological analysis, *J. Forensic Sci.*, 36, 1094, 1991.
107. **Klungel, O. H., Ensing, K., Hegge, H. F. J., Franke, J. P., and De Zeeuw, R. A.**, The potential of the X-rite reflection spectrophotometer in systematic toxicological analysis by thin layer chromatography, *J. Planar Chromatogr.*, 6, 1993, 112.
108. **Ebel, S. and Kang, J. S.**, UV/VIS spectra and spectral libraries in TLC/HPTLC. 1. Background correction and normalization, *J. Planar Chromatogr.*, 3, 42, 1990.
109. **Sunshine, I.**, Preliminary tests for drugs of abuse, *Clin. Chem.*, 34, 331, 1988.
110. **Ferrara, S. D., Tedeschi, L., Frison, G., Brusini, G., Castagna, F., Bernardelli, B., and Soregaroli, D.**, Drugs-of-abuse testing in urine: statistical approach and experimental comparison of immunochemical and chromatographic techniques, *J. Anal Toxicol.*, 18, 278, 1994.
111. **Goldberger, B. A. and Caplan, Y. H.**, Effect of glutaraldehyde (UrinAid) on detection of abused drugs in urine by immunoassay, *Clin. Chem.*, 40, 1605, 1994.
112. **Gübitz, G. and Wintersteiger, R.**, Identification of drugs of abuse by high-performance thin-layer chromatography, *J. Anal. Toxicol.*, 4, 141, 1980.
113. **O'Brien, B. A., Bonicamp, J. M., and Jones, D. W.**, Differentiation of amphetamine and its major hallucinogenic derivatives using thin-layer chromatography, *J. Anal. Toxicol.*, 6, 143, 1982.
114. **Neuninger, H.**, Nachweis und Identifizierung von Phenylethylaminen (Stimulantien und Halluzinogene), *Sci. Pharm.*, 55, 1, 1987.
115. **Willson, N. J., Kogan, M. J., Pierson, D. J., and Newman, E.**, Confirmation of Emit™ cannabinoid assay results by bonded phase adsorption with thin layer chromatography, *J. Toxicol. Clin. Toxicol.*, 20, 465, 1983.
116. **Kogan, M. J., Newman, E., and Willson, N. J.**, Detection of marijuana metabolite 11-nor-delta^9-tetrahydrocannabinol-9-carboxylic acid in human urine by bonded-phase adsorption and thin-layer chromatography, *J. Chromatogr.*, 306, 441, 1984.
117. **Kogan, M. J., Al Razi, J., Pierson, D. J., and Willson, N. J.**, Confirmation of Syva Enzyme Multiple Immunoassay Technique (EMIT™)d.a.u. and Roche Abuscreen^R Radioimmunoassay (RIA)(^{125}I) urine cannabinoid immunoassays by gas chromatographic/mass spectrometric (GC/MS) and bonded-phase adsorption/thin-layer chromatographic (BPA-TLC) methods, *J. Forensic Sci.*, 31, 494, 1986.
118. **Verebey, K., Mulé, S. J., Alrazi, J., and Lehrer, M.**, One hundred EMIT™ positive cannabinoid urine samples confirmed by BPA/TLC, RIA, and GC/MS, *J. Anal. Toxicol.*, 10, 79, 1986.
119. **Vu Duc, T.**, Additional results in the validation of the Bond Elut-THC extraction column-TLC as a confirmation method for EMIT- and RIA-positive cannabinoid urines, *J. Anal. Toxicol.*, 11, 83, 1987.
120. **Von Meyer, L.**, Zum enzymatisch-immunochemischen Nachweis des Haschischkonsums und seiner dünnschichtchromatographischen Absicherung, *Z. Rechtsmed.*, 94, 219, 1985.
121. **Sutheimer, C. A., Yarborough, R., Hepler, B. R., and Sunshine, I.**, Detection and confirmation of urinary cannabinoids, *J. Anal. Toxicol.*, 9, 156, 1985.
122. **Dextraze, P., Griffiths, W. C., Camara, P., Audette, L., and Rosner, M.**, Comparison of fluorescence polarization immunoassay, enzyme immunoassay, and thin-layer chromatography for urine cannabinoid screening, *Ann. Clin. Lab. Sci.*, 19, 133, 1989.
123. **Foltz, R. L. and Sunshine, I.**, Comparison of a TLC method with EMIT™ and GC/MS for detection of cannabinoids in urine, *J. Anal. Toxicol.*, 14, 375, 1990.
124. **Vinson, J. A. and Lopatofsky, D. J.**, A semi-automated extraction and spotting system for drug analysis by TLC I. Procedure for analysis of the major metabolite of delta^9-tetrahydrocannabinol in urine, *J. Anal. Toxicol.*, 9, 6, 1985.
125. **Meatherall, R. C. and Garriott, J. C.**, A sensitive thin-layer chromatographic procedure for the detection of urinary 11-nor-delta^9-tetrahydrocannabinol-9-carboxylic acid, *J. Anal. Toxicol.*, 12, 136, 1986.
126. **Sherma, J., Bernardo, J. E., and Higgs, M. H.**, Screening of benzoylecgonine in urine by Cyclobond solid phase extraction and high performance TLC, *J. Liq. Chromatogr.*, 11, 3135, 1988.
127. **Kogan, M. J., Pierson, D. J., Durkin, M. M., and Willson, N. J.**, Thin-layer chromatography of benzoylecgonine: a rapid qualitative method for confirming the EMIT™ urine cocaine metabolite assay, *J. Chromatogr.*, 490, 236, 1989.
128. **Funk, W. and Dröschel, S.**, Qualitative and quantitative HPTLC determination of cocaine, ecgonine, ecgonine methyl ester and benzoylecgonine, *J. Planar Chromatogr.*, 4, 123, 1991.

129. **Bailey, D. N.,** Thin-layer chromatographic detection of cocaethylene in human urine, *Am. J. Clin. Pathol.*, 101, 342, 1994.
130. **Blum, L. M., Carenzo. E. F., and Rieders, F.,** Determination of lysergic acid diethylamide (LSD) in urine by instrumental high-performance thin-layer chromatography, *J. Anal. Toxicol.*, 14, 285, 1990.
131. **Wang, S. Y. and Tham, S. Y.,** Thin-layer chromatographic and column liquid chromatographic analyses of morphine in urine via dabsylation, *J. Chromatogr.*, 381, 331, 1986.
132. **Blakesley, J., Wood, D., Howse, C., and Spencer-Peet, J.,** A simplified thin-layer chromatography system for the detection of commonly abused basic drugs, *Ann. Clin. Biochem.*, 24, 508, 1987.
133. **Nadkarni, S., Faye, S., and Hay, A.,** Experience with the use of the Toxi-Lab™ TLC system in screening for morphine/heroin abuse, *Ann. Clin. Biochem.*, 24, 211, 1987.
134. **Wolff, K., Sanderson, M. J., and Hay, A. W. M.,** A rapid horizontal TLC method for detecting drugs of abuse, *Ann. Clin. Biochem.*, 27, 482, 1990.
135. **Wolff, K., Sanderson, M. J., Hay, A. W. M., and Barnes, I.,** An evaluation of the measurement of drugs of abuse by commercial and in-house horizontal thin layer chromatography, *Ann. Clin. Biochem.*, 30, 163, 1993.
136. **Daeseleire, E., Vanoosthuyze, K., and Van Peteghem, C.,** Application of high-performance thin-layer chromatography and gas chromatography-mass spectrometry to the detection of new anabolic steroids used as growth promotors in cattle fattening, *J. Chromatogr.*, 674, 247, 1994.
137. **Gruhl, H.,** Zum Screening der Benzo-und Thienodiazepine im Urin mit dem Emit™ ST und der DC im Rahmen eines Suchtestes auf Medikamentenmissbrauch, *Z. Rechtsmed.*, 98, 221, 1987.
138. **Schütz, H.,** Modern screening strategies in analytical toxicology with special regard to new benzodiazepines, *Z. Rechtsmed.*, 100, 19, 1988.
139. **Cserháti, T. and Olajos, S.,** Separation of some benzodiazepine derivatives on CN nano-TLC plates, *Fresenius J. Anal. Chem.*, 337, 60, 1990.
140. **Klimes, J. and Kastner, P.,** Thin layer chromatography of benzodiazepines, *J. Planar Chromatogr.*, 6, 168, 1993.
141. **Abdel-Hamid, M. E., Bedair, M., and Korany, M. A.,** Identification of some β-adrenergic blockers, *Pharmazie*, 40, 494, 1985.
142. **Bonicamp, J. M. and Pryor, L.,** Detection of some beta adrenergic blocking drugs and their metabolites in urine by thin layer chromatography, *J. Anal. Toxicol.*, 9, 180, 1985.
143. **Ojanperä, I. and Ruohonen, A.,** Fast Black K salt: a visualization reagent for thin-layer chromatography of beta-adrenergic blocking drugs, *J. Anal. Toxicol.*, 12, 108, 1988.
144. **Bernhard, W., Fuhrer, A. D., Jeger, A. N., and Rippstein, S. R.,** Nachweis und die Bestimmung von Betarezeptorenblockern in Urin mittels instrumenteller Dünnschicht-Chromatographie, *Fresenius Z. Anal. Chem.*, 330, 458, 1988.
145. **Perkins, S. L. and Livesey, J. F.,** A rapid high-performance thin-layer chromatographic urine screen for laxative abuse, *Clin. Biochem.*, 26, 179, 1993.
146. **De la Vigne, U. and Jänchen, D.,** Determination of pesticides in water by HPTLC using automated multiple development (AMD), *J. Planar Chromatogr.*, 3, 6, 1990.
147. **Brose, C., Rochholz, G., Erdmann, F., and Schütz, H.,** Ein DC-Detektionsprogramm für 178 Pestizide, *Beiträge Ger. Med.*, 50, 221, 1992.
148. **Ferry, J. and Larson, R. A.,** A mixed solvent for rapid TLC analysis of phenolic compounds, *J. Chromatogr. Sci.*, 29, 476, 1991.
149. **Steinbrecher, K.,** Thin layer chromatographic identification of phenothiazine derivative drugs: interlaboratory study, *J. Assoc. Off. Anal. Chem.*, 69, 1030, 1986.
150. **Astier, A.,** Chromatographic determination of volatile solvents and their metabolites in urine for monitoring occupational exposure, *J. Chromatogr.*, 643, 389, 1993.
151. **Jain, R. and Bhatia, A.,** Thin-layer chromatographic separation of some sulpha drugs using acetoacetanilide as a coupling agent, *J. Chromatogr.*, 441, 454, 1988.
152. **Wintersteiger, R., Gübitz, G., and Hartinger, A.,** Derivatization of sympathomimetics with primary amino groups on thin-layer plates by spotting the sample spots with fluorescamine, *Chromatographia*, 13, 291, 1980.
153. **Mangalan, S., Patel, R. B., Gandhi, T. P., and Chakravarthy, B. K.,** Detection and determination of free and plasma protein-bound astemizole by thin-layer chromatography: a useful technique for bioavailability studies, *J. Chromatogr.*, 567, 498, 1991.
154. **Krauss, D., Spahn, H., and Mutschler, E.,** Quantification of baclofen and its fluoro analogue in plasma and urine after fluorescent derivatisation with benoxaprofen chloride and thin-layer chromatographic separation, *Arzneim.-Forsch.*, 38, 1533, 1988.
155. **Amin, M.,** Quantitative thin-layer chromatographic determination of dexamethasone and dexamethasone sodium hydrogen sulfate in blood and in pharmaceutical preparations, *Fresenius Z. Anal. Chem.*, 329, 778, 1988.
156. **Jain, R.,** Simplified method for simultaneous determination of diazepam and its metabolites in urine by thin-layer chromatography and direct densitometry, *J. Chromatogr.*, 615, 365, 1993.
157. **Khan, K., Paesen, J., Roets, E., and Hoogmartens, J.,** Quantitative TLC of erythromycin — application to commercial bulk samples and biological samples, *J. Planar Chromatogr.*, 7, 349, 1994.
158. **Ming, D., Chzengyie, S., Guozhu, H., Shufang, D., and Leming, L.,** Separation and quantitative determination of gentiopicroside in various biological matrices by thin layer chromatography, *J. Planar Chromatogr.*, 3, 386, 1990.

159. **Cenni, B. and Betschart, B.,** The determination of halofantrine in biological samples by high performance thin layer chromatography, *J. Planar Chromatogr.*, 7, 294, 1994.
160. **Bernhard, W., Jeger, A. N., and de la Vigne, U.,** Identification and determination of hydrochlorothiazide, *J. Planar Chromatogr.*, 2, 77, 1989.
161. **Boddy, A. V. and Idle, J. R.,** Combined thin-layer chromatography-photography-densitometry for the quantification of ifosfamide and its principal metabolites in urine, cerebrospinal fluid and plasma, *J. Chromatogr.*, 575, 137, 1992.
162. **Horne, C., Spahn, H., and Mutschler, E.,** Fluorimetric determination of levoprotiline in human plasma after thin-layer chromatographic or high performance liquid chromatographic separation, *Arzneim.-Forsch.*, 37, 1179, 1987.
163. **Liakopoulou-Kyriakides, M., Platis, F., Bala Moutsos, N., and Kortsaris, A.,** Determination of nicotine in serum by TLC-densitometry, *Anal. Lett.*, 25, 485, 1992.
164. **Warlich, R., Krauss, D., and Mutschler, E.,** Fluorimetric determination of norfloxacin in plasma and urine samples after thin-layer chromatographic separation, *Arzneim.-Forsch.*, 39, 656, 1989.
165. **Mangalan, S., Patel, R. B., and Chakravarthy, B. K.,** Detection and determination of omeprazole by HPTLC: a reliable method for the estimation of plasma levels, *J. Planar Chromatogr.*, 4, 492, 1991.
166. **Bernhard, W., de la Vigne, U., and Jeger, A. N.,** Identification and quantitative determination of oxprenolol in blood, *J. Planar Chromatogr.*, 2, 153, 1989.
167. **Ikebuchi, J., Yuasa, I., and Kotoku, S.,** A rapid and sensitive method for the determination of paraquat in plasma and urine by thin-layer chromatography with flame ionization detection, *J. Anal. Toxicol.*, 12, 80, 1988.
168. **Bauerova, K., Soltes, L., Kallay, Z., and Schmidtova, K.,** Determination of pentoxifylline in serum by high-performance thin-layer chromatography, *J. Pharm. Biomed. Anal.*, 9, 247, 1991.
169. **Ming, S., Moutian, W., and Jun, B.,** Determination of phosacetim in rabbit tissues and blood by TLC, *J. Anal. Toxicol.*, 12, 287, 1988.
170. **Guermouche, M.-H.,** Assay of pyrazinamide in human serum by HPTLC, *J. Planar Chromatogr.*, 4, 166, 1991.
171. **Detolle, S., Postaire, E., Prognon, P., Montagnier, C., and Pradeau, D.,** Quantitative over-pressure layer chromatography of quinine, quinidine, and hydroquinidine in biomedical analysis, *J. Liq. Chromatogr.*, 13, 1991, 1990.
172. **Colthup, P. V., Dallas, F. A. A., Saynor, D. A., Carey, P. F., Skidmore, L. F., and Martin, L. E.,** Determination of salbutamol in human plasma and urine by high-performance thin-layer chromatography, *J. Chromatogr.*, 345, 111, 1985.
173. **Kincaid, R. L., McMullin, M. M., Sanders, D., and Rieders, F.,** Sensitive, selective detection and differentiation of salicylates and metabolites in urine by a simple HPTLC method, *J. Anal. Toxicol.*, 15, 270, 1991.
174. **Ikebuchi, J., Suenaga, K., Kotoku, S., Yuasa, I., and Inagaki, O.,** Thin-layer chromatography with flame ionization detection for the determination of tetrodotoxin in biological fluids, *J. Chromatogr.*, 432, 401, 1988.
175. **Davis, C. M. and Harrington, C. A.,** Quantitation of thiothixene in plasma by high-performance thin-layer chromatography and fluorometric detection, *Therap. Drug Monit.*, 10, 215, 1988.
176. **Siek, T. J.,** Determination of trazodone in serum by instrumental thin-layer chromatography, *J. Anal. Toxicol.*, 11, 225, 1987.
177. **Yucang, L., Wei, W., Mingzhi, W., Yichun, P., Leming, L., and Jun, Z.,** Simultaneous determination of the residues of TNT and its metabolites in human urine by thin-layer chromatography, *J. Planar Chromatogr.*, 4, 146, 1991.
178. **Corti, P., Cenni, A., Corbini, G., Dreassi, E., Murratzu, C., and Caricchia, A. M.,** Thin-layer chromatography and densitometry in drug assay: comparison of methods for monitoring valproic acid in plasma, *J. Pharm. Biomed. Anal.*, 8, 431, 1990.
179. **Wolff, S. C. and Kovar, K.-A.,** Direct HPTLC-FTIR measurement in combination with AMD, *J. Planar Chromatogr.*, 7, 344, 1994.
180. **Wagner, J., Jork, H., and Koglin, E.,** HPTLC as a reference method in clinical chemistry: on-line coupling with spectroscopic methods, *J. Planar Chromatogr.*, 6, 446, 1993.
181. **Busch, K. L., Mullis, J. O., and Carlson, R. E.,** Planar chromatography coupled to mass spectrometry, *J. Liq. Chromatogr.*, 16, 1695, 1993.

Chapter 11

THIN-LAYER CHROMATOGRAPHY IN PHARMACEUTICAL ANALYSIS

Elena Dreassi, Giuseppe Ceramelli, and Piero Corti

CONTENTS

I. Introduction 231

II. Applications 232
    A. Alkaloids 232
    B. Antibiotics 235
    C. Compounds Containing Sulphur 237
    D. Steroids 238
    E. Compounds Containing Oxygen and Nitrogen 240
    F. Vitamins 244

References 244

I. INTRODUCTION

Planar chromatography (PC) is still an important system of analysis and is widely used for the control of pharmaceutical products, in the sense of analysis of the production cycle as well as for the detection of complex matrixes, especially in toxicological investigations. This becomes more than evident if we consider the ample scientific work conducted to acquire new techniques in this direction.[1-3] Furthermore, chromatography on thin layers is a methodology recognized by official pharmacopeias.

In recent years important methodological and instrumental modifications have signaled the positive evolution of the applications of this method. The commercialization of HPTLC plates and the availability of different types of layers have increased the possibilities of choice of the operating conditions and make it possible to elute the plates in such a way that high efficiency and reproducibility for all the zones on the plate, and between the plates, can be guaranteed. On the other hand, the availability of chemometric instruments, especially the experimental design, has given way to a further improvement of the elution phase.[4,5]

Generally speaking it may be said that the improvements concern the various steps of chromatographic separation: improvements of the stationary phases, of the application of the sample through introduction of automatic distributors, and the chromatographic development in chambers with controlled saturation.

The instrumental innovations concern, on one hand, developing techniques, such as Forced Flow Planar Chromatography (FFPC), Over Pressured Layer Chromatography (OPLC), and Rotational Planar Chromatography (RPC), and — on the other hand — methods of detection, such as the coupling of TLC with other instrumental techniques like Fast Atom Bombardment (FAB) or with the MS, or the choice between appropriate derivatization techniques.[6,7]

TLC offers various possibilities:

- UV-visible absorption
- Fluorescence emission or quenching

- Charring of organic substances
- Prechromatographic derivatization
- Postchromatographic derivatization.[8]

The derivatization reagents can be applied to the layer by five techniques:

1. Gas phase derivatization by dipping of the layer in a tank saturated with the reagent vapors (I_2 and HCl); this technique is not suitable for quantitative work on behalf of volatization of the reagent.
2. Addition of the reagent to the chromatographic solvent; of little value for the determination of quantitatives owing to the formation of a gradient of the reagent on the layer.
3. Uniform spraying of the layer by a fine spray of a sufficient quantity of reagent.
4. Dipping of the layer into the reagent.
5. Overpressure derivatization, a very recent and promising technique.

Spraying is undoubtedly the most popular technique, but dipping allows a more homogeneous distribution of the reagent, both on the surface of the layer and into the coating layer, thus permitting an enhancement of reproducibility and sensitivity. The modifications required for the transformation of a spray reagent into a dipping one depend essentially on the nature of the reagent. Spraying requires considerable skill; reproducibility among laboratories and operators is often subtle because of the possible differences in spraying equipment, spraying distance, and nature and pressure of the propellent gas. Automatic spraying partially prevents these limitations. Dipping is a rapid method requiring no particular skill and is suited to light- and oxygen-sensitive reagents. The limits of this technique consist mostly of the possibility of damaging the chromatographic layer with aggressive reagent solutions, the difficulty of obtaining a nonhomogeneous wetting, spot diffusion, elution, and chromatography of impurities from the sample matrix, which can overlap desired spots.

As far as the whole range of innovative aspects of the methodology is concerned, interesting essays and reviews have recently appeared and may be considered an essential point of reference.

II. APPLICATIONS

A. ALKALOIDS

Planar chromatography continues to be used in the qualitative analysis of alkaloids in the control phase of both pharmaceutical formulations and vegetable drugs.

Schütt and Hölzl[9] relate an identification method of *Hypericum perforatum* extracts containing naphthodianthrones by TLC using silica gel plates with fluorescence indicator and the following mobile phases: ethyl acetate–formic acid–water (30:2:3, v/v/v) and toluene–ethyl acetate–formic acid–water (50:40:5:5, v/v/v/v). After the development phase, the plates have been read at 366 nm.

Singh et al.[10] have applied a TLC method (the plates were developed in chloroform–methanol–propionic acid (72:12:10, v/v/v) and sprayed with Dragendorff's reagent and sodium nitrite) for preliminary screening of urine samples for hordenine determination, a sympathomimetic compound that gives false positives in ELISA and RIA assays. Because of the interference of hordenine with oxymorphone, hydromorphone, and apomorphine, a successive GC-MS or HPLC analysis is necessary.

Sosa et al.[11] have applied TLC (using the silica gel plates and as mobile phase acetone–light petroleum 40–60°–diethylamine (2:7:1, v/v/v), and for the visualization of spots a 1% solution of ammonium cerium(IV) sulphate in concentrated orthophosphoric acid) combined with HPLC for the identification of indolic alkaloids isolated from various *Rauwolfia* species. The originality of this work lies in the possibility of determining, with 25 μg of spotted sample, ajmaline in the presence of its stereoisomers, namely isoajmaline, sandwicine, and isosandwicine, which was impossible to determine with only one of the two methods.

Vampa et al.[12] describe a method for the qualitative and quantitative assay of glycyrrhizin in solid (tablets and granules) and liquid (extracts or liquid products) forms both in pharmaceutical formulations and food products. The method (HPTLC silica gel plates with fluorescence indicator and as mobile phase butanol–acetic acid–water (7:1:2, v/v/v) followed by fluorescence emission at 254 nm) proved valid, on the basis of precision and accuracy values and also in comparison with other techniques like HPLC and GLC referred to in the literature. The detection limit for the glycyrrhizin is 5.06 ng for deposition and it is consequently applicable to quality control of pharmaceuticals and food products containing this substance.

Ponder and Stewart[13] describe a HPTLC method used for the determination of digoxin and its related compounds: digoxigenin bisdigitoxide and gitoxin, in tablets containing digoxin. Compared to previous studies, conducted in direct phase on silica gel plates with various mobile phases containing chloroform–methanol, methyl ethyl ketone–toluene, methanol–formamide, cyclohexane–acetone–glacial acetic acid, and chloroform–methanol–formamide in different proportions, the choice of a reversed phase (plates C_{18} reversed phase) and the mobile phase containing water–methanol–ethyl acetate (50:48:2, v/v/v) has given better results, although the development time was considerably long (30 min). The quantitative determination was performed by densitometric analysis using absorbance at 218 nm for the digoxin and fluorescence at 365 nm for the other two compounds. The detection limit for digoxin is 8 ng; the other two compounds had a limit of 0.04 ng for deposit.

The review of Vetticaden and Chandrasekaran[14] contains an ample description of chromatographic and nonchromatographic techniques that were nonapplicable for the analysis of cardiac glycosides. The work underlines the function of the TLC as a preparative tool in the quantification of digoxin and derivatives.

Lavanya and Baggi[15] describe a method for the identification and confirmation of cannabinoids present in cannabis, utilizing a process of prechromatographic derivatization that later permits separation of the isomers of the derivatized products of cannabinoids. The method is based on oxidative coupling of cannabinoids with MBTH (3-methyl-2-benzthiazolinone hydrazone MBTH) using acidic ceric ammonium sulphate (CAS) as oxidant. The buffered mixture is deposited on silica gel TLC plates and the chromatography is carried out with benzene–methanol (98:2, v/v). Every cannabinoid presents a visible characteristic coloration; the single compounds may be identified more specifically through application of different wavelengths of UV light. The used compounds are Δ^9-tetrahydrocannabinol, cannabinol, cannabidiol, and cannabigerol and the respective values of the R_f are 0.69, 0.83, 0.85, and 0.77. The reproducibility of this method can be compared with other methods based on the postchromatographic derivatization, although they present a higher sensibility. One more advantage of the method described by Lavanya and Baggi consists in the noncancerogenicity of the derivatives and the stability for various days of the chromophores, no interference has been noticed. Without any doubt the derivatization in a prechromatographic phase guarantees a higher reproducibility for quantitative analysis, compared to the induction of the color after the chromatographic development. Thus it can be retained as a good alternative for the detection and identification of cannabinoids in forensic exhibits.

Funk and Dröschel[16] describe a method for the quantitative determination of cocaine and its metabolites (ecgonine, ecgonine methyl ester, and benzoylecgonine) on silica gel HPTLC plates with fluorescence indicator using water–dioxane–methanol–aqueous sodium acetate buffer at pH 8.0 (28:12:60:10, v/v/v/v) as mobile phase. Using a modified Dragendorff's reagent, the compounds may be identified with sufficient accuracy, the detection limits were 10 ng/spot for cocaine and 300 ng/spot for ecgonine, 200 ng/spot for ecgonine methyl ester, and < 100 ng/spot for benzoylecgonine. The application of thermic derivatization (260–280°C) produces intense fluorescent spots. The quantity was determined by using an excitation wavelength of 313 nm and emission of 390 nm. This technique makes it possible to obtain good quantification limits of less than 10 ng/spot for all compounds.

Duez et al.[17] report about a HPTLC method for strictosamide, an alkaloid that is found in *Nauclea latifolia*. The roots of this plant are widely used in West Africa as a traditional drug for treatment of gastrointestinal troubles and for their antipyretic and anthelmintic properties; the leaves are used as bowel function regulators. Strictosamide is supposed to be a biogenetic precursor of other alkaloids. The described technique is based on a preventive purification of the methanolic extracts with the help of chromatographic column, and a subsequent HPTLC or HPLC analysis. The quantitative determination was based on fluorodensitometric determination at 406 nm. α-Tocopherol acid succinate (at 2% in methanol) was added to the extract to prevent degradation through oxidation of the products during the chromatographic process. The chromatographic separation was performed on silica gel plates with fluorescence indicator and developed on mobile phase consisting of chloroform–acetone–methanol (65:25:10, v/v/v). A subsequent postchromatographic heating at 105°C for 24 hours gives degradation products with an intense fluorescent signal as a result, which increased the sensibility of analysis (wavelength 235 nm). The HPTLC method is further compared with a HPLC reversed phase (C_{18}) conducted with the ion-pair technique. The results for both proceedings seem similar, although a higher sensibility of the HPTLC method can be observed (detection limit for the HPTLC is 2.5 ng per spot, 11 ng injected for HPLC).

Duez et al.[8] conducted, as mentioned above, an important work that tried to put critically into evidence the different possibilities of postchromatographic derivatization on the layer. For this inquiry, various alkaloids have been taken as models:

- Cucurbitine, amino acid with anthelmintic properties, present in the leaves of *Cucurbita spp.*, analyzed with a HPTLC method using ninhydrin as reagent.
- Hyoscyamine and hyoscine, alkaloids with anticholinergic properties, present in the leaves, roots, fruits, and stem of *Datura spp.*, separated on layer and afterward visualized with *p*-dimethylaminobenzaldehyde (modificated adequately for the dipping).
- Sennosides A and B, anthraquinonic heterosides with purgative properties present in the leaves, fruits, and stem of *Senna alexandria* and *S. acutifolia* analyzed with HPTLC and visualized with two sprays and one reagent for dipping.
- Conessine, steroidal alkaloid with amoebicide properties present in the roots and stem bark of *Holarrhena floribunda*, analyzed with TLC and later visualized with the spray method and dipping.
- Reserpine and rescinnamine, alkaloids with hypotensive and sedative properties present in the root bark of *Rauwolfia vomitoria* analyzed with HPTLC methodology and subsequent transformation in fluorescence through heating. The reproducibility in this case is quite low because of the formation of interfering deterioration residues.

For every analysis the comparison between visualization spray or dipping is reported, and the latter always gives better results. Furthermore, the authors give hints on how to transform a spray reagent into a dipping one.

Corti et al.[18] have applied TLC on the analysis of reserpine and chlorthalidone in the presence of potential impurities of their pharmaceutical forms in tablets, and of reserpine, methylreserpate, chlorthalidone, and its carboxylic hydrolysis products (3-dehydroreserpine, 3-dehydromethylreserpate, 3,4,5-trimethoxybenzoic acid, 4-chloro-3-sulphonyl-benzophenoxy-2-carboxilic acid, and 3(4-chloro-3-amino-sulphonyl)phenyl-1H-isoindol-1-one) in urine. The analysis has been conducted on HPTLC silica gel plates using chloroform–methanol (2:1, v/v) as mobile phase both for the analysis of the tablets and the urine. The densitometric analysis is performed at 254 nm for all compounds; for the analysis of reserpine it is possible to increment the photodensitometric analysis using fluorescence.

Nikolova-Damyanova et al.[19] describe a method for the quantitative determination of iridoid and flavonoid glucosides in *Linaria* species in a single plate. The iridoid glucosides are antirrinoside, linarioside, 5-O-glucosylantirrinoside, and 5-O-allosylantirrinoside, while the flavonoid is acetylpectolinarin.

Zhou et al.[20] report a method for determination of allantoin in common yams (*Dioscorea opposita*). After extraction with 75% ethanol, the sample is analyzed using methanol–acetone–formic acid–water (40:2:1:6, v/v/v/v) as mobile phase and determined by TLC-scanning at exc 445 and fl 500 nm.

B. ANTIBIOTICS

Thin-layer chromatography has, for a long time, been a widely accepted methodology for the separation of multiple classes of antibiotics. During the last years numerous assays have added various revelation methods for a better quantitative determination of this methodology.

Naidong et al.[21] describe a TLC procedure on silica gel plates with fluorescence indicator and reversed phase on silanized silica gel, prepared in the laboratory and purchased, on glass and aluminum, for the identification of chlortetracycline, demeclocycline, doxycycline, metacycline, minocycline, oxytetracycline, and tetracycline conditioning the stationary phase, during the prechromatographic phase, with a 10% solution of sodium edetate, with different pH (7, 8 and 9) according to the compound and a subsequent mobile phase constituted by dichloromethane–methanol–water (59:35:6, v/v/v), and chloroform–methanol–water (74:23:3, v/v/v) for the silica gel plates, and methanol–acetonitrile–0.5 M oxalic acid (pH 2) (1:1:6, v/v/v), 0.5 M phosphoric acid (pH 2) saturated with n-butanol; 0.5 M citric acid (pH 2) saturated with n-butanol; 0.5 M oxalic acid (pH 2) saturated with n-butanol for the reversed phase. A further inquiry on the influence of humidity for the silica gel was conducted, and the best results were obtained at 79%. The reversed phase plates were analyzed under UV light at 365 nm, the ones in direct phase at 254 and 365 nm. The reversed phase analysis has given inferior results compared to the direct phase, which has furthermore been demonstrated to be practicable on plates prepared in the laboratory as well as on purchased plates.

Sekkat et al.[22] describe a simple, fast, and reliable method for the determination of various aminoglycosides, such as dibekacin, framycetin, kanamycin, netilmicin, sisomicin, tobramycin, and gentamicin in commercial formulations (injections, capsules, eye drops, solutions, and ointments). The separation was executed on silica gel plate TLC. Developing solvent system were chloroform–methanol–28% (m/m) ammonium hydroxide (2:3:2, v/v/v) for the determination of framycetin, kanamicin, and tobramycin; or the same solvents (4:3:1.5, v/v/v) for the determination of dibekacin, netilmicin, and sisomicin or 1:1:1 v/v/v for gentamicin. The R_f values obtained were, respectively, 0.15, 0.24, 0.26, 0.36, 0.60, 0.43, and 0.30, then located with ninhydrin and quantitatively determined by in situ spectrophotometer at 505 nm. The statistical analysis underlines the validity of this procedure, which presents a detection limit of 60 to 200 ng. Statistical analysis with the student's t-test did not reveal, at the 95% confidence level ($p = 0.05$), significant differences between the declared results and the recovery data.

Kang and Ebel[23] have published a procedure for the separation of tetracycline, chlortetracycline, oxytetracycline, and demechlocycline with CN HPTLC plates and a mobile phase of methanol–acetonitrile–0.5 N oxalic acid (3:1:12, v/v/v). Later they used doxycycline as an example for the quantitative assay, which is determined through densitometry at 254 nm. In the range between 400 and 900 ng per spot they found a nearly linear dependence of reflectance on the amount of substances within the spot. The results were analyzed both with linear regression as well as nonlinear regression techniques. The comparison of both results of sensibility obtained with the two procedures has revealed no significant differences between both procedures.

Naidong et al.[24] have described a TLC procedure for the identification and the purity control of chlortetracycline and demeclocycline. The silica gel plates were previously sprayed with a 10% EDTA solution (pH 8) to separate the major impurities of the two main components from the other tetracyclines and with a mobile phase of dichloromethane–methanol–water (60:35:5, v/v/v) for chlortetracycline and (59:35:6, v/v/v) for demeclocycline. The anhydrous derivatives of the two main components were separated on silanized silica gel plates with mobile phase methanol–acetonitrile–0.5 M oxalic acid (pH 2) (1:1:6, v/v/v). The authors furthermore report a comparison between the TLC and LC techniques equating the RDS values obtained through quantification (wavelength 280 nm) of the two compounds gaining values of respectively < 2% and < 1%. A further confrontation of the analytic results through t-test for the various samples compared with the two procedures did not reveal significant differences ($p = 0.05$). The detection limit of TLC is 20 ng/deposition for every compound.

Kovács-Khadady[25] has studied the behavior of some tetracyclines and derivatives on silica and reversed phase plates, impregnated with TCMA (tricaprylmethylammonium chloride), before the chromatographic analysis through overpressured layer chromatography (OPLC). It was observed that the retention on silica gel plates grew with increasing concentration of TCMA in the impregnating solution, as already reported in other works by the same author. This effect is excluded if the mobile phase contains EDTA, a factor that reduces the interaction of TCMA and tetracyclines. To obtain good chromatography the silica gel is treated with a EDTA solution after the plates have been dried. In reversed phase the retention results opposite to the reaction observed on silica gel, the retention diminishes if the TCMA is augmented. FTIR measurements have proved that the quantity of TCMA, adsorbed by the two kinds of plates, is comparable. Good separation can be obtained with TCMA 0.05M to impregnate the C_{18} layer and with a mobile phase of methanol–water (4:6, v/v).

Eneva et al.[26] describe a method for the separation of the compounds of nebramycin by TLC on silica gel with mobile phase acetone–abs. ethanol–25% ammonia (1:1:1, v/v/v), resolving the separation of apramycin and kanamycin. For quantitative determination the organic components were charred through treatment with sulfuric acid and underwent a densitometric analysis at 450 nm. Apramycin, kanamycin B, tobramycin, and carbamoyl tobramycin have thus been quantified. The methodology result is easy and reliable over a range of 2.5 to 55 µg/spot and can be used to follow processes of purification and fermentation permitting monitoring production of the nobramycin complexes and for stability studies.

Kovács-Hadady and Szilágyi[27] describe the behavior of various penicillins and cephalosporins on silica gel plates, previously impregnated with TCMA, developed with OPLC, utilizing mobile phase containing methanol and water but not TCMA. The basis of the retention is not ion-pairing, but hydrophobic interaction between the solutes and the caprylic groups of the TCMA. The retention is controlled by the surface concentration of TCMA; in fact it increases with augmenting concentration of the TCMA on the silica gel. The main advantage of the proposed procedure is the independence of the pH and the ionic force of the eluent.

Corti et al.[28] apply TLC for the determination of rifaximin and its oxidation products both for quality control of the raw material as for the pharmaceutical formulation (cream) and for the residues in the milk after the cream had been used to treat mastitis in cows. The compounds analyzed are rifaximine, 2-amino-4-methyl piridine, rifamicin O, and oxidized rifaximine. The best results have been obtained using HPTLC concentration zone silica gel plates employing a double elution as mobile phase chloroform–ethylacetate–methanol (1:9:0.5, v/v/v) (MP1) and chloroform–methanol–acetic acid (17:0.3:0.03, v/v/v) for the raw material analysis (MP2), and only the MP1 for analysis of cream and milk. The R_f value of rifaximin in the MP1 is 0.28 and the detection limits are 4.66 µg/ml for the raw material, 4.77 µg/ml for cream, and 5.20 µg/ml in the milk for the densitometric scanning at 275 nm; the other compounds have the same quantification limits.

Quintens et al.[29] describe a TLC procedure on silanized silica gel plates with fluorescence indicator of 30 cephalosporins. The results obtained with seven mobile phases buffered at pH 6.2 are satisfying and so are the results of the colorimetric reactions (sulphuric acid, sulphuric acid and formaldehyde, sulphuric acid, and nitric acid). The use of a single mobile phase, or of a combination of two mobile phases, or of one mobile phase and a colorimetric reaction enables the identification of all the cephalosporins available on the market.

Maxwell and Unruh[30] demonstrate in their work that various polycyclic ether antibiotics (lasalocid, monensin, narasin, and salinomycin) and several lipid classes can be revealed through vapor phase fluorescence technique (VPF), after heating of the compound developed on HPTLC silica gel plates both on reversed phase plates in the presence of ammonium bicarbonate with different results according to the compounds. For example, the fluorescence response of narasin and salinomycin gave a result three times greater than on silica; all lipids, except cholesterol, could be detected only on silica plates.[30] Comparison of the method with the general technique of visualization, implying spraying with a solution of vanillin, reveals that visualization through VPF may be considered a valid quantitative method for such determinations with lower sensitivity limits than the other.

C. COMPOUNDS CONTAINING SULPHUR

The TLC methodology has by now been widely recognized in various pharmacopoeias as the official method for numerous substances.

Szepesi et al.[31] have confronted two TLC methods for the sulphinpyrazone reported in the BP of 1984 and the USP of 1984. Both procedures present problems, mainly because of little stability of the main components and for the evaluation of impurities. Thus a new mobile phase chloroform–acetic acid (80:20, v/v) has been developed to which, compared to the old one (95:5, v/v), an antioxidant (2,6-D-tert-butyl-4-methylphenol, BHT) has been added, both to the eluent (0.02%) as to the sample solution (0.1%). The plates utilized are silica gel with fluorescence indicator for the official method and with concentration zone silica gel plates for the method proposed, including quantification through spectrodensitometric evaluation at 265 nm. The HPLC methodology has been indicated for the separation of the main agent from impurities, causing no decomposition of the sulphinpyrazone. The method, for reasons of accuracy and improved precision, has been indicated in the purity test.

Madelaine-Dupuich et al.[32] describe a procedure for checking the purity of N-acyl aminonaphthalene disulphonic acid derivatives, molecules with anti-human immunodeficiency virus (HIV) activity, including the use of TLC and flame ionization detection for the separation of these compounds that are difficult to analyze by conventional methods. The mobile phases utilized are composed by pure methanol and methanol–chloroform–ammonia (35:55:10, v/v/v). The TLC-FID coupling has proved to be a valid system with high performance.

Salomies[33] describes a HPTLC method for the analysis of sulfameter (1) in combination with other pharmacological sulfonamides like sulfamethoxazole (2) and sulfisoxazole (3). The advantage of the HPTLC methodology compared to HPLC methods, largely used, is that HPTLC allows a contemporaneous analysis of different samples on the same layer, with a diminished waste of solvents and is thus cheaper. The described method considers the simultaneous quantification of two sulfonamides (1+2 or 1+3) present in pharmaceutical formulations, as pure substance and in tablets. It is monodimensional and utilizes only one solvent-system chloroform–ethyl acetate (4:6, v/v). The quantitative extraction is conducted with densitometric analysis at 270 nm. To remove impurities from the layer to obtain a uniform basic line, the plates are developed two times, with methanol, before starting the analysis. This method has proved to be accurate, precise, and selective apart from being fast and cheap.

Sanchez et al.[34] propose a method for the separation of phenothiazines using TLC method on silica gel plates and mobile phase methanol–acetic acid (95:5, v/v) (it has proved to be the best alternative between the various binary and tertiary mobile phases used for the separation and for the greatest chromatic and fluorescence intensities) and subsequent detection by photochemical derivatization in six solvent systems. The qualitative and quantitative information about propiomazine, trimeprazine, promethazine, promazine, ethopropazine, acetopromazine, thioridazine, and chlorpromazine were obtained by image analysis. Images were obtained with a charge-coupled device camera (CCD) under different conditions using pseudocolor red-green-blue (RGB) and gray-scale tones and then filtering the light emitted. This is a new technique for the qualitative and quantitative determination of compounds like phenothiazines that may be separated with photochemical derivatization producing colored and fluorescent spots.

Corti et al.[35] describe a TLC method for the assay of chlorpromazine and its principal metabolites. The compounds investigated were chlorpromazine, promazine, nor1 and nor2 chlorpromazine, chlorpromazine sulfoxide, and 3-hydroxy and 7-hydroxychlorpromazine. The separation was performed using HPTLC silica gel plates with or without concentration zone, utilizing two chromatographic conditions: a single run for the layers with concentration zone with mobile phase (MP1) consisted of 72 ml water-saturated chloroform and 9.6 ml methanol having a water content of 4%, left to saturate for 12 h in the chromatographic chamber. For the layer without the concentration zone a first elution with the MP1 and an exposition for 15 min in a drying chamber saturated with formaldehyde was made. Then the layer was eluted for a second time in a mobile phase of chloroform–methanol–water (52:2:0.6, v/v/v). The best results are obtained using the concentration zone layers with densitometric detection at 254 nm and gave a sensitive detection limit of 1.0 μg/ml for chlorpromazine.

Ling et al.[36] show an application of HPTLC with fluorescence scanning densitometry for the qualitative and quantitative detection of thiols (captopril, coenzyme A, cysteamine, cysteine, glutathione, acetylcysteine), of biological and pharmacological interest, that is difficult to determine with other procedures because of their high reactivity. The mixture of thiols together with their disulfides was first derived with two reagents: ammonium 7-fluorobenzo-2-oxa-1,3-diazole-4-sulphonate (SBD-F) and 7-fluoro-4-sulphamoil-2,1,3-benzoxadiazole (ABD-F) specific for the thiols, followed by elution on HPTLC silica gel plates with a mobile phase composed of isopropionil ether–methanol–water–acetic acid (9:8:2:1, v/v/v/v). The survey is done with an excitation wavelength of 365 nm. Variations of the R_f value obtained with different pH, with the variation of the proportions of acetic acid and of water in the mobile phase, have been observed. Detection limits of the different derivatives of the thiols are about 0.2 pmol. (20–40 pg per spot); for cysteamine-SBD this limit sinks to 6.3 pg. These limits may be lowered further using surfactants (triton, paraffins or cyclodextrin solutions). The calibration curves were linear in the picogram range, with a RSD range of 1.16–3.2% of the complete procedure.

De Brabander et al.[37] describe a HPTLC method for the qualitative and quantitative determination of the residues of thyreostatic drugs in biological materials, thiouracil and analogues, propylthiouracil, phenilthiouracil, and above all methylthiouracil and tapazole. The sample for the analysis is extracted by the formation of a specific complex between the drugs and mercury ions bound to an affinity column; the qualitative and quantitative determination is based on the fluorescence induction (366 nm) of the 7-chloro-4-nitrobenzo-2-oxa-1,3-diazole (NBC-Cl) derivatives, obtained with alkaline cysteine solution after bidimensional separation on silica gel HPTLC plates, with a first mobile phase constituted of dichloromethane–methanol (98:2, v/v) and a second of dichloromethane–propionic acid (92:2, v/v). The confirmation of the suspect spots is obtained by GC-MS with positive chemical ionization. In comparison with other extraction procedures, the affinity columns with mercury ions have revealed themselves as really advantageous and furthermore allows the quantitative determination of thiouracil, methylthiouracil, and tapazole. The method of purification is fast and together with HPLC analysis and fluorescence detection very important for qualitative analysis in routine controls.

Brown and Busch,[38] who already gave evidence in other works of the possibility to apply the TLC-FAB-MS coupling for the analysis of the main components of different categories, suggest in this work the HPLC analysis on silica gel plates with a fluorescence indicator, eluted with ethyl acetate–water (100:1.5, v/v), without previous extraction, for the separation and FAB for the survey of hydrochlorothiazide (HCT), hydroflumethiazide (HFM), trichlorothiazide (TCM), chlorthalidone (CTA), amiloride hydrochloride (AMI) and furosemide (FUR). The obtained R_f values for the different compounds were HCT 0.63, HFM 0.80, TCM 0.91, CTA 0.71, AMI 0.00, and FUR 0.38. Furthermore they describe a HPTLC technique with a mobile phase ethyl acetate–methanol–ammonia (85:10:5, v/v/v) and MS for the identification of AMI and HCT in urine, obtaining three spots with R_f of 0.20, 0.33, and 0.25 respectively of AMI, HCT, and a contaminator also found in the blank urine. Detection limits of 1 to 10 ng were established for these drugs compared to the values of 10 to 50 ng obtained for the HPLC analysis followed by UV assay.

D. STEROIDS

For many years TLC has been recognized as a fundamental instrument for mainly qualitative analysis of compounds with steroidic structure. The coupling of this method with other procedures for the quantification of the single compounds after their separation on the layer, constitutes a good analytical strategy.

Vingler et al.[39] have introduced a digital autoradiographic–thin-layer chromatography method for the thorough separation of the testosterone metabolites and other androgens (androstenedione, dihydrotestosterone, androsterone, epiandrosterone, androstanedione, 3α-androstenedione, 3α-androstenediol, and 3β-androstanediol) with TLC silica gel plates containing a fluorescence indicator, with a concentration zone, applying two different methods. The first one consists of a double run of only one direction using a mobile phase of dichloromethane–diethyl ether (90:10, v/v), the second of a first double run with the same mobile phase followed by a second run at 90° with a mobile phase constituted of toluene–ethyl acetate (50:50, v/v) with a single spot and (70:30, v/v) when two tracks were to be run. Their direct quantification is performed by radioscanning. Satisfactory accuracy is obtained by both external and internal standardization.

Medina and Schwartz[40] report a method for the identification of residues of treatments with anabolic hormones on tissues of various animal species, searching for the compounds themselves and their metabolites. The main components that have been studied are estradiol, estrone, diethylstilbestrol, zeranol, zearalanone, and mycotoxins, zearalenone and zearalenol. In the first place the compounds are extracted from their biological matrices (plasma and tissue) with a clean-up procedure using solid-phase dual columns of alumina and anion-exchange resin, and after elution with a mobile phase constituted of dichloromethane–methanol–2-propanol (97:1:2, v/v/v) on silica gel TLC plates two systems of analysis are used: Fast Corinth V salt and iodine and starch. In both cases all the compounds could be found, although some of them had a not quite satisfactory sensibility limit. With Fast Corinth V the spots of zeranol and zearalenone (respectively 2 and 4 ng) are detectable, while faint spots for the phenolic estrogens were obtained. But these last compounds have a much stronger reaction if treated with iodine and starch, bringing out values for estradiol and diethylstilbestrol of 4 ng. This method can thus be considered selective for detection of zeranol and its metabolite, zearalanone, in the presence of steroidal compounds.

Datta and Das[41] propose an alternative TLC method to the official chromatographic procedures (USP XXII, BP-1988, and Indian Pharmacopoeia, 1985) for the quantitative determination of the quality control of the corticosteroid sodium phosphate salts of betamethasone, dexamethasone, and prednisolone to be found in different pharmaceutical products like eye and ear drops, lotions, and parenteral preparations, saving time and money. This method avoids all interferences found during the HPLC analysis caused by the presence of other main components within the formulations or preservatives. The separation is obtained by reversed phase ion pair TLC on silanized silica gel plates using as mobile phase aqueous 0.05M triethylamine (pH 4.2)–methanol (3:2, v/v). The analytical determination is done by scanning densitometry at 240 nm. Moreover the described method offers the possibility to utilize the same chromatogram for the identification and the assay of other main components such as chloramphenicol and methyl and propyl parabens.

Petrovic et al.[42] report the chromatographic behavior of 36 estradiol and estrone steroids on silica gel and bonded C_{18} silica gel layers analyzing the effects of various substitutes for the chromatographic behavior with several nonaqueous or aqueous-organic mobile phases. The spots are observed in UV light at 254 nm.

Smets el al.[43] describe a very quick method consisting of methanolic extraction with a Stomacher apparatus and HPTLC analysis on silica gel layers, C_{18} and CN, developed with seven different mobile phases for the qualitative and quantitative determination of different compounds with anabolic properties and their esters in injection sites in animal tissues. The plates were dipped into a 5% ethanolic solution of sulfuric acid and then heated; the spots were identified in fluorescence at 366 nm; R_f values and colors obtained were compared with those of standard compounds. In comparison with techniques currently used, as immunoassay, this method has the advantage of allowing the separation and identification of the esterificated forms of the anabolics as estrogens (stilbenes and their esters, estradiol and its esters, ethinylestradiol, and mestranol), androgens (methyltestosterone, testosterone and its esters, nortestosterone and esters, trembolone and trembolone acetate, stanozolol, boldenone, methylboldenone, and chlortestosterone) or gestagens (progesterone, medroxyprogesterone acetate, chlormadinone acetate, melengestrol acetate, and megestrol acetate). After a first monodimensional screening, the extract has to be concentrated if no characteristic spots of the different compounds can be observed. In the case of different esters within the same compound, a second elution (bidimensional) with various mobile phases has to be carried out. In case there are still doubts about the attribution of the spots, it is recommended to adopt plates of a different type other than CN or C_{18}.

Szepesi[44] has published a review containing 82 references on the role of TLC for the qualitative and quantitative analysis of steroids citing the most recent works on identification and qualification at industrial level focusing on the methodologies described in the latest editions of the United States, British, and European pharmacopeias, for identification and purity testing of steroid raw materials and formulated products.

E. COMPOUNDS CONTAINING OXYGEN AND NITROGEN

Schütz and Meister[45] report the TLC retention data of 20 commonly used β-blockers, introducing the concept of corrected R_f value. This value permits the insertion of corrections, to separate the R_f value from influences of the system (running distance, temperature, geometry of the chamber, and amount of drug applied to the chromatogram). The main components are assayed on silica gel layers with ten mobile phases and with different pre- or postchromatographic derivatizations.

Davies[46] has published a review with 283 chromatographic applications for analysis of β-blockers. Although HPLC remains the dominating procedure for the determination of those compounds, TLC, be it reversed phase or direct, has been applied especially for the preliminary screening phase for toxicological antidoping studies in sports medicine.

Betschart et al. describe a HPLC method for the determination of antimalarial drugs in biological fluids: fast, easy, cheap and semiquantitative according to the needs of an analysis in a place with little equipment, as, for example, in small laboratories or for field use, especially in developing countries.[47] The analysis is based on direct application, without the need of extraction, of a sample of urine on HPTLC plates of silica gel; after drying and developing in an open horizontal chamber, with a mobile phase formed by toluene–diethylamine–methanol (8:1:1, v/v/v), the plates were analyzed at UV at 366 nm. Chloroquine, quinine, and their principal metabolites are thus identified. Furthermore, they compared the procedures of manual and automatic application of the sample, with the Linomat, where the last obtained the better results. The detection limits for the two main antimalarics are respectively 25 and 125 ng/spot.

Abdelrahman et al.[48] have applied a TLC method as well as an HPLC method for the control of various pharmaceutical preparations containing chloroquine (ampules, tablets, and syrups).

Dreassi et al.[49] describe a TLC method for the quality control of inosiplex in its various pharmaceutical formulations. Inosiplex is a 1:3 molecular complex of inosine and p-acetilamino benzoate of 1-(dimethyl-amino)-2-propanol used as an antiviral, and sold in tablets, powders, and syrups. The authors use HPTLC silica gel concentration zone plates with a mobile phase of acetone–acetic acid (5:1, v/v) and densitometric scanning at 254 nm. The compounds investigated are inosine (R_f 0.25), p-acethylamino benzoate (R_f 0.82), p-aminobenzoic acid (R_f 0.14), and 1-(dimethylamino)-propanol (R_f 0.46). The last compound was detected after the exposition of the plate to iodine vapor; the respective detection limits for the four compounds are 0.325, 0.411, 0.025, and 4.158 µg/dep.

Zhou et al.[50] describe a TLC-scanning method for the determination of oleanolic acid in its β-cyclodextrin inclusion-complex. The plates were eluted with cyclohexane–chloroform–ethyl acetate–methanol (20:5:4:2, v/v/v/v).

Parimoo et al.[51] developed a quantitative TLC method for the determination of oxyphenbutazone and ibuprofen in the presence of paracetamol and dextropropoxyphene.

Lillsunde and Korte[52] illustrate a TLC methodology for the preliminary screening and GC-MS for the confirmation of about 300 drugs belonging to the class of barbiturates, benzodiazepines, amphetamines, phenothiazines, antidepressants, opiates, carbamates, and analgesics. The work reports extraction procedures for urine and the following chromatographic operations to determine the pharmaceutical products in the biological matrix.

Corti et al.[53] have used a TLC method for the monitoring of valproic acid in plasma after a derivatization procedure with 2-naphthacyl bromide, carrying out a comparison between TLC and

HPLC. Comparisons between the different performances of several types of stationary phases were made: HPTLC and TLC reversed phase (C_8 and C_{18}) and silica gel plates. The best results are obtained with HPTLC C_8 plates using a mobile phase of ethanol–water (1:0.4, v/v) with detection limits in plasma of 4.87 μg/ml against 3.47 μg/ml for a HPLC method. The detection on the layer was carried out with densitometric scanning at 280 nm.

Shinde et al.[54] describe a simple, fast, and precise HPTLC method for the determination of nitrendipine and nimodipine in pharmaceutical dosage, using as mobil phase toluene–ethyl acetate (1:1, v/v); for the quantitative determination a densitometric scanning at 236 for nitrendipine and 240 for nimodipine is done.

Bieganowska et al.[55] report chromatographic retention values of some basic drugs (used as therapy in Parkinson's disease) and related compounds. They suggest a reversed-phase TLC procedure with prepared silanized silica gel plates containing an appropriate concentration of ion-pairing reagents, also in the mobile phase. The spots can be observed under UV light (366 nm) or through colored reaction after spraying with Dragendorff's reagent. The retention and the separation are analyzed with the help of the variations of the pH eluent, the amount of methanol, and the type and concentration of the ion-pairing reagent. The best results were obtained when counter ion was present in the mobile phase.

Tománková and Sabartová[56] report a comparison between HPTLC and HPLC for the determination of impurities of azidothymidine (3′-azido-3′-deoxythymidine), main component with anti-HIV properties, impurities that may appear during the synthesis, Az-Tr-thymidine (1-(azido-2,3-dideoxy-5-O-trityl-β-D-threo-pentafuranosyl)thymine) (Az-Tr-T), Ms-Tr-thymidine(1-(3-O-methanesulphonyl-5-O-trityl-β-D-threo-2-deoxy-pentafuranosyl)thymine) (Ms-Tr-T), Tr-alcohol (trityl alcohol), isothymidine (1-(2-deoxy-β-D-threo-pentafuranosyl)thymine) and thymine. The HPTLC procedure is based on the separation of silica gel plates with a fluorescence indicator and two mobile phases consisting in methanol–chloroform (5:95, v/v) and (10:90, v/v); the first one enables the separation of Az-Tr-T, Ms-Tr-T, and Tr-alcohol; the second the separation of thymine and isothymidine. The densitometric determination is measured at 268 nm for isothymidine, thymine, Az-Tr-T and Ms-Tr-T and 220 nm for Tr-alcohol. Although the HPLC technique, compared to HPLTC, has proved to be more sensible, the latter one has given better results concerning speed and costs for the analysis of purity and stability in pharmaceutical products.

Vampa et al.[57] describe a HPTLC method for the investigation of the methylation rates occurring in position 1 or 3 of uracil and derivatives, in position 5 (5-nitro, 5-fluoro, 5-chloro, 5-bromo, and 5-methyl uracil), according to the amounts of product obtained and the time required to complete the reaction. The derivatives are synthesized and separated by droplet countercurrent chromatography (DCCC) and/or medium-pressure liquid chromatography; the stationary phase consists in TLC and HPTLC silica gel plates with fluorescence indicators, and the mobile phases are benzene–acetone (1:1, v/v), toluene–acetone (1:1, v/v), dichloromethane–ethyl acetate (1:1, v/v).

Aboul-Enein and Serignese[58] describe a HPTLC method on silica gel plates and fluorescence indicator with a mobile phase of chloroform–isopropyl ether–methanol–ammonia (75:25:20:1, v/v/v/v), followed by densitometric scanning at 254 nm for the quantitative determination of phenytoin contained in pharmaceutical formulations in capsules and injectables. Apart from phenytoin, the procedure indicates the hydroxylated metabolites contained in the urines, p-idroxy phenytoin and phenytoin dihydrodiol. The determined values of R_f are, respectively, 0.67, 0.37, and 0.20. The authors, like others, compare the performance of HPTLC and HPLC, and point out the fact that a lot of time and money can be saved by using the technique on layer, this procedure proving at the same time to be the better one for the quality control and the stability test.

Van Boven and Danaes[59] describe a procedure for TLC identification of azaperone (4′-fluoro-(4-(2-pyridyl)-1-piperazinyl)butyrophenone, a main component used as tranquilizer in animals, and its metabolites, in the urine and gastric contents of human patients intoxicated with this component. The procedure consisted in a TLC analysis on silica gel plates and fluorescence indicator, developed with two mobile phases of methanol–ammonia (100:1.5, v/v) and chloroform–methanol (80:20, v/v), followed by reaction on layer with acid potassium iodoplatinate solution and Mandelin reagent and a UV determination; the purity and identity of the spots was confirmed by GC and GC-MS.

Lang[60] reports a method for the determination of several drugs in biological fluids, urine, and stomach content. The pharmacological products of the group of analgesics, tranquilizers, antihistamines, and antimalarials are chloroquine, quinine, papaverine, imipramine, dibenzepine, diethazine, promethazine, chlorpromazine, aminophenazone, tripelennamine, diphenhydramine and cinnarizine. These molecules share the presence of a basic group in the structure that permits, sterically, the formation of complexes with metals. In this case the study was based on the formation of colored complexes with the ion Co^{II}. Practically, after a chromatographic run on silica gel plates with two different mobile phases of ammonia 3%–methanol (28:72, v/v) and ammonia 20%–isopropanol–ethyl acetate (6:15:79, v/v/v), blue spots can be observed after spraying with an aqueous solution of Cobalt II thiocyanate. The described method has been applied for toxicological studies and during monitoring of biological fluids of patients abusing these drugs or medically treating with them.

Makki et al.[61] describe a fast and easy method for the routine determination in plasma and suction blister fluids of 5- and 8-methoxypsoralen and of trimethylpsoralen, main components that are used in dermatology for the cure of psoriasis and vitiligo. Five different solid phases for HPTLC (silica, CN, RP_2, RP_8, RP_{18}) and various mobile phases have been examined. The best results, obtained with the use of RP_{18} plates, developed with a mobile phase of methanol–water (80:20, v/v) and with CN silica plates eluted with isobutylmethylketone–carbon tetrachloride (15:85, v/v), are derived from a statistically based study applying quite common techniques:

- Observation of the dependence of R_f value on composition solvent plotting this value against mobile phase composition.
- Observation of the variation of the difference between the R_f values of pairs of components with solvent composition.
- The use of overlapping resolution maps developed by Glajch.[62]

Petrovic et al.[63] describe a method for the separation of benzimidazoles (derivated from 2-amino-5,6-dimethylbenzimidazole, 2-aminobenzimidazole, and 5,6-dimethylbenzimidazole), with antibacterial and antifungal activity and their chromatographic behavior according to the chemical structure, on HPTLC plates, in direct phase of silica gel and in reversed phase on C_{18}, and with different mobile phases: nonpolar for the silica gel plates (hexane–1-propanol (80:20, v/v) containing three drops of aqueous ammonia; benzene–1-propanol (80:20, v/v) containing three drops of aqueous ammonia), and polar for the C_{18} (methanol–0.5 M aqueous ammonia (8:2, v/v) and acetonitrile–0.5 M aqueous ammonia (8:2, v/v)). For detection UV light at 254 nm was used.

Kincaid et al.[64] present a HPTLC method for the determination of salicylates in urines that is more sensitive and selective than Trinder's test and its modified derivatives commonly used for the screening of salicylate and its metabolites, based on the formation of colored compounds as a reaction to a ferric ion solution under moderate acidic conditions. This highly specific procedure is used in cases where low detection limits and high specificity are required, as in cases of treatment of cardiovascular or cerebrovascular diseases. The initial doses of the main component are low and the quantity of the main component and its metabolites found in the urine is about 2 to 6 mg/l. These properties are indispensable in cases of allergic attacks, asthma, and anaphylactic reactions. Furthermore it could be used for a prescreening of blood donors. The method is based on an analysis on HPTLC silica gel plates, developed with a mobile phase of benzene–acetic acid–diethylether–methanol (60:90:30:0.5, v/v/v/v), after extraction of the urine. After the plates have been dried and an analysis under UV light at 254 nm, the plates are sprayed with 5% aqueous ferric chloride, acidified with chloride acidic 0.05M as chromogen, and observed at 550 nm for confirmation and quantification. Thus it is possible to identify, together with the corresponding R_f: salicilic acid (0.70), diflunisal (0.70), aspirin (ASA)(0.67), methyl salicylate (0.67), gentisic acid (0.60), paraminosalycilic acid (0.57), and salicyuric acid (0.40). The detection limit is 1 ppm or less in comparison with more than 20 ppm for Trinder's test. Sensitivity is sufficient to give positive test results 48 hours after a single 80-mg dose of ASA by mouth or a 100-mg dose of methyl salicylate by skin injection with a muscle rub, and more than 96 hours after a 660-mg oral aspirin dose.

El Sadek et al.[65] describe a spectrodensitometric method for the determination of the components of two analgesic mixtures: paracetamol–ascorbic acid–caffeine–phenylephrine ("Rhino C") and phenazone–phenacetin–caffeine ("Caffemed"). The HPTLC method on silica gel plates with fluorescence indicator allows the separation of the pharmacologically active compounds and of the degradation products and impurities. The mobile phases used are dichloromethane–ethyl acetate–ethanol–formic acid (3.5:2:4:0.5, v/v/v/v) and dichloromethane–ethyl acetate–ethanol (5.5:1, v/v/v) for the first formulation and acetonitrile–chloroform (1:1, v/v) for the second. Wavelengths are 264 nm for ascorbic acid, 258 nm for phenazone, 274 nm for phenylephine and caffeine, 254 nm for paracetamol and phenacetin. The statistical analysis of the results obtained during the quantification phase confirms the sufficient accuracy and precision of this method, regarding quality and stability control in the pharmacological field.

Shinde et al.[66] describe an easy, precise, and rapid HPTLC method for simultaneous determination of paracetamol and diclofenac sodium in pharmaceutical preparations. The method is based on the extraction from the tablets, separation on HPTLC silica gel plates containing fluorescence indicator, developed with a mobile phase of chloroform–methanol–ammonia (10:25:0.1, v/v/v) and densitometric analysis at 286 nm with detection limits and R_f values of 50 ng and 0.75 for paracetamol, and 5 ng and 0.35 for diclofenac.

Tivert and Backman[67] apply TLC for the separation of the enantiomers of metoprolol, propanolol, and alprenolol characterized by different pharmacological activity. The separation, which is obtained on diol HPTLC plates containing a fluorescence indicator, utilizes the formation of the diastereomeric complex between the enantiomer and the chiral counter ion (N-benzyloxycarbonyl-glycyl-L-proline (ZGP)) present in the mobile phase of dichloromethane containing ZGP; the densitometric scanning is done at 280 and 300 nm. Furthermore the authors observed that the amount in ZGP and the humidity level of the plates, before the development, influence the chromatographic reproducibility and suggest keeping the plates in controlled humidity conditions. The method was applied to the analysis of a pharmaceutical formulation of metoprolol succinate.

Klimes and Kastner[68] report, in a detailed review with 50 references, of solid phases and mobile phases for the determination, also quantitative, of various benzodiazepines both from biological matrices as for pharmacological formulations.

Sioufi and Dubois[69] report, in an ample review (with 112 references), chromatographic methods that can be applied to the analysis of benzodiazepines, describing some TLC methods for quick identification of these compounds.

Weston et al.[70] report six benzodiazepines (clotiazepam, delorazepam, ethyl loflazepate, fludiazepam, haloxazolam, and oxazolam) and related benzophenones a GC-MS, a HPLC, a UV and a TLC method. For TLC analysis three different chromatographic conditions for the Rf values of the benzodiazepines and four of the benzophenones are given. The systems used for the benzodiazepines are:

1. Silica gel TLC plates pretreated with methanolic potassium hydroxide as stationary phase and a mobile phase of cyclohexane–toluene–diethylamine (75:15:10, v/v/v).
2. The same stationary phase and mobile phase of chloroform–methanol (90:10, v/v).
3. The same stationary phase and mobile phase of chloroform–acetone (80:20, v/v).

For the benzophenones the mobile phases used are toluene 100% v/v, toluene–isopropanol–ammonia (85:15:1, v/v/v), chloroform–methanol (90:10, v/v) and chloroform–acetone (80:20, v/v).

Wilson and Morden[71] demonstrate the capacities of the recent combination of off-line TLC fast atom bombardment mass spectrometry (FAB-MS) and MS-MS for the identification of various compounds with equal R_f but a different molecular weight. The compounds, polar and ionizable, were examined on silica gel HPTLC plates and developed in a mobile phase of chloroform–ethanol

(3:2, v/v). The analyzed compounds are hippuric acid, phenolphthalein and its glucuronide, paracetamol and paracetamol, sulfate. The combination FAB-MS and MS-MS enables us to obtain spectra devoid of matrix interferences containing only ions from the compounds of interest.

Scotto di Tella and Di Nunzio[72] report a method for rapid screening, permitting an orientation of the operator of toxicological studies toward particular classes of compounds. The study focuses on some atypical antidepressant drugs, of recent appearance on the market, which seem to gain importance in cases of acute voluntary and incidental intoxication. The six compounds are amineptina, minaprina, toloxatone, trazodone, and viloxazina apart from two tricyclic antidepressants like amitriptilina and nortriptilina. After extraction from plasma and urine, a bidimensional TLC on silica gel plates containing a fluorescence indicator with three different elution systems, is carried out: methanol–32% ammonia (100:1.5, v/v), cyclohexane–toluene–diethylamine (75:15:10, v/v/v), and ethyl acetate-dichloromethane–32% ammonia–methanol (80:90:15:5, v/v/v/v). The detection is obtained with seven different systems: Dragendorff's reagent, Marquis's reagent, Mandelin's reagent, Lieberman's reagent, sulphuric acid–ethanol (4:1, v/v), iodoplatinate spray, UV-VIS at 254 nm and 366 nm. This method gives semiquantitative results with detection limits of 5 µg/ml for the viloxazina and the minaprina and 2 µg/ml for the other compounds.

F. VITAMINS

Corti et al.[73] used a TLC densitometry method for the detection of vitamin B1 in the presence of hydrolytic and oxidation products to follow the process of degradation under different conditions of temperature-time treatment. HPTLC-NH_2 layers with a mobile phase consisted of methanol–water (1:0.2, v/v) and densitometric detection at 254 nm and at 366 nm as excitation wavelength for the thiochrome were used.

Diaz et al.[74] applied a TLC separation, followed by a fluorometric detection with an optic-fiber instrumentation for the separation and quantitative determination of thiamine, riboflavin, and niacin. Riboflavin shows native fluorescence, but thiamine and niacin had to be converted to fluorescent derivatives before the chromatographic analysis. To label the niacin a fluorescent tracer is used, fluoresceinamine isomer II. Thiamine is converted to fluorescent thiochrome by oxidation with potassium ferricyanide solution in aqueous sodium hydroxide. The plates used were HPTLC silica gel developed in a mobile phase of methanol–water (70:30, v/v). Detection limits and R_f values found were 7.4 ng and 0.73 for thiamine, 3.1 ng and 0.86 for riboflavin, and 2.6 ng and 0.91 for niacin. The plates were then scanned by a bifurcated optical fiber that both transmits emission radiation to the plate and collects the emission signal to the fluorometer.

Bargagna et al.[75] describe three methods of TLC, HPTLC, and HPLC for qualitative analysis of all-*trans*-retinoic acid, tretinoin, present in dermatological creams and solutions for the treatment of acne and for its anti-aging properties. This active principle isomerizes to sunlight; thus it is very important to evaluate this process in pharmaceutical preparations. The methodology is performed by a chromatographic development on HPTLC silica gel plates or reversed phase plates and successive visualization under UV light at 254 and 366 nm. For the direct phase two different mobile phases were used, according to the polarity of the retenoid compounds analyzed: petroleum ether–acetone (41:9, v/v) and cyclohexane–diethylether–acetone–acetic acid (30:20:1:0.5, v/v/v/v), and for reversed phase acetonitrile–water (45:5, v/v). Both HPLC analysis and TLC reversed phase are applied in stability tests and quality control of drugs and cosmetics containing retinoids.

REFERENCES

1. **Poole, C. F. and Poole, S. K.,** Instrumental thin layer chromatography, *Anal. Chem.*, 66, 27A, 1994.
2. **Jork, H., Keller, G., and Kocher, U.,** Application of AMD to the determination of crop-protection agents in drinking water. Part II: Limitations, *J. Planar Chromatogr.*, 5, 4, 1992.

3. Fried, B. and Sherma, J., *Thin Layer Chromatography: Practice and Applications*, 3rd ed., Marcel Dekker, New York, 1994.
4. Massart, D. L., The use of information theory in thin layer chromatography, *J. Chromatogr.*, 79, 155, 1973.
5. Massart, D. L., Vandegiste, B. G. M., Deming, S. N., Michutte, Y., and Kaufmann, K., *Chemometrics: A Textbook*, Elsevier, New York, 1988, 130.
6. Szepesi, G. and Nyiredy, S., Planar chromatography: current status and future perspectives in pharmaceutical analysis – I. Applicability, quantitation and validation, *J. Pharm. Biomed. Anal.*, 10, 1007, 1992.
7. Nyiredy, S. and Szepesi, G., Planar chromatography: current status and future perspectives in pharmaceutical analysis (Short Review)–II. Special techniques and future perspectives in planar chromatography, *J. Pharm. Biomed. Anal.*, 10, 1017, 1992.
8. Duez, P., Chamart, S., and Hanocq, M., Postchromatographic derivatization in quantitative thin layer chromatography: pharmaceutical applications, *J. Planar Chromatogr.*, 4, 69, 1991.
9. Schütt, H. and Hölzl, J., Vergleichende qualitätsuntersuchung von johanniskraut-fertigarzneimitteln unter verwendung verschiedener quantitativer bestimmungsmethoden, *Pharmazie*, 49, 206, 1994.
10. Singh, A. K., Granley, K., Misrha, U., Naeem, K., White, T., and Jiang, Y., Screening and confirmation of drugs in urine: interference of hordenine with the immunoassays and thin layer chromatography methods, *Forensic Sci. Int.*, 54, 9, 1992.
11. Sosa, M. E., Valdés, J. R., and Martinez, J. A., Determination of ajmaline stereoisomers by combined high-performance liquid and thin layer chromatography, *J. Chromatogr.*, 662, 251, 1994.
12. Vampa, G., Benvenuti, S., and Melegari, M., Determination of glycyrrhizin in pharmaceutical preparations and in licorice products by high-performance thin layer chromatography, *Acta Technol. Legis Med.*, 2, 87, 1991.
13. Ponder, G. W. and Stewart, J. T., High-performance thin layer chromatographic determination of digoxin and related compounds, digoxigenin bisdigitosside and gitoxin, in digoxin drug substance and tablets, *J. Chromatogr.*, 659, 177, 1994.
14. Vetticaden, S. J. and Chandrasekaran, A., Chromatography of cardiac glycosides, *J. Chromatogr.*, 531, 215, 1990.
15. Lavanya, K. and Baggi, T. R., An improved thin layer chromatographic method for the detection and identification of cannabinoids in cannabis, *Forensic Sci. Int.*, 47, 165, 1990.
16. Funk, W. and Dröschel, S., Qualitative and quantitative HPTLC determination of cocaine, ecgonine, ecgonine methyl ester and benzoylecgonine, *J. Planar Chromatogr.*, 4, 123, 1991.
17. Duez, P., Milcamp, A., Lompo, M., Guissou, P., and Hanocq, M., Comparison of HPTLC-fluorodensitometry and HPLC for the assay of strictosamide in the leaves, root and stem bark of *Nauclea latifolia*, *J. Planar Chromatogr.*, 7, 5, 1994.
18. Corti, P., Corbini, G., Dreassi, E., Murratzu, C., and Celesti, L., Thin layer densitometry in quantitative assay of drugs. Assay of reserpine and chlortalidone in the presence of the potential impurities of their solid pharmaceutical forms, *Pharm. Acta Helv.*, 65, 222, 1990.
19. Nikolova-Damyanova, B., Ilieva, E., Handjieva, N., and Bankova, V., Quantitative thin layer chromatography of iridoid and flavonoid glucosides in species of *Linaria*, *Phytochem. Anal.*, 5, 38, 1994.
20. Zhou, B., Luo, S., and Cai, H., Determination of allantoin in common yam (*Dioscorea opposita*) by TLC-scanning, *Zhongcaoyao*, 25, 132, 1994.
21. Naidong, W., Cachet, T., Roets, E., and Hoogmartens, J., Identification of tetracyclines by TLC, *J. Planar Chromatogr.*, 2, 424, 1989.
22. Sekkat, M., Fabre, H., Simeon, De Bouchberg, M., and Mandrou, B., Determination of aminoglycosides in pharmaceutical formulations — I. Thin layer chromatography, *J. Pharm. Biomed. Anal.*, 7, 883, 1989.
23. Kang, J. S. and Ebel, S., Identification and quantitation of tetracycline antibiotics by cyanophase HPTLC, *J. Planar Chromatogr.*, 2, 434, 1989.
24. Naidong, W., Hauglustaine, C., Roeets, E., and Hoogmartens, J., Assay and purity control of chlortetracycline and demeclocycline by TLC: a comparison with LC, *J. Planar Chromatogr.*, 4, 63, 1991.
25. Kovács-Hadady, K., OPLC study of the retention behavior of several tetracyclines on thin layers impregnated with tricaprylmethylammonium chloride, *J. Planar Chromatogr.*, 4, 456, 1991.
26. Eneva, G., Nikolova-Damyanova, B., Spassov, S., and Haimova, M., Determination of nebramycin components by TLC and densitometry, *J. Planar Chromatogr.*, 3, 232, 1990.
27. Kovács-Hadady, K., and Szilágyi, J., Retention behavior of several natural and semisynthetic penicillins and cephalosporins by means of OPLC, using silica gel impregnated with TCMA, *J. Planar Chromatogr.*, 4, 194, 1991.
28. Corti, P., Corbini, G., Dreassi, E., Politi, N., and Montecchi, L., Thin layer chromatography in the quantitative analysis of drugs. Determination of rifaximine and its oxidation products, *Analysis*, 19, 257, 1991.
29. Quintens, I., Eykens, J., Roets, E., and Hoogmartens, J., Identification of cephalosporins by thin layer chromatography and color reactions, *J. Planar Chromatogr.*, 6, 181, 1993.
30. Maxwell, R. J. and Unruh, J., Comparison of induced vapor phase fluorescent responses of four polycyclic ether antibiotics and several lipid classes on RP-18 and silica gel HPTLC plates, *J. Planar Chromatogr.*, 5, 35, 1992.
31. Szepesi, G., Gazdag, M., Pap-Sziklay, Zs., and Végh, Z., Problem of semiquantitative TLC methods prescribed in the United States and British pharmacopoeias for the purity testing of sulphinpyrazone, *J. Pharm. Biomed. Anal.*, 4, 123, 1986.

32. **Madelaine-Dupuich, C., Azema, J., Escoula, B., Rico, I., and Lattes, A.,** Analysis of N-acyl aminonaphthalene sulphonic acid derivatives with potential anti-human immunodeficiency virus activity by thin layer chromatography and flame ionization detection, *J. Chromatogr.*, 653, 178, 1993.
33. **Salomies, H.,** Quantitative HPTLC of sulfonamides in pharmaceutical preparations, *J. Planar Chromatogr.*, 6, 337, 1993.
34. **García Sánchez, F., Navas Díaz, A., and Fernández Correa, M. R.,** Image analysis of photochemically derivatized and charge-coupled device-detected phenothiazines separated by thin layer chromatography, *J. Chromatogr.*, 655, 31, 1993.
35. **Corti, P., Dreassi, E., Politi, N., and Valoti, M.,** Thin layer chromatography in the quantitative analysis of pharmaceuticals. Comparison of layers in the analysis of a complex mixture of phenotiazine derivatives, *Pharm. Acta Helv.*, 66, 329, 1991.
36. **Lin Ling, B., Baeyens, W. R. G., Del Castillo, B., Imai, K., De Moerloose, P., and Stragier, K.,** Determination of thiols of biological and pharmacological interest by high-performance thin layer chromatography and fluorescence scanning densitometry, *J. Pharm. Biomed. Anal.*, 7, 1663, 1989.
37. **De Brabander, H. F., Batjoens, P., and Van Hoof, J.,** Determination of thyreostatic drugs by HPTLC with confirmation by GC-MS, *J. Planar Chromatogr.*, 5, 124, 1992.
38. **Brown, S. M. and Busch, K. L.,** Direct identification and quantitation of diuretic drugs by fast atom bombardment mass spectrometry following separation by thin layer chromatography, *J. Planar Chromatogr.*, 4, 189, 1991.
39. **Vingler, P., Kermici, M., and Krien, P.,** Direct quantitative digital autoradiography-thin layer chromatography of 3α,3β- and 5α-reduced and 17β-dehydrogenated androgens derived from testosterone metabolism, *J. Chromatogr.*, 571, 73, 1991.
40. **Medina, M. B. and Schwartz, D. P.,** Thin layer chromatographic detection of zeranol and estradiol in fortified plasma and tissue extracts with Fast Corinth V, *J. Chromatogr.*, 581, 119, 1992.
41. **Datta, K. and Das, S. K.,** Densitometric quantification of corticosteroid sodium phosphate salts in parenteral preparations or eye and ear drops after reversed phase ion pair TLC, *J. Planar Chromatogr.*, 6, 204, 1993.
42. **Petrovic, S. M., Acanski, M., Pejanovic, V. M., and Petrovic, J. A.,** Retention behavior of estradiol and estrone derivatives in normal and reversed phase thin layer chromatography, *J. Planar Chromatogr.*, 6, 29, 1993.
43. **Smets, F., De Brabander, H. F., Bloom, P. J., and Pottie, G.,** HPTLC of anabolic compounds in injection sites, *J. Planar Chromatogr.*, 4, 207, 1991.
44. **Szepesi, G.,** The role of thin layer chromatography in steroid analysis, *J. Planar Chromatogr.*, 5, 396, 1992.
45. **Schütz, H. and Meister, T.,** Thin layer chomatographic screening program for commonly used β-blockers, *Arzeim.-Forsch.*, 40, 651, 1990.
46. **Davies, C. L.,** Chromatography of β-adrenergic blocking agents, *J. Chromatogr.*, 531, 131, 1990.
47. **Betschart, B., Sublet, A., and Steiger, S.,** Determination of antimalarial drugs under field conditions using thin layer chromatography, *J. Planar Chromatogr.*, 4, 111, 1991.
48. **Abdelrahman, A. N., Karim, E. I. A., and Ibrahim, K. E. E.,** Determination of chloroquine and its decomposition products in various brands of different dosage forms by liquid chromatography, *J. Pharm. Biomed. Anal.*, 12, 205, 1994.
49. **Dreassi, E., Celesti, L., Ceramelli, G., Savini, L., and Corti, P.,** Thin layer chromatography in pharmaceutical quality control. Assay of inosiplex in different pharmaceutical forms, *Pharm. Acta Helv.*, 67, 341, 1992.
50. **Zhou, Z. and Dong, Z.,** Determination of oleanolic acid in its β-cyclodextrin inclusion-complex by TLC-scanning, *Zhongguo Yiyuan Yaoxue Zazhi*, 14, 29, 1994.
51. **Parimoo, P., Bharathi, A., and Shajahan, M.,** Determination of oxyphenbutazone and ibuprofen in the presence of paracetamol and dextropropoxyphene in dosage forms by quantitative thin layer chromatography, *Indian Drugs*, 31, 139, 1994.
52. **Lillsunde, P. and Korte, T.,** Comprehensive drug screening in urine using solid-phase extraction and combined TLC and GC/MS identification, *J. Anal. Toxicol.*, 15, 71, 1991.
53. **Corti, P., Cenni, A., Corbini, G., Dreassi, E., Murratzu, C., and Caricchia, A. M.,** Thin layer chromatography and densitometry in drug assay: comparison of methods for monitoring valproic acid in plasma, *J. Pharm. Biom. Anal.*, 8, 431, 1990.
54. **Shinde, V. M., Desai, B. S., and Tendolkar, N. M.,** Selective determination of nitrendipine and nimodipine in pharmaceutical dosage by high performance thin layer chromatography, *Indian Drugs*, 31, 119, 1994.
55. **Bieganowska, M. L., Petruczynik, A., and Doraczynska-Szopa, A.,** Ion pair, reversed phase thin layer chromatography of some basic drugs and related pyridine derivatives, *J. Planar Chromatogr.*, 5, 184, 1992.
56. **Tománková, H. and Sabartová, J.,** Determination of impurities in azidothymidine by HPTLC densitometry and HPLC, *Chromatographia*, 29, 562, 1990.
57. **Vampa, G., Benvenuti, S., and Pecorari, P.,** Determination of the reactivity of uracil derivatives with respect to methyl iodide by high-performance thin layer chromatographic densitometry, *J. Chromatogr.*, 604, 261, 1992.
58. **Abdoul-Einen, H. Y. and Serignese, V.,** Thin layer chromatographic (TLC) determination of phenytoin in pharmaceuticals formulations and identification of its hydroxylated urinary metabolites, *Anal. Lett.*, 27, 723, 1994.
59. **Van Boven, M. and Daenens, P.,** Analysis and identification of azaperone and its metabolites in human, *J. Anal. Toxicol.*, 16, 33, 1992.

60. **Láng, K. L.,** Some observations on the determination of organic bases with cobalt(II) thiocyanate for the identification of synthetic pharmacological products, *Microchem. J.*, 41, 191, 1990.
61. **Makki, S., Thanh, T. T., Chinkarenko, T., and Guinchard, C.,** Optimizing the one-dimensional planar chromatographic separation of the principal psoralens used in therapy, *J. Planar Chromatogr.*, 4, 213, 1991.
62. **Glajch, J. L., Kirkland, J. J., Squire, K. M., and Minor, J. M.,** Optimization of solvent strength and selectivity for reversed-phase liquid chromatography using an interactive mixture-design statistical technique, *J. Chromatogr.*, 199, 57, 1980.
63. **Petrovic, S. M., Acanski, M., Perisic-Janjic, N. U., and Vlaovic, D.,** Separation and retention behavior of some newly synthesized benzimidazoles in normal and reversed phase thin layer chromatography, *J. Planar Chromatogr.*, 4, 475, 1991.
64. **Kincaid, R. L., McMullin, M. M., Sanders, D., and Rieders, F.,** Sensitive, selective detection and differentiation of salicylates and metabolites in urine by a simple HPTLC method, *J. Anal. Toxicol.*, 15, 270, 1991.
65. **El Sadek, M., El Shanawany, A., Aboul Kier, A., and Rücker, G.,** Determination of the components of analgesic mixtures using high performance thin layer chromatography, *Analyst*, 115, 1181, 1990.
66. **Shinde, V. M., Tendolkar, N. M., and Desai, B. S.,** Simultaneous determination of paracetamol and diclofenac sodium in pharmaceutical preparations by quantitative TLC, *J. Planar Chromatogr.*, 7, 50, 1994.
67. **Tivert, A. M. and Backman, Å.,** Separation of the enantiomers of β-blocking drugs by TLC with a chiral mobile phase additive, *J. Planar Chromatogr.*, 6, 216, 1993.
68. **Klimes, J. and Kastner, P.,** Thin layer chromatography of benzodiazepines, *J. Planar Chromatogr.*, 6, 168, 1993.
69. **Sioufi, A. and Dubois, J. P.,** Chromatography of benzodiazepines, *J. Chromatogr.*, 531, 459, 1990.
70. **Weston, S. I., Japp, M., Partridge, J., and Osselton, M. D.,** Collection of analytical data for benzodiazepines and benzophenones, *J. Chromatogr.*, 538, 277, 1991.
71. **Wilson, I. D. and Morden, W.,** Application of thin layer chromatography-mass spectrometry to drugs and their metabolites: advantages of tandem MS-MS, *J. Planar Chromatogr.*, 4, 226, 1991.
72. **Scotto di Tella, A. and Di Nunzio, C.,** Separazione in cromatografia su strato sottile bidimensionale di otto antidepressivi, *Riv. Tossicol. Sper. Clin.*, 18, 93, 1988.
73. **Corti, P., Caricchia, A. M., Franchi, G., Lencioni, E., Murratzu, C., and Corbini, G.,** Contrôle densitométrique des produits de dégradation de la vitamine B_1 et calcul des facteurs d'activation, *Ann. Pharm. Fr.*, 47, 117, 1989.
74. **Navas Díaz, A., Guirado Paniagua, A., and García Sánchez, F.,** Thin layer chromatography and fiber-optic fluorimetric quantitation of thiamine, riboflavin, and niacin, *J. Chromatogr.*, 655, 39, 1993.
75. **Bargagna, A., Mariani, E., and Dorato, S.,** TLC, HPTLC, and HPLC determination of *cis*- and *trans*-retinoic acids, retinol and retinyl acetate in topically applied products, *Acta Technol. Legis Med.*, 2, 75, 1990.

Chapter 12

PLANAR CHROMATOGRAPHY APPLICATIONS IN VETERINARY TOXICOLOGY

H. M. Stahr

CONTENTS

I. Introduction 250

II. Applications 251
   A. Pesticide Analysis—Insecticides 251
      1. Carbamates 251
         a. Sample Preparation 251
         b. Planar Chromatography 252
      2. Organophosphorus (OP) Insecticides 252
         a. Sample Preparation 252
         b. Planar Chromatography 253
      3. Organochlorine (OC) Insecticides 253
         a. Sample Preparation 253
         b. Planar Chromatography 253
      4. Pyrethroid Insecticides 253
         a. Sample Preparation 254
         b. Planar Chromatography 254
      5. General Considerations 255
         a. TLC Separation 255
         b. Confirmation of Identity 255
   B. Rodenticides–Avicides 255
      1. Rodenticide Anticoagulants 255
         a. Sample Preparation 255
         b. Planar Chromatography 255
      2. Avicides and Alkaloids 256
         a. Sample Preparation 256
         b. Planar Chromatography 256
      3. General Conclusions 256

III. Food and Feed Additives 257
   A. Vitamins A, E, and D 257
      1. Nutrient Analysis 257
         a. Extraction 257
         b. Planar Chromatography 257
   B. Selenium 258
      1. Sample Preparation 258
      2. Planar Chromatography 258
   C. Ionophores 258
      1. Monensin 258
         a. Sample Preparation 258
         b. Planar Chromatography 258

        2. Other Ionophores 259
            a. Sample Preparation 259
            b. Planar Chromatography 259
        D. Sulfa Drugs 259
            1. Sample Preparation 259
            2. Planar Chromatography 259
        E. Antibiotics/Antimicrobials 259
            1. Sample Preparation 259
            2. Planar Chromatography 260

    IV. Mycotoxins 260
        A. Sample Preparation 260
        B. Planar Chromatography 261
            1. Classic Mycotoxins 261
            2. Fumonisins 262

References 264

I. INTRODUCTION

Pathologists and veterinary toxicologists have a natural affinity for planar chromatography. They are trained to use their visual senses to observe and their deductive powers to interpret what they see.

The separation process itself can be controlled so that solvent strength, planar surface, and visualization methods can be used to determine the most difficult analytes in the most challenging matrices. The goal is to identify an unknown chemical that has caused a harmful (fatal) effect on a biological system (animal). Even if the identity of the substance is not known, its effect can be used with screening tests to try to find the source of the problem. Carbamate, organophosphorus, organochlorine, and pyrethroid pesticides can be grouped and analyzed to try to select the compound of concern if a pesticide is suspected, but the specific chemical involved is not known. Identifying the substance is critical to finding the source and preventing further harm from the exposure.

Alkaloids can be grouped to determine if one of these central nervous system–directed toxic chemicals is the cause of the symptoms observed by the diagnostician. Our toxicology group has multiple alkaloid screens, one for economic poisons (strychnine, avitrol, nicotine, and caffeine) and one for mycotoxin tremorgens (aflatrem, penitrems, roufortine, and lysergic acid-like compounds such as ergotamines). Ionophores (monensin, narasin, lasalocid, and salinomycin) are used as feed additives and antimicrobials and can be screened by one sample preparation and chromatography analysis. The same concept can be followed for mycotoxins (zearalenone, ochratoxin, aflatoxins, T-2 toxin, and vomitoxin), rodenticides (warfarin, brodificoum) and drugs (barbiturates, amphetamines, antimicrobials sulfas, and tetracyclines).

These qualitative assessments are best done by a combination of selective cleanups, multiple planar chromatography surfaces, and solvent systems to allow the greatest differentiation of the analytes. With planar chromatography surfaces that vary from silica and alumina to C_{18} reverse phase and polymer, separations of analytes ranging from pesticides to heavy metals are possible.

Quantitative TLC can be done once the identity is established by direct densitometric analysis or removing the bands by scraping and elution followed by spectroscopy. Instrumental analysis is essential to speciate unknown analytes and to confirm that a suspected carbamate or drug is really that substance.

In legal cases, the TLC plate may be photographed for the record. The spectroscopic data become invaluable documentation to show in courtrooms, usually with a contentious audience and many months after the submission of the original case. Planar chromatography allows screening, confirmation, and quantitation depending on the gravity and economics of the case.

Other means of confirmation are available as well. These may not be as esoteric, but are much less expensive. Compounds that inhibit acetylcholinesterase (ACHE) can be detected by removing bands from the TLC plate, exposing the bands to ACHE, and then determining the activity of the ACHE.

In a similar manner, bands that are suspected to be toxic can be removed and tested for toxin by bioluminescent bacteria (Microtox®), or a genotoxic effect (Mutatox®), or tissue cultures sensitive to drugs, hormones, or other biological effects being studied. These tests cost tens of dollars rather than hundreds of dollars, and the capital investment is much less than that of modern instrumentation.

Derivatization allows selective confirmation of analytes with fluorescamine or *o*-phthalaldehyde for amines; dansyl or ester functions for acids; (TCBI) for carbamates; anisaldehyde for organophosphates; dimethylaminobenzaldehyde for amines and sulfas; or dimethylaminocinnaminaldehyde for sulfas or arsenicals.

With the right combination of derivatization techniques and selective chromatography, the identity of the suspected substance can be established to the satisfaction of the most exacting analyst. An illustration is the procedure developed by Rottinghaus[16] and described below. This procedure uses fluorescence detection, chemical hydrolysis, and rechromatography and detection of the hydrolyzate product of fumonisins. These simple steps make very selective analysis possible in complex samples.

The compactness, facilitation, and dependability of TLC allowed FSIS/USDA to use this technique to control sulfa residues in the meat industry (S.O.S. test). Many veterinary clinics obtained the equipment to do this test in the field. It allows preharvest testing and residue abatement in livestock before the residues become a problem at the slaughterhouse.

Veterinary toxicology in the field, in the clinic, and in the diagnostic lab is well served by thin-layer chromatography. Rapid screens for clinical cases or acute poisonings are possible. Quantitative, documented data for legal or regulatory situations are obtainable.

Unfortunately, thin-layer chromatography is still underutilized. Veterinarians should be taught TLC techniques to increase their competence to make them better analysts, and to train them in deductive reasoning and problem solving concepts that are essential to the successful investigator. For this purpose, we have detailed techniques used in veterinary toxicology at our institution.

II. APPLICATIONS

A. PESTICIDE ANALYSIS – INSECTICIDES
1. Carbamates[1]
a. Sample Preparation

Carbamates are relatively labile chemicals. Strong bases are to be avoided to prevent losses due to chemical reactions. The solubility in petroleum solvents is also limited. Sample extraction is normally done by acetonitrile or methanol to solubilize the carbamate pesticide. A small amount of water allows actively absorbed pesticides to be released into the solvent. The principle devised by Mills et al.[2] of defatting by hexane extraction is then used. Partitioning into methylene chloride ($MeCl_2$) is accomplished by adding water to the methanol or acetonitrile (with 2% salt to assist in the transfer of analyte) and an equal volume of methylene chloride to receive the pesticide. The $MeCl_2$ is concentrated, and an aliquot representing 1 g of sample is spotted from a minimum of solvent on the TLC plates. The alternative procedure is a solid phase extraction (SPE) concentration step. A portion of the sample representing 5 g of sample is diluted with water to make it < 20% acetonitrile (ACN) or methanol (MeOH). This solution is then passed through a SPE C_{18} cartridge. The carbamate will remain on the SPE column. It may be removed by elution with $MeOH–H_2O$

(75:25) and then partitioned into $MeCl_2$ or eluted directly from the SPE cartridge with ethyl acetate (ETAC) if the extract is clean.

b. Planar Chromatography

The extract is concentrated and spotted on TLC plates. Two kinds of plates are used routinely for analysis of carbamates: C_{18} reverse phase TLC (RPTLC) plates and silica gel normal phase TLC (NPTLC) plates.

The solvent used to develop the RPTLC plates is ethanol–water–acetic acid (65:35:1) (Table 12.1). The carbamates may be visualized by fluorescence quenching (FQ) or by fluorescence (Fl) after hydrolysis and fluorescamine derivatization. The sensitivity is 10 ng by FQ or ng by Fl. By spraying the plate with TCBI, fractions in micrograms can be detected. The color of the bands is blue for most carbamates.[1]

TABLE 12.1
Carbamate Pesticides

Carbamates	R_f Values	
	NP	RP
Carbofuran	0.37	0.43
Carbaryl	0.45	0.40
Methomyl	0.09	0.00
Baygon®	0.44	0.44
Temik®	0.20	0.00
Temik Sulfone	0.01	0.00
Landrin®	0.47	0.32

Note: NP = Normal phase high performance silica gel (Whatman); toluene–ethyl acetate (4:1). RP = Reverse phase C_{18} (Whatman); ethanol–water–acetic acid (65:35:1 + 0.5% sodium chloride).

Alternatively, NPTLC may be used with channeled preadsorbent (Whatman) and unchanneled (Merck) silica gel plates. The preadsorbent plates are useful for spotting urine, body fluids, or other samples that are difficult to handle. The development solvent used for carbamates is toluene–ethyl acetate (4:1) (see Table 12.1). The same visualization method is used for either silica gel plate.

A strategy used in our laboratory is to visualize the bands on a phosphorus-containing plate under UV light to determine if separation is complete (FQ). If bands are not indicative of optimum separation, a second or third development is done before the chemical conversion process is used to visualize and help confirm the identity of the analyte.

The combination of fluorescence quenching/fluorescence derivatization/TCBI and aniseldehyde are used as multiple detection methods so that optimum separation and identification are possible.[3]

Thiocarbamates, anilines, and similar compounds may be separated and, if convertible to amines, can be detected by this technique. Hydrolysis and reaction products of carbamates are often detectable by these techniques. A profile of stomach contents is helpful to use to detect poisonings by ten of the parent compounds in very low concentrations.

2. Organophosphorus (OP) Insecticides
a. Sample Preparation

The extraction of organophosphorus (OP) pesticides is possible using the procedure under carbamates above. The differences are that some OPs are more soluble in petroleum solvents, and these can be used for selective partitioning or elution from silica or Florisil® columns. The OP insecticides are substantially less labile than the carbamate insecticides and may, therefore, be subjected to more active adsorbents and higher temperature gas chromatography analysis. It is still preferable to use a polar partition solvent to receive the analytes. This assists in reducing the lipid residues that are co-partitioned and increases the likelihood of not "missing" a relatively less soluble OP substance.

b. Planar Chromatography

After the partition and cleanup of extracts, they are concentrated and spotted on silica gel plates and developed with heptane–dimethylformamide (90:10) (Table 12.2). The same detection strategy is used, fluorescence quenching to screen for possible residues, then color development to confirm. The OP insecticides give characteristic colors with anisaldehyde spray reagent and heat. They are not persistent, though, so for densitometric determination they should be oversprayed with TCBI, which produces a stable colored band for quantitation.

TABLE 12.2
R_fs of Organophosphorus Pesticides Relative to Methyl Parathion

Pesticide	R_f^*
Counter®	2.83
Thimet®	2.67
Dyfonate®	2.50
Diazinon	1.33
Malathion	0.67
Methyl parathion	1.00

* Normal phase Merck silica gel 60; heptane–dimethylformamide (90:10).

3. Organochlorine (OC) Insecticides

a. Sample Preparation

OC insecticides are covered third, not because they are unimportant, but because the order of concern right now is carbamate-organophosphorus, chlorinated hydrocarbons, and pyrethroid insecticides for diagnostic toxicologists.

The organochlorine insecticides, polychlorinated biphenyls (PCBs), and chlorinated carboxylic acid esters all lend themselves to the aforementioned extraction, defatting, and cleanup by selective absorption by normal or reverse phase column chromatography.[2] They have the advantage that heat volatization (Sweepco distillation) can also be used for cleanup, since they are relatively stabile.[5] The PCBs can be eluted from silica or Florisil® with petroleum ether, which makes speciation and analysis much more readily done.[1] After extraction, defatting, partitioning, and cleanup, concentration is done with nitrogen effusion, and the extracts are spotted on TLC plates along with appropriate standards.

b. Planar Chromatography

NPTLC and RPTLC are used to screen for the OCs by comparing the observed bands with standard compounds. Often, the two or three most common, or the compounds with extreme migration rates, are chosen to represent a whole series of possible analytes. See Table 12.3 for R_f values.

The bands are made visible if silver ions are impregnated into the absorbent layer before development[6] (e.g., Analtech Ag–containing plate), and after development of the plate, it is irradiated with shortwave UV to reduce the Ag; chlorine-containing compounds are detected as dark spots. Alternatively, $AgNO_3$ may be sprayed on the plate post-development.[1]

4. Pyrethroid Insecticides

This group of insecticides is less analyzed than the other groups. They are used extensively and probably should be of more concern. They are as toxic to warm-blooded animals as the chlorinated hydrocarbons and are now used more frequently as insecticides.

TABLE 12.3
R_f Values of Common Chlorinated Pesticides Relative to Aldrin

Pesticide	R_f
Chlordane	0.81[a]
o,p-DDT	0.79
o,p-DDE	0.79
o,p-DDD	0.50
Hexachlorobenzene	1.11
2,4,6-TCP	0.19
PCP	0.13
2,4D-Butylester	0.45
2,4D-Acid	0.00
BHC	0.31
Lindane	0.47
Aldrin®	1.00
Dieldrin®	0.59
PBB	1.00
Arochlor-1254®	1.00[a]
Toxaphene	0.75[a]
Heptachlor®	0.91
Endrin®	0.66

[a] Gives a broad band
* Analtech silica gel plates impregnated with $AgNO_3$; heptane-ethyl acetate (99:1)

a. Sample Preparation

As with the OC hydrocarbons, pyrethroids can be extracted with acetonitrile, the extract defatted, partitioned into hydrocarbons or methylene chloride, concentrated, and cleaned up by absorption chromatography on normal and reverse phase sorbents.

b. Planar Chromatography

The extracts can be analyzed by HPTLC and RPTLC as for the chlorinated hydrocarbons. The pesticides can be visualized by fluorescence quenching on layers containing fluorescent indicator.

The R_f values of the common pyrethroid insecticides are shown in Table 12.4.

"Natural" pyrethrins turn rose to purple and give multiple spots with anisaldehyde spray reagent. The two main spots turn purple. This differentiates them from synthetic pyrethroids.

TABLE 12.4
Thin-Layer Chromatography of Pyrethroids

Compound	Reverse Phase Thin-Layer Chromatography (RPTLC) R_f	Normal Phase Thin-Layer Chromatography (NPTLC) R_f
cis-Permethrin	0.70	0.57
trans-Permethrin	0.60	0.50
ASANA®	0.50	0.43
Pyrethrin		
(Natural)	0.30	0.33
(2 Large Bands)	0.25	0.40

Visualization — Short wave UV light. Plates RPTLC-KC18F Whatman; heptane mobile phase. Plates NPTLC-Merck Silica Gel 60; heptane–ethyl acetate (9:1 v/v) mobile phase.

5. General Considerations
a. TLC Separation
The same conventions used with other analytes are used with insecticides. The smallest possible spotting volume is used to assure the greatest resolving power of the process. Multiple developments in the same direction can resolve interferences from co-extracted materials. A weaker solvent may be used if the bands of interest are migrating too rapidly to allow another development. With reverse phase plates, a complete change of solvent, such as methanol–methylene chloride (65:35), makes ionophores (e.g., monensin) move to R_f O.5. A more typical solvent, ethanol–water–acetic acid (65:35:1), does not cause monensin to move from the origin. This is an excellent cleanup device. First, move other compounds up the plate in the latter solvent, then redevelop in the former solvent to move the rumensin. Extraction of a band and respotting on another plate has been used to improve an imperfect separation caused by band proximity to the top of the plate.[3]

b. Confirmation of Identity
The extraction of the analytes so that spectral confirmation can be made is strongly recommended. The lack of definitive speciation from observation of band or peak position is a serious concern. The lack of speciation in gas chromatography of residues has resulted in multiple columns being used to determine retention times. The same principle can be used with planar chromatography. Multiple separations using different plate surfaces and different development solvents are highly recommended. Removing bands and making ultraviolet, infrared, and mass spectral analysis is even more recommended for certainty in assigning identities of the substances.

B. RODENTICIDES–AVICIDES
Currently, the most popular rodenticides used are anticoagulants, alkaloids (strychnine), and organophosphorus insecticides. The primary avicide used is avitrol. These became favored agents to use to kill pets, so they are used in panels to screen for possible poison agents.[1]

1. Rodenticide Anticoagulants
The principal anticoagulants used presently are brodificoum and bromodiolone. There are still diphacinone, chlorphacinone, and warfarin available on farms and probably other compounds like pival pindone in agricultural communities and practice.

The compounds usually persist for perhaps a day in sufficient quantities to measure in the liver of exposed animals, or in the case of acute death, stomach contents. If the animal has lived several days and then succumbed, it will be necessary to use the most sensitive means to detect residues.[7]

a. Sample Preparation
Baits need an acid/solvent extraction to dissolve the coumarin compounds. Defatting this solvent is recommended. Baits usually are concentrated enough so that merely concentrating the solvent and analyzing a sample equivalent to 0.1 gram will suffice to identify it.[5]

Stomach contents are handled like a complex sample according to the procedures used for insecticides, i.e., extraction with acetonitrile/acid, defatting, partitioning into methylene chloride, concentration, and absorption column cleanup using silica gel, Florisil®, or RP silica SPE column cleanup.

Liver samples are extracted with 2% citric acid in methylene chloride using celite as a filter aid. Once the extract is separated, it can be treated like the other sample extracts with appropriate cleanup (above). Felice[8] has published a particularly good cleanup procedure.

b. Planar Chromatography
RPTLC and NPTLC may be used to screen for the anticoagulants (see Table 12.5). Usually all the possible anticoagulants are used as reference standards in both reverse and normal phase systems. Warfarin and diphacinone represent the extremes in R_f for anticoagulants as a group, so that they may also be used as surrogates to rule out particular commercial anticoagulants. Both fluorescence of the compounds and fluorescence quenching are used to screen for possible rodenticides.

TABLE 12.5
R_f Values of Rodenticides (Normal Phase)

Rodenticide	UV	Relative R_f^*	R_f
Raucumin®	Blue	1.2	0.80
Chlorofacinone®	Yellow	0.25	0.19
Brodifacoum®	Blue	1.5	0.90
Fumarin®	Pink	1.0	0.77
Pival®	Gold	1.0	0.78
Warfarin®	Blue	1.0	0.75
Dicoumarol	Blue	0.3	0.27
Diphacinone®	Gold	0.3	0.28
Difenacoum®	Blue	1.4	0.85
Coumachlor®	Blue	1.0	0.80

Note: Silica gel; toluene–ethyl acetate–acetone (3:2:1)
* Relative to warfarin

The coumarins all fluoresce and absorb in the ultraviolet region. This characteristic can be used to speciate and quantitatively analyze for these compounds. TLC bands have been observed from liver sample extracts that resemble commercial rodenticides; when analyzed by mass spectroscopy, however, they were found to be naturally occurring coumarin compounds. Spectral confirmation is recommended for all unknown bands.

2. Avicides and Alkaloids

The most commonly used alkaloid for rodent killing is strychnine. It is still available at garden supply outlets. Avitrol® is used to cause avoidance of an area by birds. These two substances are pesticides to which pets are exposed in abundance. Nicotine is used as a tranquilizer and caffeine is a normal, frequent food constituent to which pets are exposed.

a. Sample Preparation

The bait, stomach contents, or liver can be extracted by solubilizing the analyte in acetonitrile and acid (dilute).[1] The extract can be defatted with solvent (when acidic) without loss of alkaloid. Then the extract is made basic and the free base is extracted into organic solvent. Care must be taken not to lose Avitrol® or nicotine-volatile alkaloids and to cause decomposition of strychnine and other more complex alkaloids. Urine and blood samples may be used also to determine exposure in acute toxicity.

b. Planar Chromatography

Reverse phase and normal phase TLC are used as analytical tools for screening for alkaloids. A panel of alkaloids is spotted with the sample extract and the plate is developed (Table 12.6). Fluorescence quenching is used before visualizing the bands with color-forming reagents so additional developments can be done if desired. Iodine fumes may also be used as a visualization technique. The fumes may be removed and spectral confirmation done on the bands. After the analyst is content with the separation, the bands may be visualized with iodoplatinate spray. The bands may be removed from the TLC plates, and UV spectral, mass spectral, and other techniques can be used to confirm and further analyze extracts.

3. General Conclusions

Normally, no alkaloid bands are found in stomach contents, hence detection of alkaloids is suspicious. Other methods of confirmation are recommended for doubtful matches of R_f values or other uncertainties. There are many natural alkaloids that may be present and may or may not be a significant factor in a suspected intoxication.

TABLE 12.6
Alkaloids R_f on Planar Chromatography

Alkaloid	R_f	Color with Iodoplatinate
Strychnine	0.4	Purple
Nicotine	0.8	Blue
Caffeine	0.6	—
Avitrol	0.1	Green

Note: Stationary phase: Merck silica gel 60; mobile phase: chloroform–methanol (8:2 v/v)

III. FOOD AND FEED ADDITIVES

A. VITAMINS A, E, AND D

1. Nutrient Analysis

Veterinary medicine has become more concerned about deficiency in the fat-soluble vitamins A, E, and D than in poisonings because lack of these vitamins is of particular concern in the growth and development of young animals and for all livestock in developing a normal immune response and healthy physiological process.

a. Extraction[5]

The usual samples that are chosen for fat-soluble vitamin analysis are blood, serum, or plasma, liver is used from deceased animals. The procedure used is alcohol-hexane extraction of the serum or liver, separation of the solvent, removal of solvent, concentration of the extract, and TLC. Cleanups used are absorption column chromatography with silica gel or Florisil® when the vitamins are eluted with hydrocarbon-alcohol or SPE, or C_{18} column reversed phase cleanup when a polar (methanol) solvent is used to elute the vitamins.

b. Planar Chromatography

Normal phase and reverse phase thin-layer chromatography are used with the extracts and standards. After development, fluorescence quenching is used to observe the plate and standards for speciation and an estimate of the concentration. Vitamin E can be visualized by phosphomolybolic acid spray, and at a glance the situation of over-supplementation or deficiency can be ascertained. Table 12.7 shows the R_f values of vitamins and detection colors.

TABLE 12.7
Table Planar Chromatography of Fat-Soluble Vitamins

Vitamin	R_f^*NPTLC	R_f^{**}RPTLC	R_f^*RPTLC	Phosphomolybic Acid	Anisaldehyde
A	0.15	0.25	0.50	Dark Blue	Blue
E	0.40	Origin	0.60	Dark Blue	Green
K	0.70	Origin	0.75	Dark Blue	Green
D_3	0.20	Origin	0.55	Dark Blue	Blue

* Heptane–chloroform (50:50)
** Ethanol–water–acetic acid (65:35:1)
Note: RPTLC = Whatman KC18F; NPTLC = Merck Silica gel 60 with fluorescent indicator

Vitamin D is a special case because the levels are only easily detectable in feeds or in poisonings. A fluorescence derivitization can be used for Vitamin D_3, since it contains a hydroxy group. Vitamin K can be detected by these same techniques to enhance its sensitivity.[1,9]

B. SELENIUM

Selenium levels are important in antioxidant levels to protect physiological process and function.

1. Sample Preparation

The tissue sample needs to be treated to free the selenium for analysis. An acid digestion[1] with nitric acid–magnesium nitrate followed by a muffle furnace removal of residual nitrate may be used. The sample is then solubilized in hydrochloric acid, which dissolves and reduces selenium, which is complexed with diaminonapthalene forms for analysis. The complex can be extracted with cyclohexane and analyzed by fluorescence spectrometry, atomic absorption, or planar chromatography.

2. Planar Chromatography

The extract can be analyzed on normal and reverse phase thin-layer chromatography. The greatest sensitivity has been achieved by NPTLC using plates coated with liquid paraffin to enhance fluorescence.[10] Sub-picogram quantities may be detected.

With reverse phase TLC and normal fluorescence visualization, nanogram levels are readily detected. The absence of a band at 1/100 of a µg/ml equivalent indicates a deficiency. Quantitative TLC is readily done with coated normal phase plates or RPTLC (C_{18}) plates. The band stability in air, if not on RPTLC plates, or without paraffin coating is not good.[10] The bands fade like those of polynuclear aromatic hydrocarbons in ambient air.

The R_f of piazselenole (the diaminonapthene complex of Se) is about 0.5 in ethanol–water–acetic acid (65:35:1, v/v) with C_{18} toluene–ethyl acetate (4:1, v/v) on normal phase plates.

C. IONOPHORES

1. Monensin

a. Sample Preparation

Feed samples, rumen contents, and stomach contents all may be extracted with a polar solvent like 90% methanol or acetonitrile–10% water. Experience has shown the need to make multiple extractions to achieve complete recovery of monensin. Three extractions provide quantitative recovery from feeds without urea. Only 50% of the monensin is recoverable by this same procedure if the sample has urea in it. Once the monensin is solubilized, the sample can be defatted with a hydrocarbon solvent, decolorized, partitioned into methylene chloride or ethyl acetate–ethanol 95:5, then concentrated for chromatography.

Cleanup using absorption column chromatography with Florisil® is quantitatively possible. Using solid phase C_{18} column cleanup, it is not quantitatively possible to recover monensin, and sample spiking or standard additions must be done to get quantitative results.

b. Planar Chromatography

Normal phase thin-layer chromatography may be done using chloroform–acetone–isopropanol (85:10:5) or ethyl acetate–water (97:3) as developing solvent. Monensin may be visualized by use of 5% vanillin–sulfuric acid or anisaldehyde spray. Tenths of a microgram are visualized with vanillin spray, and the intensity of the color relative to background and other colors improves on standing overnight. The anisaldehyde color is more discernable especially under longwave ultraviolet light, where nanogram levels are discernable. This color is less stable, but may be restored by reheating the plate. The respective R_f values for rumensin in the above development solvents are 0.3 and 0.5.

Reverse phase thin-layer chromatography may also be done using C_{18} Whatman plates and developing with ethanol–water–acetic acid (65:35:1). Monensin does not move appreciably from the origin ($R_f = 0$). Once all interfering substances are moved up the plate, methanol–methylene chloride (65:35) can be used for development, and the rumensin will move up the plate ($R_f = 0.5$) and be detectable with the color reagents mentioned above.

2. Other Ionophores
a. Sample Preparation
The same procedure used for monensin can be used for salinomycin and lasalocid. Levels encountered are lower in diagnostic laboratory samples, so problems are multiplied.

b. Planar Chromatography
The same developing solvents and detection methods can be used for the other ionophores. Lasalocid has the characteristic of fluorescing so it can be detected under long wavelength UV light with use of spray reagent.

The solvent ethyl acetate–water (97:3) separates four ionophores in one development. Multiple developments may be useful in RPTLC analysis of ionophores using C_{18} layers and ethanol–water–acetic acid (65:35:1 v/v) mobile phase (Table 12.8).

TABLE 12.8
Separation of Ionophores with Normal Phase TLC

Compound	R_f
Monensin	0.4
Salinomycin	0.5
Norasin	0.6
Lasalocid	0.7

Silica gel; ethyl acetate-water (97:3).[19]

D. SULFA DRUGS
1. Sample Preparation
Feeds may be extracted with methanol. An aliquot of the extract equivalent to 2 grams is diluted with water to make the methanol level < 10%, the extract is placed onto a C_{18} Sep Pak SPE column and eluted off with methylene chloride, concentrated, and spotted on TLC. Urine may be eluted directly through the SPE column, washed, and eluted with methylene chloride. The eluate is then concentrated for TLC analysis. Tissues may be extracted and processed like feeds.

2. Planar Chromatography
The concentrated extracts are spotted on NPTLC or RPTLC plates. The R_f 0.5 for 6-sulfaquinoxaline on NPTLC (Merck silica gel) developed in 3–2–1 toluene–ethyl acetate–acetone (v/v/v) is shown in Table 12.9. The sulfas may be visualized by spraying with base and then fluorescamine to produce fluorescent bands of the sulfas. Nanogram levels may be detected in this way. The band may be oversprayed with DMAB spray (dimethylaminobenzaldehyde), which produces yellow bands from sulfa drugs, or DMAC (dimethylaminocinnaminaldehyde), which produces intense red bands. The DMAC-sprayed sulfas are roughly as visible as the fluorescamine-sprayed sulfas.

Reverse phase TLC may also be used. Whatman C_{18} plates developed with ethanol–water–acetic acid (65:35:1) gives a separation with the sulfas near the top of the plate.

Whatman also makes a multidimensional plate with C_{18} on one side and normal phase silica on the remaining plate surface. Developing in both media separates all known sulfa drugs. Straight ethylacetate may be used as the development solvent for normal phase TLC.

E. ANTIBIOTICS/ANTIMICROBIALS
1. Sample Preparation
Extractions are best done with acidic solvents containing methanol or acetonitrile and water. This allows extraction of most antibiotics, which are acidic and chelate calcium. Basic extractions are done where solubility can be enhanced for amine functional compounds. Concentration and cleanup can be combined using a solid phase cartridge with the C_{18} silica or polymer partition. Elution from

TABLE 12.9
Planar Chromatography Separation of Antimicrobial Agents

Compound	R_fs on C_{18} RPTLC 65:35:1[a]	Silica NPTLC 85:15[b]	DMAB Fluorescent Color	Anisaldehyde Color
Chlortetracycline	0.13	0.73	+	Orange yellow
Beta-tetracycline	0.17	0.73	+	Orange yellow
Tetracycline	0.23	0.82	+	Orange yellow
Oxytetracycline	0.55	0.95	+	Orange yellow
Chloramphenicol	0.71	0.77		Yellow
Penicillin	0.78	0.88		Yellow
Lincomycin	0.82	0.95		Yellow
Sulfaquinoxaline	0.51	0.55	Yellow	
Arsanilic acid	1.00	1.00		
Hydroxy nitro				
Furazolidone	0.096	0.066	Yellow	Brown
Nitrofurazone	0.73	0.89	Yellow	Brown

[a] Ethanol-H_2O-HAC (65:35:1); C_{18}
[b] Ethanol-H_2O-NH_4OH (85:15:1); EDTA-treated silica

C_{18} silica can be best accomplished with EDTA in the water portion of the elution solvent and using EDTA-saturated water in the cartridge preparation. Where methanol, then water,[1] are used to activate the cartridge to overcome the effect of alkaline earth ion chelation, EDTA-saturated water is used to prepare the cartridge.

2. Planar Chromatography

Normal phase silica and reverse phase (C_{18}) thin-layer chromatography can be used for antibiotics. Table 12.9 lists the R_f values of some frequently encountered antibiotics in veterinary samples.[3] The same precautions discussed above are necessary with normal phase plates and antibiotics that chelate calcium ions. EDTA is sprayed or dipped as applied to the plates, and the water in the developing solvent phase is saturated with EDTA. This latter device may be used also with C_{18} RPTLC plates to sharpen bands and give optimum sensitivity. Confirmation of antibiotics by TLC-FAB/MS is an excellent way of identifying compounds that are labile and difficult to confirm by usual techniques.[18]

IV. MYCOTOXINS

Fungal metabolites are the most analyzed group of substances in veterinary practice after antibiotics and antimicrobials. They can be found everywhere. The propagules are ubiquitous. It is only required that conditions of culture are adequate for microbial growth and toxin production.

Most of the mycotoxins that were of interest prior to 1991 were fat soluble, and basic polar-nonpolar extraction, partition, concentration, cleanup, and TLC were a possible route for analysis. When the South Africans discovered fumonisins,[13,14] the situation changed. These compounds are water soluble and very polar, with their carboxylic acid and amine functional groups. They require solid phase concentration and cleanup, and official methods are yet to be developed. The South Africans[15] have developed an extraction and cleanup procedure, and Rottinghaus[16] has developed a planar chromatography analysis procedure.

A. SAMPLE PREPARATION

Classical mycotoxins are the most commonly found mycotoxins. Aflatoxin, zearalenone, zearalenol, vomitoxin, T-2 toxin, and ochratoxin may be extracted from feeds with acetonitrile–water (80:20), defatted with petroleum ether, decolorized with ferric gel, partitioned into methylene chloride, and concentrated for analysis by TLC.[1]

Tissues can be analyzed by extraction with methylene chloride containing 1% citric acid, followed by concentration of the solvent and analysis of extracts by TLC.[5] Cleanup of extracts[1] can be done by C_{18} or silica gel SPE column cleanup. The first elution solvent with C_{18} SPE columns is water, followed by 25% methanol–water, which will elute vomitoxin. Aflatoxin, T-2 toxin, acetylated vomitoxin, zearalenone, and zearalenol may then be eluted with 75% methanol–water. Ochratoxin is eluted with 75% methanol–water plus 1% acetic acid. The toxins may be partitioned into methylene chloride from the methanol–water eluates by adding water to dilute the alcohol to less than 10% of the solution, and then the methylene chloride is concentrated for analysis.

Two percent water added to silica gel or Florisil® allows a column cleanup for mycotoxins using the elution order[1]:

1. Petroleum ether
2. Chloroform–methanol (97:3)
3. Chloroform–methanol (90:10)
4. Chloroform–acetic acid (99:1)
5. Chloroform–methanol (80:20)

The toxins are eluted in the order:

Fraction (2) aflatoxin, zearalenone
(3) T-2 toxin, diacetoxyscirpenol, zearalenol
(4) Ochratoxin
(5) Vomitoxin, ergotamines

Basic mycotoxins — ergotamines, penitrems, aflatremus, routfortine, and tremogens — can be extracted into acidic acetonitrile–water, defatted with petroleum ether, extracted with methylene chloride, pH adjusted to 9–12, and partitioned into methylene chloride. The extract is concentrated and can then be analyzed by TLC. If a cleanup is required, the silica column described above or SPE using methanol–water (75:25) plus 1% NH_4OH to elute and tremogen fraction may be used.

Acidic mycotoxins — Rubratoxins and citrinins, for example, can be analyzed like ochratoxin.

Fumonisin sample preparation.

Preparing fumonisin samples requires a different approach. The extraction requires acetonitrile–water (50:50) to extract the fumonisins. Blending for 5 minutes or shaking 30 minutes is favored. Solid phase concentration is essential. Either ion exchange or C_{18} SPE may be used. There is considerable acceptance of both SPE procedures. If ion exchange is used, the sample is placed on a SPF ion exchange column and eluted.[15] If C_{18} SPE is used, the sample is put on the column with the aliquot made up to 70% water. The elution is done with a solvent containing greater than 80% methanol or acetonitrile. Usually acetonitrile is favored to prevent methylation of the fumonisins. Concentration of the eluate is then done prior to analysis.[17]

B. PLANAR CHROMATOGRAPHY
1. Classic Mycotoxins

Normally, extracts are spotted on normal phase and reverse phase (C_{18}) thin-layer chromatography plates. A screening procedure used in our laboratory involves spotting extracts on normal phase TLC plates and developing in toluene–ethyl acetate–acetone (3:3:1, v/v).[1] The fluorescent mycotoxins can be observed under longwave and shortwave UV light. Zearalenone and zearalenol are most visible under shortwave UV (254 nm). Aflatoxin, ochratoxin, and $AlCl_3$-derivatized vomitoxin all are most visible under longwave UV (360 nm).

After the first development in the (3:2:1) solvent, aflatoxin, zearalenone, and zearalenol should be detectable. Multiple developments may be done to sharpen the bands and improve the separation. Chloroform–acetone (90:10) is excellent also with pure samples and aflatoxins B1, B2, G1, and

G2. Aflatoxin M1, M2, and hydroxyaflatoxins require development in a more polar solvent; chloroform–acetone–propanol (85:10:5) is often used. The ochratoxin zone may be seen after redevelopment over one quarter of the plate in the 3:2:1 solvent containing 1% acetic acid. The band is visible two centimeters above the origin. The plate may be sprayed with anisaldehyde reagent, and visible spots will appear for zearalenone, zearalenol, T-2 toxin, and vomitoxin. The respective zones under UV light are orange, purple, and yellow; the colored bands fluoresce and increased sensitivity is observed. Aflatoxins form fluorescent zones with the spray, providing further confirmation of identity.

By using a reverse phase separation as well, a further confirmation of identity is produced. Aflatoxins reverse their order: M2 and M1, G2, G1, B2, B1. Ochratoxin and citrinin are observed near the top of the plate. Vomitoxin and polar substances are near the top of the plate, and T-2 toxin and other more nonpolar toxins are below them.

Tremogens[1] can be analyzed by reverse phase TLC. Ethanol–water–NH_4OH (85:10:5) is used as a developing solvent. Normal phase separations can be done with solvents described above under antibiotics that chelate calcium. Tremogens turn purple with dimethylaminobenzaldehyde spray.

Rubratoxins do not fluoresce and require sulfuric acid (10% methanol) spray and heating to observe fluorescence under long wavelength UV light. The compounds turn brown to black depending on the heating process or fluoresce under UV light.

Other mycotoxins such as penicillic acid, patulin, cyclic esters, and trichothecenes can be analyzed by planar chromatography, and their R_f values are shown in Tables 12.10–12.12.

TABLE 12.10
Typical R_f Values of Selected Mycotoxins on Silica Gel (Adsorbosil 5)

Mycotoxin	Toluene–EtAC–90% HCOOH (6:3:1)	C_6H_6–MeOH–AcOH (24:2:1)
Citrinin	0.16–0.48	0–0.20
Luteoskyrin	0–0.47	0–0.23
Nivalenol	0–0.02	0–0.01
Butenolide	0.10	0.03
Kojic acid	0.16	0.03
Aflatoxin G_2	0.17	0.13
Nivalenol acetate (fusarenone X)	0.19	0.09
Aflatoxin G_1	0.23	0.14
Aflatoxin B_2	0.26	0.20
Aflatoxin B_1	0.31	0.23
Diacetoxyscirpenol	0.33	0.24
Aspertoxin	0.35	0.13
T-2 toxin	0.36	0.27
Patulin	0.41	0.21
Penicillic acid	0.47	0.22
Gliotoxin	0.53	0.39
Ochratoxin A	0.55	0.35
Zearalenone	0.78	0.42
Sterigmatocystin	0.85	0.75

From Scott, P., *Advances in Thin-Layer Chromatography*, Touchstone, J. C., Ed., John Wiley & Sons, New York, 1982. With permission.

2. Fumonisins

These compounds are ubiquitous, are found where corn is found, are extremely toxic to horses, producing leukoencephalomalacia, and are carcinogens. They will be regulated as soon as established levels of concern are developed.

Here again, two systems are used for analysis: normal phase and reverse phase. Normal phase planar chromatography on silica gel requires a strong aqueous development solvent,[16] chloro-

TABLE 12.11
R_f Values of Miscellaneous Mycotoxins for Normal Phase TLC on Silica Gel

Toxin	R_f	Solvent	Visualization	Limit of Detection
Sterigmatocystin	0.85	6:3:1 Toluene–ethyl acetate–formic acid	Aluminum chloride	0.1 µg
	0.75	24:2:1 Benzene–methanol–acetic acid	Yellow fluorescence[a]	
Alternariol	0.6	90:10 Chloroform–acetone	Blue fluorescence[a]	0.5 µg
Alternariol methyl ether	0.7		Blue fluorescence[a]	
Penicillic acid	0.5	3:2:1:1 Toluene–ethyl acetate–acetone–acetic acid	Anisealdehyde Blue fluorescence[a]	0.5 µg
Patulin	0.4	90:10 Chloroform–acetone	10% Pyridine in methanol	0.1 ng
Moniliformin	0.6	3:2 Chloroform-methanol	Fluor. quenching[a]	0.1 µg
Dicumarol	0.2	90:10 Chloroform-acetone	Ammonia[a] Fluorescence	0.5 µg

[a] Long wavelength UV light.

TABLE 12.12
R_f Values of Mycotoxins of Diagnostic Significance

Toxin	R_f RP C_{18} silica gel: 65:3:1 + 0.5% Sodium chloride Ethanol–water–acetic acid	R_f NP Silica gel: 3:2:1:1 Toluene–ethylacetate– acetone-acetic acid	Visualization	Detection Sensitivity
Citrinin	0.78	0.20	Long wavelength UV, yellow fluorescence (acid fumes)	0.2 µg
Ochratoxin A	0.68	0.64	Long wavelength UV, blue fluorescence	0.1 µg
Penitrem A	0.35 85:15:1 Ethanol–water–ammonium hydroxide	0.70 90:10 Chloroform-methanol	Blue color dimethylamino benzaldehyde spray	1–2 µg
Aflatrem	0.20	0.80	Blue color dimethylamino benzaldehyde spray; blue fluorescence, long wavelength UV	1–2 µg
Slaframine	0.3 85:15:1 Ethanol–water–ammonium hydroxide	0.5 90:10 Chloroform-methanol	Iodiometric	1.0 µg
Zearalenone	0.2	0.69	Long wavelength UV fluorescence and short wavelength UV fluorescence	0.1 µg
Zearalenol	0.3	0.60	Aluminum chloride, long wavelength UV fluorescence	0.1 µg
Rubratoxin	0.8	0.40	20% sulfuric acid/methanol long wavelength UV green fluorescence: short wavelength UV fluorescence quenching	0.5 µg 0.2 µg

form–methanol–acetic acid (60:30:10). Reverse phase TLC can be done using a variety of solvents, including ethanol–water–acetic acid (65:35:1) or methanol–1% aq. KCl (80:20).[16]

The fumonisins may be visualized by anisaldehyde spray, producing purple spots microgram levels of the compounds. Rottinghaus[16] developed the fluorescamine detection system, which gives nanogram sensitivity.

REFERENCES

1. **Stahr, H. M., Ed.,** *Analytic Methods in Toxicology,* John Wiley and Sons, New York, 1991.
2. **Stahr, H. M.,** Planar chromatography applications in analytical toxicology, in *Critical Reviews in Analytical Chemistry,* Zielenski, W. L., Jr., Ed., CRC Press, Boca Raton, FL, 1987.
3. **McMahon, B. M. and Harkin, N. F.,** *Pesticide Analytical Manual,* 3rd ed., FDA, USH & HS, Washington, DC, 1994.
4. **Sherma, J.,** *Manual of Analytical Quality Control for Pesticides and Related Compounds,* USEPA, Office of Research and Development, Health Effects Research Lab, Research Triangle Park, NC, 1976.
5. **Cunniff, P.,** *Official Methods of AOAC,* 16th ed., AOAC, Arlington, VA, 1995.
6. **Getz, M.,** *Paper and Thin Layer Chromatography,* Heydon, London, 1980.
7. **Mount, M. C. and Feldman, B. F.,** Mechanism of rodenticide toxicosis in the dog and its therapeutic implications, *Am. J. Vet. Res.,* 44, 2009, 1983.
8. **Chalermachackit, T., Felice, L. J., and Murphy, M. J.,** *J. Anal. Tox.,* 17, 56, 1993.
9. **Madden, U. A. and Stahr, H. M.,** Reverse phase thin layer chromatography assay of Vitamin K in bovine liver, *J. Liq. Chromatogr.,* 16, 2825, 1993.
10. **Frank, W., Kerler, R., Schiller, J. T., and Dammann, V.,** Prechromatographic derivitization of samples for HPTLC, in *Instrumental High Performance Thin Layer Chromatography,* Kaiser, R. E., Ed., Eburton Institute for Chromatography, Bad Durkheim, Germany, 1982.
11. **Stahr, H. M., Imerman, P. J., and Schrunk, D.,** An improved analysis for monensin and feeds, *J. Diagn. Invest.,* 8, 140, 1994.
12. **Armstrong, D.,** private communication, 1992, Pittsburgh Conference.
13. **Blzuidenhout, S. C., Gelderbloom, W. C. A., Gonst Allman, C. Horahet, R. M., Marasas, W. F. O., Spiteller, G., and Vleggar, R.,** Structure elucidation of the fumonisins, mycotoxins from *Fusarium moniliforme, J. Chem. Soc. Comm.,* 743, 1988.
14. **Marasas, W. F. O., Kellerman, T. S., Gelderbloom, W. C. A., Coetzer, J. A. W., Thiel, P. F., and VandeLuzt, J. J.,** Leukoencephalomalacia in a horse induced by fumonisin B, isolated from *Fusarium moniliforme, Ondespoort J. Vet. Res.,* 55, 197, 1988.
15. **Shephard, G. S., Sydenbessn, E. E., Thiel, P. G., and Gelderbloom, W. C. A.,** Quantitative determination of fumonisins b_1 and b_2 by high-performance liquid chromatography with fluorescence detection, *J. Liq. Chromatogr.,* 13, 2077, 1990.
16. **Rottinghaus, G. E., Coatney, C. E., and Minor H. C.,** A rapid sensitive thin layer chromatography procedure for the detection of fumonisin B1 and B2, *J. Vet. Diagn. Invest.,* 4, 326, 1992.
17. **Murphy, P. A., Ross, P. F., Rice, L. G.,** *J. Agric. Food Chem.,* 41, 263, 1993.
18. **Hisao, O., Yoshitomo, I., Jarnko, H., Katauyoshi, M., Hareda, K. I., Suzuki, M.,** Determination of tetracycline antibiotics in milk by liquid chromatography and thin layer chromatography/fast atom bombardment mass spectrometry, *J. AOAC Int.,* 77, 891, 1994.
19. **Owles, P. J.,** Identification of monensin, narasin, salinomycin and lasalocid in pre-mixes and feeds by thin-layer chromatography, *Analyst,* (Cambridge, UK) 109, 1331, 1984.

INDEX

A

p-Acetylamino benzoate, 240
Acetylpectolinarin, 234
Acheta domestica (house cricket), 87
Aconitine, 38
Aconitum kusnezoffi, 38
Additives, food, 183–184
Adenosine derivatives, 15
Adhesins, 20, 25
β-Adrenergic-blocking agents, 223, 240
Aeshna cyanea (dragonfly), 80
Aflatoxins
　food analysis, 171, 174, 188
　veterinary toxicology, 250, 260–262
Aglycones, flavonoid TLC, 40–41
Ajmaline, 15
Aldehydes and ketones, 164
Aliphatic thiols, derivatization, 13
Alkaloids
　forensic toxicology, 223
　pharmaceutical analysis, 232–235
　plant, 15, 35–39
　　Catharanthus, 36
　　ergot, 36–37
　　indole, 36
　　insect interactions, 99
　　miscellaneous classes of, 38–39
　　opium, 37–38
　　pyrrolizidine, 38
　　quinoline and isoquinoline, 37
　　quinolizidine, 37
　　tropolone, 38
　vertebrate skin secretions, 111, 117–118
　veterinary toxicology, 250, 256
Allantoin, pharmaceutical analysis, 234
Aloprenolol, 243
Alumina layers, 3, *see also* specific applications
Aluminum alloys, 3
Amiloride hydrochloride, 238
Amines
　derivatization, 251
　food analysis, 188–189
　insect, 90–92
Amino acids, 4
　clinical chemistry, 141–142
　food analysis, 183
　parasite, 61, 66
Amino layers, 3
Aminophenazone, 242
Amitriptiline, 244
Ammonium thiolacetate, 13
Amphetamines, 250
　forensic toxicology, 220–221
　pharmaceutical analysis, 240

Amphibian skin secretions, 110–111, 114, 126
Anabolic steroids, forensic toxicology, 223
Analgesics, pharmaceutical analysis, 240, 242
Anas platyrhynchos (mallard ducks), 109
Anise, 43
Anopheles, 73
Anopheles stevensi, 61
Anthocyanins, 39, 42–43
Antibiotics, 4, 144
　food analysis, 187–188
　pharmaceutical analysis, 235–237
　veterinary toxicology, 250, 259–260
Anticarsia gemmatilis, 80
Anticircular development, 12, 136
Anticoagulants, veterinary toxicology, 250, 255–256
Anticonvulsants, 144
Antidepressants, 240, 244
Antihistamines, 242
Antimalarials, 242
Antioxidants
　food analysis, 170–171, 183
　pharmaceutical analysis, 237
　veterinary toxicology, 258
Antirrinoside, 234
Apigenins, 41
Apomorphone, 232
Araschinia levana, 93
Arctic beetle (*Pytho americanus*), 77
Aromatic-*N*-glycosides, 4
Arrhenatherum, 41
Arseniacals, 251
Ascaris, 52
Asclepias, 45
Asialo GM1, 26
Atropa belladona, 39
Auport's reagent, 37
Autographa californica, 87
Automated Multiple Development System, 9–11, 136
Automatic Development Chamber (CAMAG), 9, 10, 213
Avicides–alkaloids, 256
Azadirachtin, 95–96, 97
Azaperone, 241
Azidothymidine, 241
Azoxyglycosides, 99

B

Bacterial contamination of food, 188–189
Bacteriology, 19–28
　methods, 21–23
　overlay methods for receptor identification, 23–27
　　peroxidase-based assays, 25–27
　　preparation chromatograms, 25
　　radiolabel assays, 27
　protocols, 27–28

sample preparation, 20–21
species and strain identification, 23
Bacteroides, 23
Baicalein/baicalin, 3
Baljet reagent, 44–45
Barbiturates, 4, 240, 250
Benzimidazoles, 242
Benzoate, 7, 171, 172, 183–184
Benzodiazepines and derivatives, 4, 143, 197
    forensic toxicology, 223
    pharmaceutical analysis, 240, 243
Benzophenones, 243
N-Benzyloxycarbonyl-glycyl-L-proline (ZGP), 243
Berberine, 39
Beta-blockers, 223, 240
Betaine, 39
Bifidobacterium, 23
Bile acids, clinical chemistry, 142
Biogenic amines
    food analysis, 170, 188–189
    insect, 90–92
Biological fluids, *see* Forensic toxicology
Biological samples, *see* Clinical chemistry
Biomphalaria glabrata, 56–58, 64–66
    neutral lipids, 60
    phospholipids, 61, 63
Birds
    alkaloids, 117–118
    avicides, 255–256
    ceramides and glucosylceramides, 115–116
    skin secretions, 109, 110, 114
Bis(2-ethylhexyl)phosphate (HDEHP), 198
Bismuth, 4
Blood analysis, human, *see* Clinical chemistry; Forensic toxicology
Blowfly (*Calliphora erythrocephala*), 93
Blowfly (*Calliphora vicinia*), 80
Blowfly (*Lucilia cuprina*), 89
Blowfly (*Phormia terraenovae*), 86–87
Body fluids, *see* Clinical chemistry; Forensic toxicology
Bombyx mori, 77
Borneol, 44
Bufadienolides, 44, 45
Bufogenins, 114, 123
Bull snake (*Pituophis melanoleucus*), 117
Buprenorphine, 14, 143
Butenolids, 262

C

Cabbage butterfly (*Pieris brassicae*), 73–74, 80
Caffeic acid, 3
Caffemed, 243
Calendula officinalis, 40
Calliphora erythrocephala (blowfly), 93
Calliphora vicinia (blowfly), 80
Candida albicans, 20
Canids, skin secretions, 119
Cannabinoids, 143
    forensic toxicology, 221
    pharmaceutical analysis, 233

Captopril, derivatization, 13
Carbamates, 14
    derivatization, 251
    environmental analysis, 157–159
    pharmaceutical analysis, 240
    veterinary toxicology, 251–252
Carbaryl, 14, 157
Carbohydrates
    clinical chemistry, 142–143
    food analysis, 173–175, 182–183
    insect, 72–80
    parasite, 56, 61, 66
Cardenolides, 44, 96–97
    insect effects, 95
    visualization systems, 99
Cardiac glycosides, 44–45
Carnitine, 39
Carotenoids
    food analysis, 170
    insect, 94–95
    parasite, 56, 59, 61, 66–67
    pharmaceutical analysis, 244
Catecholamines, 3
Catharanthus roseus, 15, 36
Cavia porcellus (guinea pig), 119
Celastrus rasthornianus, 44
Cellulose, 3, *see also* specific applications
    flavonoids, 40
    food analysis, 175, 176
    parasitology, 52
Centrifugal development, 137
Ceramides, vertebrate skin secretions
    pig, 125–126
    TLC system and detection method, 115–116
Cerebrosides, 148
Chamomilla, 43
Chamomilla recutita, 43
Chelidonium alkaloids, 38
Chelisonium majus, 38
Chicken (*Gallus domesticus*), 116–117
Chiral plates, 3
Chlamydia trachomatis, 24
Chloramphenicol, 239
Chloroquine, 242
Chlorpromazine, 144, 237, 242
Chlorthalidone, 238
Cholesterol
    insect, 83
    vertebrate skin secretions, 114
Cholesteryl sulfate, 117
Choline, 39
Choristoneura fumiferana, 82
Choristoneura orae, 82
Cinchona, 37
Cinnamins, 44
Cinnarizine, 242
Circular development
    clinical samples, 136
    principles of TLC, 12
Citric acid, food analysis, 171
Citrus oils, 43

Index

Civet cat, 123
Cleome, 41
Clinical chemistry, 131–148
   applications, 141–148
      amino acids, 141–142
      bile acids, 142
      carbohydrates, 142–143
      drugs, 143–144
      lipids, 144–145
      phospholipids, 145–146
      porphyrins, 147
      prostaglandins, 146
      steroid hormones, 147–148
   comparison of TLC with other techniques, 140–141
   historical aspects of TLC, 132
   methodology, 132–140
      development and detection, 137–138
      materials, 132–134
      sample application, 137
      sample preparation, 135–137
      specimen collection and storage, 134–135
      validation of results, 138–140
   principles of TLC, 132
   recent developments in plates, 140
Clostridium botulinum, 24
Clostridium difficile, 24
Clotiazepam, 243
Cocaine, 143
   forensic toxicology, 221–222
   pharmaceutical analysis, 233
   plant science applications of TLC, 39
Cockroach (*Nauphoeta cinerea*), 83
Colchicum alkaloids (colchicine), 38
Colchicum ritchii, 38
Column chromatography
   bacterial lipids, 21
   food analysis, 174
   parasitology, 59
   vertebrate skin secretions, 112–113
Concentrating zone, 8
Conessine, 234
Convallaria majalis, 44
Corallus, 121, 122
Corticosteroids, 239
Coumarins, 44, 255–256
Cricket (*Gryllus bimaculatus*), 77
Cryptocercus puntulatus (wood cockroach), 77
Cryptococcus neoformans, 20
Cucurbita, 14
Cucurbitine, 234
Culex pipiens, 80
Cumin, 43
Cupressus sempervirens, 41
Cyanidine, 42
Cycasin, 99
α-Cyclodextrin, 4, 221
β-Cyclodextrin, 3, 4, 221, 240
Cycnia tenera (Dogbane Tiger Moth), 95

D

Danaus plexippus (monarch butterfly), 95
Dansylation, 4, *see also* Derivatization; specific applications of TLC
Datura, 234
Datura stramonium, 39
Dehydroacetic acid, 7
Delorazepam, 243
Densitometry, 14, *see also* specific applications
   clinical samples, 137
   food analysis, 178
Derivatization, 178, 180, 189, *see also* specific applications of TLC
   pharmaceutical analysis, 232
   postchromatographic, 12–13
   sample, 6
   for veterinary toxicology, 251
Dermacentor variabilis (dog tick), 81–83
Desacetyl pieristoxin B, 96
Detection, 12–13, *see also* specific applications
   bacterial lipids, 22
   clinical samples, 137
   plant products
      alkaloids, 35
      cardiac glycosides, 44–45
      flavonoids, 41–42
Detection limits, 14
Development, 8–12, *see also* specific applications of TLC
   anticircular, 12
   circular, 12
   clinical samples, 136–137
   distance of, 8
   linear ascending, 8–10
   linear horizontal, 10–12
Dextropropoxyphene, 240
Dialysis, clinical samples, 136
Diaprepes abbreviatus, 96
Diatomaceous earth, 3
Dibenzipine, 242
Dicoumarol, 263
Diester waxes, vertebrate, 124
Diethazine, 242
Diflubenzuron, 7
Diflunisal, 242
Digitalis, 44
Digitalis lanata, 7, 45
Digitalis purpurea, 7
Digoxigenin bisdigitoxide, 233
Dimethylsilozane, 3
Diol layers, 3
Dioscorea opposita (yams), 234
Diphenhydramine, 242
Documentation, 13
Dog, 119
Dogbane Tiger Moth (*Cycnia tenera*), 95
Dog tick (*Dermacentor variabilis*), 81–83
Dopamine, 90–92
Dragonfly (*Aeshna cyanea*), 80
Drosophila, 81, 93, 94

Drosophila melanogaster, 89, 92
Drugs, *see also* Pharmaceutical analysis
   clinical chemistry, 143–144
   forensic toxicology, 218–223
   veterinary toxicology, 259–260
Dynamic modification, 4
Dysdercus, 93

E

Ecdysones, 86–89
Echinostoma caproni, 56, 58, 60, 61, 63–65
Echinostoma trivolvis, 52, 56, 59, 63–64, 66–67
   amino acids, 61
   neutral lipids, 60
   phospholipids, 60
   pigments, 61–62
Eleutherococcus, 45
Elution strength, 4–5, *see also* specific applications
Enantiomeric separations, 4, 243
Entomology, 72–100
   biogenic amines, 90–92
   carbohydrates, 72–80
   lipids
      ecdysones, 86–89
      techniques, 80–86
      terpenoids, 89–90
   pigments, 92–95
   plant toxins, 95–100
Environmental analysis, 153–166
   general principles, 154–156
   monitoring, 156–164
      aldehydes and ketones, 164
      inorganic pollutants, 164
      military and warfare agents, 162–163
      pesticides, 156–159
      phenols, 161–162
      polychlorinated biphenyls, 162, 163
      polycyclic aromatic heterocycles, 160–161
      polycyclic aromatic hydrocarbons, 159–160
      trinitrotoluene, 163
   as research technique, 164–165
Environmental samples, 4
Epicrates (tree boas), 121, 122
Equolides, 114, 123
Ergot alkaloids, 36–37
Erigeron canadensis, 7
Escherichia coli, 24
Essential oils, 43, 44
Ethyl loflaxepate, 243
Eucalyptus, 40, 41, 43
Eugenol, 44
Eumaeus atala, 96
Exoenzyme S, 24
Extraction, *see* Liquid-liquid extraction; Liquid-solid extraction; Solid-phase extraction
   bacterial lipids, 20
   sample preparation, 7

F

Fatty acids
   bacterial, 22
   food analysis, 175
   insect, 80, 84, 86
   vertebrate skin secretions, 109, 111, 117
Ferulic acid, 3
Fibrial proteins, 24
Fimbriae, 20
Flavonoid glucosides, pharmaceutical analysis, 234
Flavonoids, 3, 7
   pharmaceutical analysis, 234
   plant analysis, 39–42
Flavor compounds, 7
Fludiazepam, 243
Fluorescence detection, 12–14, *see also* specific applications
Folyella agamae, 62
Food analysis, 7, 170–189, 251
   antibiotics, 236
   protocols, 181–189
      additives, 183–184
      composition, 181–183
      contaminants, 184–188
      decomposition, 188–189
   sample preparation, 172–174
   sampling, 171–172
   techniques, 174–179
   validation of method, 179–181
   veterinary toxicology, 257–260
      antibiotics–antimicrobials, 259–260
      ionophores, 258–259
      selenium, 258
      sulfa drugs, 259
      vitamins A, E, and D, 257–258
Forced flow technique, 10–12
Forensic toxicology, 193–224
   analytic techniques, 194, 195
   chromatographic systems, 198–214
      evaluation of systems, 198–209
      R[f] correction, 209–210
      TIAFT system, 210–212
      Toxi-Lab system, 212
      UniTox system, 202–209, 212–214
   drugs of abuse, 218–223
   extraction, 197–198
   identification, 214–218
   miscellaneous substances, 223–224
   toxicologically relevant substances, 194
   utility of TLC, 194–196
Fox, 119
FTIR, 1, 223
Fumonisins, 250, 260–264
Fungal toxins, *see* Aflatoxins
Furosemide, 238
Fusobacterium, 23

G

Gallus domesticus (chicken), 116–117
Gangliosides
   bacterial, 21

clinical chemistry, 148
Gas contamination, 5
Gas phase, 9
Gentisic acid, 242
Gitoxin, 233
Glucosylceramides
    pig, 125–126
    TLC system and detection method, 115–116
Glutathione, derivatization, 13
Glycerol, insect, 77–79
3-Glycidooxipropyl, 4
Glycolipids
    bacterial, 25
        indicator dyes and standards, 22
        sample preparation, 21
    insect, 73–74, 80
    parasite, 61
    vertebrate skin secretions, 116–117
Glycols
    food analysis, 174
    insect, 77–79
Glycosidase, bacterial, 23
Glycosides, 7
    bacterial, 23
    flavonoid, 41
Glycosphingolipids
    bacterial, 20
    bacterial toxin receptors, 27
    bacterial toxins and, 24
    granulocyte, 7
    insect, 80, 83
    parasite, 61
Glycyrrhizin, 233
Gossypol, 174
Granulocyte glycosphingolipids, 7
Grayanotoxins, 96, 98, 99
Gryllus bimaculatus (cricket), 77
Guaphaliinae, 41
Guinea pig (*Cavia porcellus*), 119

H

Haloxazolam, 243
Heavy metals, 4, 155, 164
Helicobacter pylori, 24
Heliothis zea, 81, 82
Helisoma trivolvis, 56, 59, 61–62, 66–67
Helleborus, 44
Helminths, *see* Parasitology
Hemolymph, 66, *see also* Entomology
Hemolytic saponins, 45
Hemophilus influenzae, 24
Heroin, 143
High-performance liquid chromatography, see HPLC
High-performance TLC, see HPTLC
Hippuric acid, 244
Holarrhena floribunda, 234
Homobatrachotoxin, 117–118
Homo-cysteine, 13
Hordenine, 232
Horse, 114, 117, 123
House cricket (*Aceta domestica*), 87

HPLC, 4
    clinical chemistry applications, comparison with TLC, 141
    environmental analysis, 165–166
    forensic toxicology, 196
    TLC as pilot method, 15, 165–166
HPTLC
    alkaloids, 234
    bacterial lipids, 21
    clinical chemistry, 140, 144–145
    dimethylsiloxane-treated, 3
    environmental analysis, 158
    food analysis, 176
    insect amines, 90–92
    insect lipids, 85
    insect polyols, 78
    linear horizontal development, 10
    mobile phase, 5
    parasitology, 58, 63
    pharmaceutical analysis, 238–242, 244
    plates, 2–3
H-separation chamber, 10
Humans, *see* Clinical chemistry; Forensic toxicology
Human skin secretions, 124, 145
Humic acids, 4
Hydrastinine, 39
Hydrochlorothiazide, 238
Hydroflumethiazide, 238
Hydromorphone, 232
Hydrophobicity, 15–16
5-(Hydroxymethyl)-2-furfuralcans, 14
Hymenolepis, 52, 54–55, 57
Hyoscyamine, 39, 234
Hyoscyamus niger, 39
Hypericum perforatum, 232
Hypoaconitine, 38

I

Ibuprofen, 240
Iguana iguana, 118
Illcium henryi, 40, 41
Image analysis, 237
Imipramine, 242
Immunoassay, 218–220
Inachis io, 80
Indole alkaloids, 36
Infrared spectra, 13
Inorganic compounds, 4
    environmental analysis, 155, 156, 164
    veterinary toxicology, 258
Inosine, 240
Inosiplex, 240
Insecticides, *see* Pesticides
Ionophores, 258–259
Ipomoea batata, 42
Iridoid glucosides, 234
Isoajmaline, 232
Isoborneol, 44
Isoflavonoids, *Streptomycetes*, 41
Isoquinoline alkaloids, 37
Isorhamnetin, 41

Isosandwichine, 232
Isothymidine, 241

J

Juvenile hormones, 89

K

Kaempferol, 41
Kalmia latifolia (mountain laurel), 96
Keckiella, 40
Kieselguhr plates, *see also* specific applications
　food analysis, 175, 176
　insect polyols, 78
Klebsiella pneumoniae, 24
Krim system, 212–216, 220

L

Lactic acid, food analysis, 171
Lactobacillus, 24
Lactones, visualization systems, 99
Lactosyl ceramide-binding bacteria, 23–24
Lasalocid, 259
Lastenia californica, 41
Lavandula, 43
Lead, 4
Leishmania amazonensis, 61
Leucochloidiomorpha constantiae, 52, 55, 56
Libacedries, 41
Libocedrus, 41
Limonoids, 7, 43
Linari, 234
Linarioside, 234
Linear ascending development, 8–10
Linear horizontal development, 10–12
Lipids, *see also* specific classes of lipids
　bacterial, *see* Bacteriology
　clinical chemistry, 144–145
　food analysis, 174, 182
　　pesticide extraction, 185
　　sample preparation, 173
　insect, 73–74
　　ecdysones, 86–89
　　techniques, 80–86
　　terpenoids, 89–90
　parasite, 53–56, 60–61, 64–65
　sample preparation, *see* specific applications
　vertebrate skin secretions, 113, *see also* Skin secretions, vertebrates
　　column chromatography, 112–113
　　reptile, 124–125
　　TLC systems and detection methods, 113–117
Lipocedrus, 40
Lipophilicity, 15–16
Liquid-liquid extraction
　clinical samples, 136
　environmental samples, 154
　food analysis, 174
　forensic toxicology, 197–198
Liquid-solid extraction
　clinical samples, 136
　forensic toxicology, 198
Littorina saxatilis rudis, 62
Liver, 145, 196–197
Lizards, 118, 124–125
Locusta migratoria, 87
LSD–lysergide, 222
Lucilia cuprina (blowfly), 89
Lutein, 56
Luteolin, 41
Lysergide, 222

M

Macaca fascicularis (monkey), 118
Macrozamin, 99
Magnesium, 3
Mallard ducks (*Anas platyrhynchos*), 109
Malvinine, 42
Mammals, skin secretions, 106, 107, 115–116
Manduca sexta (tobacco hornworm), 81, 83
Mangold system, 60
Mannitol, insect, 77–79
Matricaria chamomilla, 7
Mean list length (MLL) method, 198, 209
Melissa officinalis, 43
Mentha, 43
Mercaptopropionylglycine, 13
Merck Tox Screening System, 214
Mercury, 4
Mesaconitine, 38
Methoxymorpholinodoxorubicin hydrochloride, 14
5-Methylcytosine, 3
Methyl tryptophan, 4
Metoprolol, 3, 243
Microbial toxins
　aflatoxins, *see* Aflatoxins
　food analysis, 171, 188
Microbiology, *see* Bacteriology
Microphallus similis, 62
Microtox, 251
Midazolam, 4
Military and warfare agents, 162–163
Milkweed bug (*Oncopeltus fasciatus*), 95
Minaprine, 244
Minocarboxykato cobalt (III) complexes, 3
Minoxidil, 4
Mobile phases, 4–5, *see also* specific applications
Mobile R chamber, 10
Molecular parameter determination with adsorptive and reversed-phase TLC, 15–16
Mole (*Scalopus aquaticus*), 118
Monarch butterfly (*Danaus plexippus*), 95
Monensin, 258
Moniliformes dubius, 58
Monkey (*Macaca fascicularis*), 118
Monoterpenes, 43
Monothioglycerol, 13
Mountain laurel (*Kalmia latifolia*), 96
Mucin, 21
Mutatox, 251
Mycobacterium, 23

Mycoplasma, 23
Mycotoxins
    food analysis, 171
    veterinary toxicology, 250, 260–264
Myrcene, 43
Myricetin, 41

N

Narcotic analgesics, 197
Nauclea latifolia, 233–234
Neem tree (*Azadirachta indica*), 95–96
Neisseria gonorrhoeae, 24
Neoglycolipids, bacterial, 21
Nerium oleander, 44
Neutral lipids
    bacterial
        indicator dyes and standards, 22
        sample preparation, 21
    insect, 81–84
    parasite, 52–57, 60, 64
Newt (*Taricha*), 118, 123
Niacin, 244
Nimodipine, 241
Nippostrongylus brasiliensis, 57
Nitrendipine, 241
Nitrogen-containing pharmaceutical compounds, 240–244
Nonnutritive sweeteners, 184
Nonylphenyl ethylene oxide oligomers, 3
Norepinephrine, insect, 90–92
Nortriptiline, 244
Nuclear polyhedrosis virus, 87
Nucleic acid derivatives, 148

O

Ochratoxins, 250, 260, 261, 263
Octopamine, insect, 90–92
Olea europea, 7, 40, 41
Oleanolic acid, 240
Ommochromes, 93
Oncopeltus fasciatus (milkweed bug), 95
Opiates
    forensic toxicology, 222–223
    pharmaceutical analysis, 240
    plant products, 37–38
Organic acids, 7
    food analysis, 171, 183
    urinary, 148
Organochlorine compounds, *see also* Pesticides
    environmental toxicology, 157, 159
    veterinary toxicology, 253, 254
Organophosphorus compounds, *see also* Pesticides
    environmental toxicology, 157, 159
    veterinary toxicology, 252–253
Oroxylin-A, 3
Overpressure TLC, 10–11, 39
    clinical samples, 136–137
    forensic toxicology, 224
Oxazolam, 243
Oxygen-containing compounds, pharmaceutical analysis, 240–244
Oxymorphone, 232
Oxyphenbutazone, 240

P

Papain, 197
Papaverine, 242
Papaver somniferum, 37–38
Para-amino benzoic acid, 240
Paracetamol, 240, 244
Paracetamol–ascorbic acid–caffeine–phenylephrine, 243
Paraminosalicylic acid, 242
Parasitology, 51–67
    Biomphalaria glabrata, 64–66
    Echinostoma caproni, 64–65
    Echinostoma trivolvis, 66–67
    material for, 56–58
    quantification, 62–63
    sample preparation, 58–60
    TLC systems and detection, 60–62
    validation of results, 63–64
Paratenuisentis ambiguus, 62
Pelargonidine/pelargonine, 42
Penicillins, 4, *see also* Antibiotics
Peroxidase-based assays, bacterial lipid-binding proteins, 25–26, 28
Pesticides, 15
    derivatization, 251
    environmental analysis, 155–159
    food analysis, 171, 184–187
    forensic toxicology, 223
    veterinary toxicology, 251–255
        carbamates, 251–252
        organochlorine compounds, 253, 254
        organophosphorus compounds, 252–253
        pyrethroids, 253–254
Petaurus breviceps (sugar glider), 114
Pharmaceutical analysis, 14, 15, 231–244
    alkaloids, 232–235
    antibiotics, 235–237
    instrumentation, 231–232
    oxygen- and nitrogen-containing compounds, 240–244
    steroids, 238–240
    sulfur-containing compounds, 237–238
    vitamins, 244
Pharmacology, clinical chemistry, 143–144
α-Phellandrene, 43
Phenazone–phenacetin–caffeine (Caffemed), 243
Phenolphthalein, 244
Phenols, 161–162
    environmental analysis, 155–156
    forensic toxicology, 223
    plant
        plant–insect interactions, 98
        visualization systems, 99
Phenothiazines
    forensic toxicology, 223
    pharmaceutical analysis, 237, 240
Phenyldimethylsiloxane, 3
2-Phenyl-γ-benzopyrone, 39
Pheophorbide, 7

Pheromones
  insect, *see* Entomology
  parasite, 52
Phormia terraenovae (blowfly), 86–87
Phospholipids
  bacterial, 20, 22
  clinical chemistry, 145–146
  insect, 80–81, 83, 86
  parasite, 60–61, 64–65
    quantitation, 62–63
    silica gel plates or sheets, one-dimensional, 53–57
  vertebrate skin secretions, 114–115
Phytosphingosine, 115
Pieris brassicae (cabbage butterfly), 73–74, 80
Pigments
  food analysis, 170–171
  insect, 92–95
  parasite, 56, 59, 61, 66–67
  plant, 42–43
Pig (*Sus scrofa*) skin secretions, 125–126
Pilot method for HPLC, 15
Pimpinella anisum, 43
Pinenes, 43
Pitohui, 117–118
Pituophis melanoleucus (bull snake), 117
Plants, 7, 14–15
  advantages of TLC, 34
  alkaloids, 35–39, 232–235
  anthocyanins, 42–43
  cardiac glycosides, 44–45
  essential oils, 43, 44
  flavonoids, 39–42
  food analysis, 170
  pharmaceutical products, 232–235
  saponins, 45–46
  toxin-insect interactions, 95–100
Plasmodium berghei, 56, 61, 73
Polanisia trachysperma, 41
Polarity, mobile phase, 3
Pollutants, *see also* Pesticides
  environmental analysis, *see* Environmental analysis
  food analysis, 185
Polyamide plates, 3, *see also* specific applications
Polychlorinated biphenyls, 162, 163, 185, 253
Polycyclic aromatic heterocycles, 160–161
Polycyclic aromatic hydrocarbons, 159–160
Polygonum glabrum, 44
Polyols, insect, 77–80
Porphyrins, clinical chemistry, 147
Preadsorbent zone, 8
Precision, clinical chemistry, 139–140
Preservatives, 7, 171
Programmed multiple development, 136
Promethazine, 242
Propionibacterium, 24
Propranolol, 3, 243
Propyl parabens, 239
Prostaglandins, 146
Protein, food analysis, 173, 174, 181–182
Protozoan parasites, 58, *see also* Parasitology
Pseudomonas aeruginosa, 21, 24
Pteridine pigments, insect, 93
Puffer fish, 123

Pyrethroids
  environmental analysis, 157–159
  veterinary toxicology, 253–254
Pyrrochorid bugs, 93
Pyrrolizidine alkaloids, 38, 96, 99
Pytho americanus (Arctic beetle), 77

Q

Qualitative and quantitative evaluation, 13–14, *see also* specific applications
Quantitative structure–activity relationship, 15–16
Quaternary alkaloids, 39
Quaternary ammonium antiseptics, 3
Quercetin, 41
Quinine, 242
Quinoline alkaloids, 37
Quinolizidine alkaloids, 37

R

Radiolabel-based assays, bacteriology, 27
Rawolfia, 232
Rawolfia cubana, 36
Rawolfia vomitoria, 234
Raymond-Marthoud reagent, 45
Reproducibility, 13–15, *see also* specific applications
Reptiles, *see* Snakes
Reptiles, skin secretions, 106, 108, 111, 115–116, 124–125
Rescinnamine, 234
Reserpine, 234
Resolution, 5–6, *see also* specific applications
Retention factor, 5
Retention mechanism, 6
Retention parameters, 5–6, *see also* specific applications
Retinoids, pharmaceutical analysis, 244
Reversed-phase TLC, *see* RPTLC
R[f] correction, 209–210
R[f] zones, 214
Rhino C, 243
Rhododendron, 96
Riboflavin, 244
Rodenticides, veterinary toxicology, 250, 255–256
Rosa, 41, 43
Rosmarinus, 43
RP-18, 2–3
RP system, 212–216
RPTLC
  enantiomeric separations, 4
  environmental samples, 155
  food analysis, 175
  forensic toxicology, 197
  molecular parameter determination, 15–16
  parasite, 52
  plant products, flavonoids, 41
  retention mechanism, 6
  veterinary toxicology
    anticoagulants, 255
    pesticides, 251
    vitamins, 257
Rumensin, 258
Ruminococcus, 23

S

Salicylates, 242
Salicyluric acid, 242
Salinomycin, 259
Salvia, 43
Sample application, 7–8, *see also* specific applications
Sample preparation, 6–7, *see also* specific applications
Sample volumes, 8
Sandwich chambers, 9
Sandwichine, 232
Sanguinarine, 39
Saponins, 45–46
Scalopus aquaticus (mole), 118
Schistosoma, 52, 55–56, 61
Scopolamine, 39
Scutellaria radix, 3
Seawater analysis, 4
Sediment analysis, 165
Sedum sediform, 40
Selecta Sol chamber, 12
Selenium, veterinary toxicology, 258
Semiquantitative determinations, 14
Sena alexandria, 234
Senna acutifolia, 234
Sennosides, 234
Sensitivity and reproducibility, 14–15, *see also* specific applications
Serotonin, insect, 90–92
Serpentine, 39
Sesquiterpene lactones, 98, 99
Sesquiterpenes, 44
Shiga toxins, 24
Shigella, 24
Short Bed Continuous Development, 10
Sideritis sipylea, 44
Silica, 2–3, *see also* specific applications
Silica gel
    flavonoid TLC, 40
    food analysis, 175, 176
    parasitology, 52
Skin lipids, 52, 145
Skin secretions, vertebrates, 106–127
    chromatographic systems and detection techniques, 113–118
        alkaloids, 117–118
        ceramides and glucosylceramides, 115–116
        nonpolar lipids, 113
        phosphatides, 114–115
        polar lipids, 114–117
        sphingosines, 116
        steryl glycosides, 116–117
        sulfolipids, 117
    material collection and preservation, 106–111
        epidermis and epidermal derivatives, 106–110
        exocrine gland secretions, 110–111
        storage and preservation methods, 111
    protocols, 124–126
        ceramides and acylglucosylceramides from pig, 125–126
        lipids, reptile, 124–125
        tetrodotoxin, 126
    quantification, 118–121
    sample preparation, 111–113
    systematics, 121
    validation of results, 121–124
Skipski system, 60
Snakes
    scent gland sampling, 111–112
    skin secretions, 117, 124–125
    systematics, 121
Soil analysis, 165
Solid-phase extraction, 7
    environmental samples, 154
    food analysis, 174
    veterinary toxicology, 251
Solvents
    bacterial lipids, 22
    plant products, alkaloids, 35–36
Sorbate, 7
Sorbents, impregnation of, 4
Sorbic acid, 172
Sorbitol, insect, 77, 78, 79
S.O.S. test, 251
Southern armyworm (*Spodoptera eridania*), 81
Spectrophotometers, 14
Spectroscopy, 1
Sphingosines, vertebrate skin secretions, 116
Spodoptera eridania (southern armyworm), 81
Squalene, 114
Stable fly (*Stomoxys calcitrans*), 80
Staphylococcus aureus, 24
Stationary phases, 2–4, *see also* specific applications
Stepwise elution, 5
Stereoisomers, essential oils, 44
Steroidal alkaloids, 234
Steroids, 3, 144
    clinical chemistry, 147–148
    forensic toxicology, 223
    pharmaceutical analysis, 238–240
    vertebrate skin secretions, 114
Steroid saponins, 45–46
Sterols
    bacterial, 22
    food analysis, 173
    parasite, 52–57
    vertebrate skin secretions, 117
Steryl esters, vertebrate, 123
Steryl glycosides, vertebrate skin secretions, 116–117
Stomoxys calcitrans (stable fly), 80
Streptococcus pneumoniae, 24
Streptomycetes isoflavonoids, 41
Strictosamide, 233–234
Strophanthus, 44
Strychnine, 256, 257
Subtilisin, 197
Sugar alcohols, insect, 77–80
Sugar glider (*Petaurus breviceps*), 114
Sugars, food analysis, 175, 182
Sulfa drugs
    derivatization, 251
    forensic toxicology, 223
    pharmaceutical analysis, 237
    veterinary toxicology, 250, 259
Sulfinpyrazone, 237

Sulfolipids, vertebrate skin secretions, 117
Sulfonamides, 144
Sulfur-containing compounds
  pharmaceutical analysis, 237–238
  skin secretions, 117
Surface-modified silicas, 2–3
Sus scrofa (pig) skin secretions, 125–126
Sympathomimetics
  forensic toxicology, 223
  hordenine, 232
Systematics, vertebrate skin secretions, 121

T

T-2 toxin, 250, 260–262
Tapazole, 238
Taricha (newt), 118, 123
Tarichatoxin, 123
Techniques and instrumentation, 1–16, *see also* specific types of studies
  detection of solutes, 12–13
  development, 8–12
    anticircular, 12
    circular, 12
    linear ascending, 8–10
    linear horizontal, 10–12
  molecular parameter determination with adsorptive and reversed-phase TLC, 15–16
  as pilot method for HPLC, 15
  qualitative and quantitative evaluation, 13–14
  sample application, 7–8
  sample preparation, 6–7
  sensitivity and reproducibility, 14–15
  theoretical aspects of TLC, 2–6
    mobile phases, 4–5
    retention mechanism, 6
    retention parameters, 5–6
    stationary phases, 2–4
Temperature, development chamber, 9
Terpenoids
  insect, 89–90
  plant-insect interactions, 95–98
    azadirachtin, 97
    cardenolides, 96–97
    grayanotoxins, 98
    sesquiterpene lactones, 98
Tetracyclines, 4, 187–188, *see also* Antibiotics
Tetrahydrofuran, 3
Tetrodotoxin, 118, 123, 126, 224
Thebaine, 14
Theoretical aspects of TLC, 2–6
  mobile phases, 4–5
  retention mechanism, 6
  retention parameters, 5–6
  stationary phases, 2–4
Thermobia domestica, 82
Thiamine, 244
Thiazides, 238
Thioglycolic acid, derivatization, 13
Thiols, derivatization, 13
Thiouracil, 238

Thorium nitrate, 4
Thymine, 241
Thymus vulgaris, 43
Thyroid drugs, 238
TIAFT system, 210–212
Ticks, 81–82
Tin, 4
Tobacco hornworm (*Manduca sexta*), 81, 83
Tocopherols
  food analysis, 170
  veterinary toxicology, 257
Toloxatone, 244
Toxi-Lab system, 212, 214, 220–222
Toxins
  alkaloids, bird skin secretions, 117–118
  bacterial, , 224–225
  food analysis, 171, 188
  pharmaceutical analysis, 244
  plant–insect interactions, 95–100
    azadirachtin, 97
    cardenolides, 96–97
    grayanotoxins, 98
    sesquiterpene lactones, 98
  vertebrate, 122–124
  warfare agents, 156, 162–163
Tranquilizers, pharmaceutical analysis, 241, 242
Trazodone, 244
Tree boas (*Epicrates*), 121
Trehalose, insect, 74, 76–78
Tretinoin, 244
Triacylglycerols
  insect, 85
  parasite, 52
  vertebrate skin secretions, 111
Triatoma infestans, 81
Tributylamine, 4
Tributyl phosphate, 4
Tricaprylammonium chloride, 4
Trichinella spiralis, 57
Trichlorothiazide, 238
Trichoplusia ni, 83
Tricyclic antidepressants, 144
Trigonelline, 39
Trinitrotoluene, 163, 224
Tripelennamine, 242
Triterpene saponins, 45
Trollius macropetalus, 41
Tropane alkaloids, 39
Tropolone alkaloids, 38
Trypsin, 197
Tryptophan, 4
Tubocurarine, 39
Turtles, 111, 124
Twin-trough chamber, 9
Two-phase triangular development, 137

U

U-chamber, 12
Ulmus, 40
Ultrafiltration, clinical samples, 135

Index

UniTox System, 202–209, 212–214
Ureaplasma, 23
Urginea maritima, 45
Urginea scilla, 44
Urine, *see* Clinical chemistry; Forensic toxicology
UV spectra, 13, 41, *see also* specific applications
UV-VIS, 1

V

Valproic acid, 240–241
Vancomycin, 4
Vario chamber, 10
Verotoxins, 24, 27
Vertebrate skin secretions, *see* Skin secretions, vertebrates
Veterinary toxicology, 249–263
  avicides–alkaloids, 256
  food and feed additives, 257–260
    antibiotics–antimicrobials, 259–260
    ionophores, 258–259
    selenium, 258
    sulfa drugs, 259
    vitamins A, E, and D, 257–258
  insecticides, 251–255
    carbamates, 251–252
    organochlorine compounds, 253, 254
    organophosphorus compounds, 252–253
    pyrethroids, 253–254
  mycotoxins, 260–264
  rodenticides, anticoagulant, 255–256
Vibrio cholerae, 24
Viguiera, 40
Viloxazine, 244

Visual pigment chromophores, insect, 95
Vitamin K, 7
Vitamins
  pharmaceutical analysis, 244
  veterinary toxicology, 257–258
Vomitotoxin, 250, 260, 261

W

Wagner system, 60
Warfare agents, 156
Water analysis, 4, 164
Water hyacinth weevil (*Neochetina eichornia*), 96
Wax esters, vertebrate, 123, 124
Wogonin, 3
Wolf, 119
Wood cockroach (*Cryptocercus puntulatus*), 77

X

Xanthine derivatives, 15
Xanthophylls, parasite, 59

Y

Yams (*Dioscorea opposita*), 234
Yeasts, 20
Yersinia, 24

Z

Zearalenone, 250, 260–263
ZGP (*N*-benzyloxycarbonyl-glycyl-L-proline), 243
Zirconia, 3